Curriculum for Wales

Mathematics and Numeracy

Mastering Mathematics

FOR 11–14 YEARS

Linda Mason, Jonathan Agar, Laszlo Fedor

BOOK 3

Photo credits

p10 © steeve-x-art/Alamy; p60 © stocksolutions/Adobe Stock; p76 left © GLandStudio/Adobe Stock, middle © khuruzero/Adobe Stock; p101 © NASA/Roscosmos; p116 © Marco Uliana/Adobe Stock; p294 © nicoletaionescu/Adobe Stock

Acknowledgments

Every effort has been made to trace all copyright holders, but if any have been inadvertently overlooked, the Publishers will be pleased to make the necessary arrangements at the first opportunity.

Although every effort has been made to ensure that website addresses are correct at time of going to press, Hodder Education cannot be held responsible for the content of any website mentioned in this book. It is sometimes possible to find a relocated web page by typing in the address of the home page for a website in the URL window of your browser.

This book is based on material written for and published in *Key Stage 3 Mastering Mathematics: Book 3*, Second Edition (978 1 3983 0841 1) by Sophie Goldie, Luke Robinson and Heather Davis, with Series Editor Steve Cavill. The publisher would like to thank them for permission to re-use their work in the present volume.

Hachette UK's policy is to use papers that are natural, renewable and recyclable products and made from wood grown in well-managed forests and other controlled sources. The logging and manufacturing processes are expected to conform to the environmental regulations of the country of origin.

Orders: please contact Hachette UK Distribution, Hely Hutchinson Centre, Milton Road, Didcot, Oxfordshire, OX11 7HH. Telephone: +44 (0)1235 827827. Email education@hachette.co.uk Lines are open from 9 a.m. to 5 p.m., Monday to Friday. You can also order through our website: www.hoddereducation.co.uk

ISBN: 978 1 3983 4447 1

© Hodder & Stoughton Limited 2022

First published in 2022 by
Hodder Education,
An Hachette UK Company
Carmelite House
50 Victoria Embankment
London EC4Y 0DZ

www.hoddereducation.co.uk

Impression number   10 9 8 7 6 5 4 3 2 1

Year   2026 2025 2024 2023 2022

All rights reserved. Apart from any use permitted under UK copyright law, no part of this publication may be reproduced or transmitted in any form or by any means, electronic or mechanical, including photocopying and recording, or held within any information storage and retrieval system, without permission in writing from the publisher or under licence from the Copyright Licensing Agency Limited. Further details of such licences (for reprographic reproduction) may be obtained from the Copyright Licensing Agency Limited, www.cla.co.uk

Cover photo © Musicman80/Adobe Stock

Illustrations by Aptara, Inc.

Typeset in India by Aptara, Inc.

Produced by DZS Grafik, Printed in Bosnia & Herzegovina

A catalogue record for this title is available from the British Library.

Contents

The curriculum vi

How to use this book.. vii

1 Powers and indices..1
   1.1 Index notation 2
   1.2 Standard form 9
   1.3 Prime factorisation..14
   Review exercise...19

2 Fractions 22
   2.1 Fractions review 22
   2.2 Mixed numbers 30
   2.3 Multiplying and dividing mixed numbers...35
   Review exercise...39

3 Accuracy 42
   3.1 Significant figures...43
   3.2 Approximating 46
   3.3 Accuracy 49
   Review exercise...53

Consolidation 1 55

4 Percentages 59
   4.1 Percentages review...59
   4.2 Using multipliers...65
   4.3 Appreciation and depreciation 70
   4.4 Reverse percentages...72
   Review exercise...78

5 Using measures..81
   5.1 Calculations involving time 81
   5.2 Speed 86
   5.3 Dimensions of a calculation 92
   5.4 Converting between metric units of area and volume........................95
   Review exercise...100

Consolidation 2 104

6 Using ratio and proportion and percentage 107
6.1 Ratio and proportion review 108
6.2 Profit and loss 113
6.3 Simple and compound interest 116
Review exercise 119

7 Equations, expressions, formulas and inequalities 121
7.1 Solving equations review 122
7.2 Using the laws of indices 129
7.3 Expanding brackets 132
7.4 Formulas 138
7.5 Solving inequalities 145
Review exercise 149

Consolidation 3 152

8 Graphs 155
8.1 The gradient of a line 157
8.2 The equation of a straight line 163
Review exercise 171

9 Real-life graphs 174
9.1 Distance–time graphs 175
9.2 Reading from real-life graphs 184
Review exercise 192

10 Transformations 197
10.1 Reflection and translation 197
10.2 Rotation 209
10.3 Enlargement 215
Review exercise 223

Consolidation 4 228

11 Prisms and cylinders 233
11.1 Volume of a prism 233
11.2 Volume of a cylinder 239
11.3 Surface area 244
Review exercise 251

12 Trigonometry 254
- 12.1 Similarity 254
- 12.2 Finding a missing side 260
- 12.3 Finding a missing angle 268
- 12.4 Solving problems 271
- Review exercise 275

13 Real-life finance 278
- 13.1 Income, expenditure and budgets 278
- 13.2 Foreign currency exchange 282
- 13.3 Taxation on goods and services 285
- Review exercise 289

Consolidation 5 291

14 Working with data 296
- 14.1 Hypothesis testing and questionnaires 297
- 14.2 Grouped frequency tables 302
- 14.3 Displaying grouped data 310
- 14.4 Scatter diagrams 320
- Review exercise 328

15 Probability 335
- 15.1 Probability space diagrams 335
- 15.2 Venn diagrams 342
- 15.3 Combined events 347
- Review exercise 355

Consolidation 6 360

Glossary 367

The curriculum

A Curriculum for Wales

The Curriculum for Wales has been developed in Wales, by practitioners for practitioners, bringing together educational expertise and wider research and evidence. Our resources are designed to reflect the Welsh context and to help develop your identity as a citizen of Wales and the world.

We have worked in partnership with University of Wales Press to produce this resource. They have reviewed it to make sure it is tailored to the new curriculum and explores Welsh culture and heritage in an authentic way. Find out more about University of Wales Press by visiting their websites www.uwp.co.uk and www.gwasgprifysgolcymru.org

Our authors have a wealth of experience teaching, examining and working in education in Wales:

- **Laszlo Fedor** has 20 years' experience of working in state-funded schools in Wales. He teaches Mathematics to children of all abilities in secondary and sixth-form education.
- **Jonathan Agar** has been teaching in South Wales for over a decade, including time spent as head of department, assistant headteacher and principal examiner. He has a particular interest in pupils' misconceptions in Mathematics and has completed a Research Masters in Mathematics Education.
- Series Editor **Linda Mason** has many years of experience as teacher, adviser, curriculum consultant and principal examiner in Wales, including work across North Wales and with WJEC.

We would also like to thank the teachers from schools across Wales who helped to plan and review this title, including:

- Llanishen High School, Cardiff
- Cefn Hengoed Community School, Swansea
- Bishop Hedley High School, Merthyr Tydfil

How to use this book

▶ How to get the most from this book

Hodder Education's Mathematics resources support the learning and experience of Mathematics for years 7–9 and comprise:

- three textbooks covering the Wales National Curriculum for ages 11–14
- Boost online content.

Our Book 3 material is split into 15 chapters, and each chapter comprises two, three or four units. In total there are 49 units in the book. The material across all three books, and the scheme of work, is designed to be used whenever the teacher feels it is appropriate for the class; for example, some content in Book 2 or even Book 1 may be suitable for some teaching in Year 9. Similarly, our scheme of work is designed to be flexible.

The book contains indication of five proficiencies: **conceptual understanding**, **communication using symbols**, **fluency**, **logical reasoning** and **strategic competence**.

These five proficiencies are intertwined, so no individual proficiency is developed in isolation. Consequently, in general, many of the proficiencies could be highlighted in activities, examples and exercises throughout the book.

As an aid, the best fit or a principle proficiency is flagged as guidance only, to raise awareness of a particular proficiency for the learner. We have chosen to highlight good examples of conceptual understanding and communication using symbols as relevant alongside various mathematical explanations, activities and worked examples. Fluency, logical reasoning and strategic competence are highlighted in relation to individual questions in our exercises, reviews and consolidation sections. All of these indicators are intended as guidance to aid the learner in understanding their own proficiency development.

Fluency

Logical reasoning

Strategic competence

Conceptual understanding

Communication using symbols

In summary, the five proficiencies capture a learner's developing understanding of the multi-faceted nature of their learning.

- **Conceptual understanding** allows learners to develop their ability to connect ideas through increasing depth of knowledge. Understanding the way in which concepts are connected aids learner development.
- Through progression in **communication using symbols**, learners develop understanding of conventions and abstract representation.
- With experience, learners will develop **fluency** in remembering facts, relationships and techniques.
- Learners develop **logical reasoning**, including justification and proof, in understanding the relationship between concepts.
- In developing **strategic competence**, learners show independence in applying ideas within a problem, and recognise mathematical structure.

All exercise questions that relate to finance are indicated with this symbol.

Each **chapter** includes:

- *Coming up* – a list of learning objectives that will be tackled in the chapter
- a *Starter* problem – either an activity or a puzzle – to engage the students with a new topic and designed to be used before the first lesson, or at the start of the first lesson in that topic

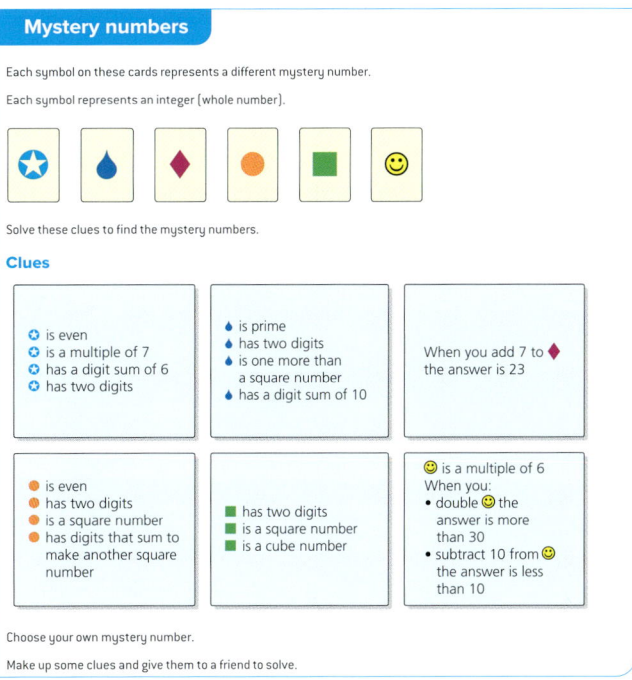

- activities, investigations and whole-class discussion points
- a *Review exercise* at the end of the chapter; this encompasses all of the units covered in the chapter.

Each of the **individual units** within the chapter includes:

- a *Skill checker* – simple diagnostic questions to test basic understanding in preparation for the unit

Skill checker

Match together the calculations which have the same answers.

$60 - 6 \times 3$	$5 + 4^2$	$18 - 3^2$	$16 - 2 \times 3$	$7 \times 8 + 50 \div 2$
$1 + 4 \times 2$	$7 \times 9 \div 3$	$7 \times (2 + 4)$	$2 \times 3 + 4$	$9 \times (4 + 5)$

- a clear and detailed explanation of the topic
- plenty of worked examples with solutions
- a focus on fluency, with a carefully structured approach that takes into account cognitive load theory
- helpful hints and guidance on misconceptions and pitfalls to watch out for
- *Now try these* exercise questions, which:
  ▸ develop conceptual understanding and communication using symbols
  ▸ are split up into three bands of increasing demand: Band 1 questions are for those students who are working towards age-related expectations, Band 2 are for those at age-related expectations, and Band 3 are for those working beyond age-related expectations. Most students will engage with Band 2 questions, and either Band 1 or Band 3, depending on which is most appropriate
  ▸ are carefully calibrated to enable the whole class to understand each question and answer before moving on
  ▸ give the opportunity to apply skills, including working systematically, modelling, breaking problems down into stages, visualising, working backwards, and trial and improvement

  Non-calculator questions are indicated.

- a list of key words (highlighted in the text). These are fully explained in a glossary at the back of the book.

There are six sets of consolidation questions throughout the book, each of which appears after a sequence of two or three chapters. These are designed to cover approximately half a term's work.

The book encourages learners to use physical equipment (manipulatives) and representations as well as visual and abstract representations, for example using cards, bar model diagrams and physical number lines to aid the development of understanding.

Opportunities to link with Science and Technology, Humanities, Expressive Arts, Health and Well-being, Languages, and Literacy and Communication teaching and learning are included in the *Cross-curricular activities* panels. Scattered throughout the books are examples that we hope will encourage exploration of historical Welsh mathematicians and contexts.

All answers are provided online at **www.hoddereducation.co.uk/MasteringMathematicsWales** and are freely accessible. You can also find an editable course planner here, with lesson suggestions and time built in for consolidation, assessment and application lessons. A suggestion for how content in these resources can be mapped to the Wales curriculum's What Matters statements and Progression Steps has been included in an editable format, to enable schools to create their own structure, as well as a full set of links to other areas of the curriculum across all subjects.

1 Powers and indices

Coming up...
- Index notation and the laws of indices
- Using standard form
- Writing integers as a product of prime factors
- Finding the least common multiple and highest common factor of two numbers

The tower of Hanoi

This puzzle is called 'The tower of Hanoi'.
You have 3 poles and 4 coloured discs of different sizes set on one of the poles.

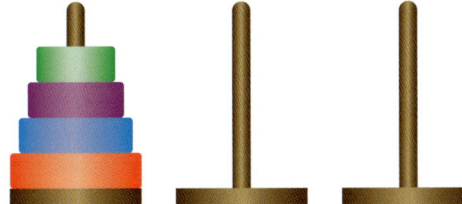

The aim is to move the discs onto one of the other poles.

There are two rules for moving the discs.

Rule 1: You can only move one disc at a time

Rule 2: You must never put a larger disc on top of a smaller disc

① Make your own Tower of Hanoi.
   - Cut out 4 circles of card of different diameters to use as your discs.
   - Make 3 large square bases for your towers.
   - Put the circles on one of your towers in order of size.
② What is the smallest number of moves needed to move all your discs to another tower?
③ Investigate further for different numbers of discs.

Find a rule for the smallest number of moves, m, needed to move d discs from one tower to another.
Find out about the history and myths of the Tower of Hanoi puzzle.

1.1 Index notation

Skill checker

Make a copy of this cross-number and then solve the clues.

Across
1. $7^3 - 5^3$
4. $3^2 + 4^2$
5. 2×3^3
6. $4^2 \times \sqrt{36}$
10. $2 \times \sqrt{100}$
11. $\dfrac{5^2 \times 4^3}{\sqrt{4}}$

Down
1. 5^2
2. $10^3 - 5^3$
3. $\sqrt{256}$
4. $6^2 - \sqrt{100}$
7. 3×6^3
8. $\dfrac{\sqrt{10\,000}}{2}$
9. $\dfrac{6^3}{2^3}$
10. 5×2^2

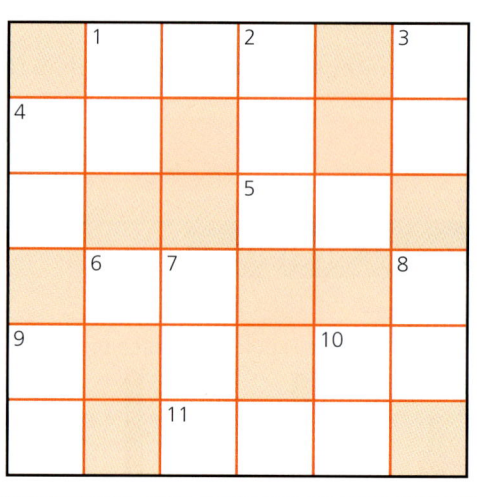

▶ Using indices

Remember that 5×5 is written as 5^2. *(Say 5 squared.)*

You say that 5 is the **base** and 2 is the **index** or **power**.

In the same way, repeated multiplications can be written using **index notation** like this:

$\underbrace{5 \times 5 \times 5}_{3 \text{ fives}} = 5^3$ *(Say 5 cubed.)*

$\underbrace{5 \times 5 \times 5 \times 5}_{4 \text{ fives}} = 5^4$ *(Say 5 to the power of 4.)*

Note

You have a power button on your calculator; it may look like $x^\square$ or x^y. Make sure you know how to use it: press `5` `x^☐` `4` `=` and check you get an answer of 625.

You can have any number as the base.

For example, $\underbrace{7 \times 7 \times 7 \times 7 \times 7 \times 7 \times 7 \times 7}_{8 \text{ sevens}} = 7^8$

Check you can use your calculator to work out 7^8. You should get 5 764 801.

In general, $\underbrace{a \times a \times a \times \ldots \times a \times a}_{n \text{ times}} = a^n$

The next examples show you how to multiply and divide numbers written using index notation.

Communication using symbols

1 Powers and indices

Worked example

Write $3^5 \times 3^4$ as a single power of 3.

Solution

$3^5 \times 3^4$
$= (3 \times 3 \times 3 \times 3 \times 3) \times (3 \times 3 \times 3 \times 3)$
$= 3^9$

> There are 5 threes multiplied together multiplied by 4 threes multiplied together which makes 9 threes multiplied together.

> The powers have been added: 5 + 4 = 9.

Worked example

Write $3^6 \div 3^4$ as a single power of 3.

Solution

$3^6 \div 3^4 = \dfrac{3^6}{3^4}$

$= \dfrac{3 \times 3 \times \cancel{3} \times \cancel{3} \times \cancel{3} \times \cancel{3}}{\cancel{3} \times \cancel{3} \times \cancel{3} \times \cancel{3}}$

$= 3 \times 3$
$= 3^2$

> It is easier to write this as a fraction.

> If you multiply by 3 and then divide by 3 it cancels out, so 4 of the 3s cancel from the top and bottom.

> The powers have been subtracted; 6 − 4 = 2.

In the examples the base was 3, but the same would be true for any base.
Here are the rules for multiplying and dividing powers.

$$a^m \times a^n = a^{m+n} \qquad a^m \div a^n = a^{m-n}$$

For example, $9^3 \times 9^8 = 9^{3+8} = 9^{11}$

and $\dfrac{4^{12}}{4^7}$ or $4^{12} \div 4^7$ is $4^{12-7} = 4^5$

> Take care! The bases must be the same − you can't combine powers if the bases are different. There is no way to simplify $3^7 \times 5^4$ any further.

Sometimes powers involve brackets like in this next example.

Worked example

Write $(3^2)^4$ as a single power of 3.

Solution

$(3^2)^4$ means $\underbrace{(3^2) \times (3^2) \times (3^2) \times (3^2)}_{4 \text{ times}}$

So $(3^2) \times (3^2) \times (3^2) \times (3^2) = (3 \times 3) \times (3 \times 3) \times (3 \times 3) \times (3 \times 3)$
$= 3^8$

> There are 2 threes multiplied together 4 times.

> The powers have been multiplied; 2 × 4 = 8.

$(3^2)^4$ is sometimes called a power of a power. Again, the example would have worked for any base, not just 3.
Here is the rule for a power of a power.

$$(a^m)^n = a^{m \times n}$$

For example, $(11^4)^5 = 11^{4 \times 5} = 11^{20}$

3

Activity

a Work out these powers of 2:

    i 2^1 **ii** 2^2 **iii** 2^3

    iv 2^4 **v** 2^5 **vi** 2^6

b Using only your answers to part **a**, complete these sums.

The first one has been done for you.

You can only use each number once!

You are writing each number as a sum of powers of 2.

    i   21 = [16] + [4] + [1]

    ii   7 = ☐ + ☐ + ☐

    iii   10 = ☐ + ☐

    iv   31 = ☐ + ☐ + ☐ + ☐ + ☐

    v   51 = ☐ + ☐ + ☐ + ☐

c The table shows how you can write the numbers 1 to 8 as sums of powers of 2.

▶ 1 means that the power of 2 is needed.

▶ 0 means that the power of 2 is NOT needed.

▶ You don't need to write in any zeros **before** the first 1.

Number	Powers of 2					
	32	16	8	4	2	1
1						1
2					1	0
3					1	1
4				1	0	0
5				1	0	1
6				1	1	0
7				1	1	1
8			1	0	0	0

5 = 4 + 1

7 = 4 + 2 + 1

Carry on the table for the numbers up to 32.

Can all numbers be written using powers of 2?

d The table in part **c** is used to write numbers in **binary**.

In binary, 7 is written as 111 and 8 is 1000.

Convert these binary numbers back to ordinary numbers.

    i   100000   **ii**   101010

    iii   1101011   **iv**   10000000

Each binary digit is called a 'bit'. The number 10000000 uses 8 bits, and 8 bits is called a byte. A kilobyte is 1024 bytes and a megabyte is roughly 1 million bytes. Binary numbers are used in computers to store information. Binary is a very powerful tool as data can be represented as strings of 0s and 1s which are represented as 'on/off' signals.

e Find out more about binary numbers and how they are used in computing.

Can every number be represented as a binary number?

How do you think text is converted to binary?

1 Powers and indices

▶ Powers of 1 and 0

Activity

① Complete these.
  a Use your calculator to work out
      i $3^7 \div 3^6$ ii $4^8 \div 4^7$ iii $9^{11} \div 9^{10}$
  b Use the laws of indices to write these as a single power.
      i $3^7 \div 3^6 = 3^{\square}$ ii $4^8 \div 4^7 = 4^{\square}$ iii $9^{11} \div 9^{10} = 9^{\square}$
  c What do you notice?
      Write down the value of 56^1.

② Complete these.
  a Use your calculator to work out
      i $2^9 \div 2^9$ ii $5^8 \div 5^8$ iii $19^3 \div 19^3$
  b Use the laws of indices to write these as a single power.
      i $2^9 \div 2^9 = 2^{\square}$ ii $5^8 \div 5^8 = 5^{\square}$ iii $19^3 \div 19^3 = 19^{\square}$
  c What do you notice?
      Write down the value of 56^0.

In the activity you found that
- any number to the power 1 is itself
- any number divided by itself is 1. ← *The only exception to this rule is 0.*

Using indices these are written as

$$a^1 = a \qquad a^0 = 1$$

▶ Negative powers

Look at this pattern.

$$2^3 = 2 \times 2 \times 2 = 8$$
$$2^2 = 2 \times 2 = 4$$
$$2^1 = 2$$
$$2^0 = 1$$
$$2^{-1} = \frac{1}{2^1} = \frac{1}{2}$$
$$2^{-2} = \frac{1}{2^2} = \frac{1}{4}$$
$$2^{-3} = \frac{1}{2^3} = \frac{1}{8}$$

(each step $\div 2$)

You can carry on the pattern so $2^{-8} = \dfrac{1}{2^8}$ and $2^{-n} = \dfrac{1}{2^n}$

You could have used any base so you can say

$$a^{-n} = \frac{1}{a^n}$$ ← *A negative power means '1 over'.*

> **Worked example**
>
> Write 10^{-3} as a decimal.
>
> **Solution**
>
> $10^{-3} = \dfrac{1}{10 \times 10 \times 10}$
>
> $= \dfrac{1}{1000}$ ← Find $1 \div 1000$.
>
> $= 0.001$

▶ Roots

Remember that you can take roots to 'undo' squaring and cubing.

For example, $3^2 = 9$ so $\sqrt{9} = 3$ ← Say the **square root** of 9 is 3.

and $5^3 = 125$ so $\sqrt[3]{125} = 5$ ← Say the **cube root** of 125 is 5.

In the same way you can take roots to 'undo' higher powers.

For example, $4^6 = 4096$

so $\sqrt[6]{4096} = 4$ ← Say the **sixth root** of 4096 is 4.

Communication using symbols

> You have a root button on your calculator; it may look like □√☐ or ˣ√☐.
> You may have to use the SHIFT key or 2nd key to reach it. Make sure you know how to use the root button: press 6 √☐ 4096 = and check you get an answer of 4.

> **Watch out!**
> 9 has two square roots, +3 and −3, as $(-3) \times (-3) = 9$ and $3 \times 3 = 9$.
> However, the symbol $\sqrt{}$ means the positive square root only, so $\sqrt{9} = 3$.

> **Worked example**
>
> Calculate
>
> a $5^4 + 4^5$
>
> b $\sqrt[5]{343} - \sqrt[6]{64}$
>
> **Solution**
>
> a $5^4 + 4^5 = 625 + 1024$
>
> $\qquad = 1649$
>
> b $\sqrt[5]{243} - \sqrt[6]{64} = 3 - 2$
>
> $\qquad = 1$
>
> *Watch out! You can't combine sums and differences into a single power because the base number is different, so you just need to use your calculator to work these out.*

1.1 Now try these

Band 1 questions

1. Write each expression as a power of 2.
   a $2 \times 2 \times 2 \times 2$
   b $2 \times 2 \times 2 \times 2 \times 2 \times 2 \times 2$
   c $2 \times 2 \times 2 \times 2 \times 2 \times 2 \times 2 \times 2 \times 2 \times 2 \times 2 \times 2$
   d $2 \times 2 \times 2 \times 2 \times 2 \times 2$

2. Write each expression as a single power.
   a $7 \times 7 \times 7$
   b $5 \times 5 \times 5 \times 5 \times 5$
   c $14 \times 14 \times 14 \times 14 \times 14 \times 14 \times 14$
   d $12 \times 12 \times 12 \times 12 \times 12 \times 12$
   e $8 \times 8 \times 8 \times 8 \times 8 \times 8 \times 8$
   f $33 \times 33 \times 33 \times 33 \times 33 \times 33 \times 33 \times 33 \times 33$

3. a Calculate the difference between 3^2 and 2^3.
   b Find the value of $3^3 + 4^2 + 2^4$.

4. 5^4 can be written in several ways.
   Here are some.

   $5 \times 5 \times 5 \times 5$ $\quad$ $5^2 \times 5^2$ $\quad$ $5^3 \times 5^1$

   Write 4^5 in as many different ways as you can.

5. Copy and complete these.
   a $4^2 = \square$
   b $\square^3 = 125$
   c $2^\square = 32$

Band 2 questions

6. Write the correct symbol <, > or = between each pair of numbers.
   a $2^5 \square 6^2$
   b $2^7 \square 5^3$
   c $4^3 \square 3^4$
   d $2^9 \square 8^3$
   e $10^3 \square 2^{10}$
   f $12^5 \square 3^8$

7. The expressions on these 12 cards can be matched into six pairs.
   All the missing numbers are the same.

   $4^\square$ $\quad$ $3 \times \square$ $\quad$ 9 $\quad$ 32 $\quad$ $\square^3$ $\quad$ 6 $\quad$ 10 $\quad$ $\square \times 5$ $\quad$ 16 $\quad$ $3^\square$ $\quad$ 8 $\quad$ $\square^5$

   a Match the cards into pairs.
   b What is the missing number?

8. Write each of these as a single power.
   a $4^5 \times 4^3$
   b $6^{12} \times 6^4$
   c $5^9 \times 5^3$

9. Write each of these as a single power.
   a $3^5 \div 3^3$
   b $7^{14} \div 7^8$
   c $8^9 \div 8^5$
   d $\dfrac{2^7}{2^3}$
   e $\dfrac{10^8}{10^3}$
   f $\dfrac{20^6}{20}$

10. Write each of these as a single power.
    a $(6^2)^3$
    b $(2^4)^5$
    c $(13^6)^3$

11
  a Write down the value of $49 \div 49$.
  b Write 49 as a power of 7.
  c Write $7^2 \div 7^2$
      i as a power of 7   **ii** as a number.
  d What is the value of 7^0?
  e What are the values of
      i 2^0 **ii** 3^0 **iii** 17^0?
  f State a general value for the power zero.

12 Write each of these calculations as a single power.
  a $(13^7)^8$ **b** $7^9 \div 7^2$ **c** $4^5 \times 4^9$
  d $3^6 \div 3^4$ **e** $(12^4)^6$ **f** $11^3 \times 11^4 \times 11^5$

13 Work out the value of these. Give each of your answers as a decimal.
  a 2^{-1} **b** 10^{-1} **c** 5^{-1} **d** 4^{-1} **e** 100^{-1} **f** 8^{-1}

14 Jerin says that $3^4 \times 2^2$ is equal to 6^6.
  Show that Jerin is wrong.
  What mistake has she made?

Band 3 questions

15 Work out the value of these. Give each of your answers as a decimal.
  a 2^{-2} **b** 10^{-2} **c** 5^{-2} **d** 2^{-3} **e** 10^{-3} **f** 5^{-3}

16 Work out the value of these.
  a $\sqrt[5]{243}$ **b** $\sqrt[7]{128}$ **c** $\sqrt[4]{625}$

17 Write these as a single power of 5.
  a $5^3 \times 5^{-2}$ **b** $5^3 \div 5^{-2}$ **c** $(5^{-2})^3$
  d $5^{-4} \times 5^8$ **e** $5^{-1} \times 5$ **f** $5^{-2} \div 5^{-2}$

18 Find the missing digits.
  a $7^{\square} = 7^5 \div 7^2$ **b** $7^3 = 7^3 \times 7^{\square}$ **c** $(7^{\square})^4 = 7^8$ **d** $7^6 \div 7^2 = 7^{\square} \times 7^3$

19 Write each of these as a single power.
  a $\dfrac{3^5 \times 3^4}{3^6}$ **b** $(2^4 \times 2^2)^3$ **c** $\dfrac{(2^2)^3}{(2^4)^2}$

  d $\dfrac{6^{-7}}{6^{-3}}$ **e** $(5^{-3} \times 5^4)^5$ **f** $\dfrac{9^3 \times 9^2}{9^{-2} \times 9^{-1}}$

20
  a Priya says that $(5^3)^2$ is the same as $(5^2)^3$.
      Show that Priya is right.
  b Show that $(a^m)^n$ is the same as $(a^n)^m$.
  c Is it **always true, sometimes true or never true** that $a^m \div a^n$ is the same as $a^n \div a^m$?
      Explain your answer fully.

1 Powers and indices

1.2 Standard form

Skill checker

① Work out these powers of 10.
  a 10^1
  b 10^2
  c 10^3
  d 10^4
  e 10^5
  f 10^6

② a Write down 1 million as
    i   an ordinary number
    ii a power of 10.
  b Write down 1 billion as
    i   an ordinary number
    ii a power of 10.
  c A googol is 10^{100}.
    Aeron writes a googol down as an ordinary number.
    How many zeros should follow the 1?

③ a Copy and complete this pattern.

   $2.7 \times 10 = \boxed{}$

   $2.7 \times 10^2 = 2.7 \times 100 = \boxed{}$

   $2.7 \times 10^3 = 2.7 \times 1000 = \boxed{}$

   $2.7 \times 10^4 = 2.7 \times \boxed{} = \boxed{}$

   $2.7 \times 10^5 = 2.7 \times \boxed{} = \boxed{}$

   $2.7 \times \boxed{} = 2.7 \times \boxed{} = 2\,700\,000$

  b Work out the value of 2.7×10^9.
     Which of these is the correct way to say this number?

   | 2.7 billion | 27 hundred million |
   | 27 thousand million | 2 thousand 7 hundred million |

▶ Using standard form for large numbers

You often see headlines in newspapers that involve large numbers. For example, there are more than 400 billion plastic toy bricks in the world or there are 8.9 million school children in the UK.

In newspapers, large numbers are usually written using the words 'billion' and 'million' rather than being written out in full like this:

  400 000 000 000 Lego bricks or 8 900 000 school children

In Maths and Science, large numbers are usually written using powers of 10.

For example, 4×10^{11} Lego bricks or 8.9×10^6 school children.

Standard form is a way of writing down large numbers without writing down all the zeros.

A number is in standard form when it is written as

  a number between 1 and 10 multiplied by a power of 10

In symbols this is written as

  $A \times 10^n$

- A can be any number from 1 up to 10 (but not 10). ← $1 \leq A < 10$
- n must be an integer.

Cross-curricular activity

This painting is called 'Salvator Mundi' and it was painted by Leonardo da Vinci.

It was bought for over $450 million dollars, making it one of the world's most expensive paintings (at the time of writing, this is still the case!).

How would you write this figure in standard form?

Find out about other expensive paintings and the artists who painted them.

Worked example

The Pacific Ocean has a surface area of 168 000 000 km². Write this number in standard form.

Solution

168 000 000 in standard form is $1.68 \times 10^?$ ← *A* is always between 1 and 10.

Use a place value diagram to help you work out what the power of 10 should be.

H M	T M	M	H Th	T Th	Th	H	T	O	.	t	h
								1	.	6	8
1	6	8	0	0	0	0	0	.			

$168\,000\,000 = 1.68 \times 10\,000\,000$
$ = 1.68 \times 10^8$ ← You have to multiply 1.68 by ten 8 times to get 168 000 000.

So the Pacific Ocean has a surface area of 1.68×10^8 km²

You don't need to draw a place value diagram each time. The digits move 8 places so that the decimal point is now between the 1 and the 6, so you multiply by 10^8.

$168\,000\,000. = 1.68 \times 10^8$ ← Count the arrows. They tell you the power of 10.

▶ Using standard form for small numbers

Activity

① Complete this pattern.

$\div 10 \begin{pmatrix} 10^2 = 100 \\ 10^1 = 10 \\ 10^0 = \underline{} \\ 10^{-1} = 0.1 \end{pmatrix} \div 10$
$10^{-2} = \underline{}$
$10^{-3} = \underline{}$
$10^{-4} = \underline{}$

② a Work these out.
    i 2×10^{-1} 2×10^{-2} 2×10^{-3}
    ii 84×10^{-1} 84×10^{-2} 84×10^{-3}
    iii 7.9×10^{-1} 7.9×10^{-2} 7.9×10^{-3}

  b Complete each statement.

Multiplying by 10^{-1} is the same as dividing by _____ once

Multiplying by 10^{-2} is the same as dividing by 10 _____

Multiplying by 10^{-3} is the same as dividing by 10 _____

1 Powers and indices

Worked example

A flea weighs around 0.000 087 kg.

Write this number in standard form.

Solution

> A is always between 1 and 10.

0.000 087 in standard form is $8.7 \times 10^?$

Use a place value diagram to help you work out what the power of 10 should be.

O	.	t	h	th	t th	h th	m
8	.	7					
0	.	0	0	0	0	8	7

$0.000\,087 = 8.7 \div 10 \div 10 \div 10 \div 10 \div 10$
$= 8.7 \times 10^{-5}$

So the flea weighs 8.7×10^{-5} kg.

> Remember: dividing by 10 is the same as multiplying by 10^{-1}.
> You have to divide 8.7 by ten five times to get 0.000 087.
> Dividing by 10 five times is the same as multiplying by 10^{-5}.
> You don't need to draw a place value diagram each time.
> You have moved the digits 5 places to get the decimal point between the 8 and the 7, so you multiply by 10^{-5}.
> $0.000\,087 = 8.7 \times 10^{-5}$

▶ Converting numbers from standard form

You also need to be able to convert from standard form back to ordinary numbers.

Remember when the power of 10:
- is positive, then the number is BIG
- is negative, then the number is SMALL.

Worked example

Convert these numbers from standard form to ordinary numbers.

a 5.67×10^4

b 3.08×10^{-6}

Solution

a 5.67×10^4 means you multiply 5.67 by 10 four times.

The digits move 4 places.

56 700.

So $5.67 \times 10^4 = 56\,700$

> A positive power means the number is big!

b 3.08×10^{-6} means you divide 3.08 by 10 six times.

The digits move 6 places.

0.00000308

So $3.08 \times 10^{-6} = 0.000\,003\,08$

> A negative power means the number is small!

Curriculum for Wales **Mastering Mathematics: Book 3**

> **Activity**
>
> ① Are there any tall buildings or structures in Wales? Write down their heights in standard form.
> ② a What is Wales's current estimated population in standard form?
>   b How many times bigger is China's population? Write this in standard form.

> **Maths in context**
>
> Standard form, also known as scientific notation, was invented by Muhammad Al-Khwarizmi. He was a Persian mathematician and astronomer in the 9th century. Al-Khwarizmi was also responsible for introducing the 10 digits that we still use today (0, 1, 2, 3, 4, 5, 6, 7, 8, 9).
> Why do you think standard form is important in the world of an astronomer?

1.2 Now try these

Band 1 questions

Fluency

① Write each of these numbers as a power of 10.
  a 1 000 000
  b 100 000 000
  c 10
  d 0.01
  e 0.001
  f one hundred thousand
  g one thousand million
  h one ten thousandth

② Write each of these as an ordinary number.
  a 10^5
  b 10^{10}
  c 10^{-1}
  d 10^{-2}
  e 10^{-3}

③ Work out these multiplications. Part **a** has been answered for you.
  a $2.6 \times 10^3 = 2.6 \times 1000 = 2600$
  b 4×10^2
  c 4.8×10^5
  d 1.3×10^4
  e 2.4×10^7
  f 9.3×10^6

Strategic competence

④ Copy and complete these. Fill in the missing numbers.
  a $\square \times 10^2 = 500$
  b $\square \times 10^3 = 3000$
  c $6 \times 10^\square = 6\,000\,000$
  d $\square \times 10^2 = 350$
  e $4.2 \times 10^\square = 42\,000$
  f $\square \times 10^5 = 450\,000$

Band 2 questions

Fluency

⑤ These numbers are in standard form. Write them as ordinary numbers.
  a 2×10^3
  b 7×10^6
  c 4.2×10^5
  d 7.1×10^7
  e 8.6×10^9

⑥ Write these numbers in standard form.
  a 200
  b 5000
  c 7 000 000
  d 3600
  e 7 200 000

⑦ Work out these multiplications.
  a 3×10^{-1}
  b 4×10^{-2}
  c 5×10^{-3}

⑧ Copy and complete these.
  Fill in the missing numbers.
  a $4 \times 10^\square = 0.4$
  b $\square \times 10^{-3} = 0.006$
  c $3 \times 10^\square = 0.03$

9 Write these as ordinary numbers.
  a The length of a human chromosome is 5×10^{-6} m.
  b The CN tower in Toronto is 5.53×10^2 m tall.
  c The mass of an electron is 9.11×10^{-31} kg.

10 Write these numbers in standard form.
  a The distance between the Earth and the Moon is 239 000 miles.
  b A £5 note is 0.000 22 m thick.
  c Quartz fibre has a diameter of 0.000 001 m.

Band 3 questions

11 a Write the correct inequality symbol < or > between each of these pairs of numbers.
  i 3×10^4 ☐ 3×10^5 ii 4.6×10^6 ☐ 5.8×10^6
  iii 7×10^9 ☐ 5.3×10^9 iv 6×10^5 ☐ 1 million
  b Explain how you can compare the sizes of numbers written in standard form.

12 Write these numbers in order, starting with the smallest.
  4.56×10^5 3.4×10^4 563 000 7.4×10^6 820 000

13 Correct each of the pieces of homework below.
  For each question say who has got it right and explain where the other has gone wrong.

  Seren
  Write 3800 in standard form
  3.8×10^2
  Write 0.000 006 78 in standard form
  6.78×10^{-6}
  Write 5×10^7 as an ordinary number
  50 000 000
  Write 9.6×10^{-5} as an ordinary number
  9 600 000
  Put in order of size, small to big:
  2.6×10^4, 2.9×10^2, 2.7×10^{-3}
  2.6×10^4, 2.9×10^2, 2.7×10^{-3}
  Double 6.7×10^5
  13 400 000 000

  Bryn
  Write 3800 in standard form
  3.8×10^3
  Write 0.000 006 78 in standard form
  6.78×10^{-8}
  Write 5×10^7 as an ordinary number
  500 000
  Write 9.6×10^{-5} as an ordinary number
  0.000 096
  Put in order of size, small to big:
  2.6×10^4, 2.9×10^2, 2.7×10^{-3}
  2.7×10^{-3}, 2.9×10^2, 2.6×10^4
  Double 6.7×10^5
  1.34×10^6

14 a i   Explain how to work out 2000×300 in your head.
      ii Now write your answer in standard form.
  b i   Write the answer to $(2 \times 10^3) \times (3 \times 10^2)$ in standard form.
      Look carefully at your answer.
      ii How is the number worked out?
      iii How is the power of 10 worked out?
  c Work out
      i $(4 \times 10^2) \times (2 \times 10^3)$ ii $(1.2 \times 10^4) \times (3 \times 10^2)$
  d Explain how to multiply numbers when they are written in standard form.
  e Use your method to work out $(3 \times 10^{-2}) \times (4 \times 10^{-3})$.
      Make sure you write your answer in standard form.

Strategic competence

15 The problems below can be solved either by multiplying or by dividing.
Choose the correct operation for each one and then answer the question.

a A mouse weighs 1.5×10^{-2} kg.
An owl eats 1000 mice in a year.
What weight of mice is this?

b The speed of sound is 3.3×10^2 metres per second.
How far does sound travel in an hour?

c A grain of salt weighs 2×10^{-5} grams.
How many grains of salt are there in a 750 gram packet?

d A packet of 500 sheets of paper is 55 mm thick.
How thick is each sheet of paper?

e The average number of clover leaves in a square metre of lawn is 1.5×10^3.
Estimate the number of clover leaves in a park with 5×10^4 m² of lawn.

16 The galaxy that the Earth is in is called the Milky Way.
The Milky Way is about 120 000 light years across.
The Earth is around 27 000 light years from the centre of the Milky Way.
A light year is the distance that light can travel in a year and it is 9.46×10^{15} metres.

a Imagine a spaceship that can travel at half the speed of light.
It sets off from the Earth towards the centre of the Milky Way. How long would it take?

b How many generations of astronauts do you think it would take?

c Assuming an original crew of 20 people, how many people do you think would arrive at the centre of the Milky Way?

Logical reasoning

17 Which of these numbers are not in standard form? Why not?
Where possible re-write the number so it is in standard form.

a 0.3×10^4 b 7×10^{-3} c 10.1×10^{-4}
d $4 \times 10^{0.5}$ e 10^9 f 9.9×10^{100}

1.3 Prime factorisation

Skill checker

① Look at the numbers in this box.

9	7	51	36	20
16	37	2	27	24
11	21	100	1	42
8	64	6	12	5

Write down all the numbers from the box that are

a square
b cubes
c factors of 36
d prime
e multiples of 6
f factors of 100 and prime.

② Write down the first ten prime numbers.

Remember

Prime numbers are numbers that have exactly 2 factors: 1 and the number itself.

▶ Writing numbers as a product of prime factors

Remember that the **factors** of a number divide into it exactly.
For example, the factors of 12 are 1, 2, 3, 4, 6 and 12
A **prime factor** is a factor of a number that is also prime.
The prime factors of 12 are 2 and 3.

> **Remember**
> 1 is not a prime number.

Activity

Show that every number between 2 and 20 is either prime or can be made by multiplying prime numbers together.

Number	Answer
2	2 (prime)
3	3 (prime)
4	2 × 2
5	5 (prime)
6	2 × 3
7	

In the activity you found that every number up to 20 is either prime or can be made by multiplying together prime numbers. In fact, every whole number above 1 is either prime or can be made by multiplying together prime numbers.

This is known as the Fundamental Theorem of Arithmetic.

For example, $60 = 2 \times 2 \times 3 \times 5$
$ = 2^2 \times 3 \times 5$

> This is called a **product** of prime factors.
> Remember: product means multiply.

The next example shows you how to use a factor tree to help you write a number as a product of prime factors.

Worked example

Write 1540 as a product of prime factors.

Solution

Using a factor tree:

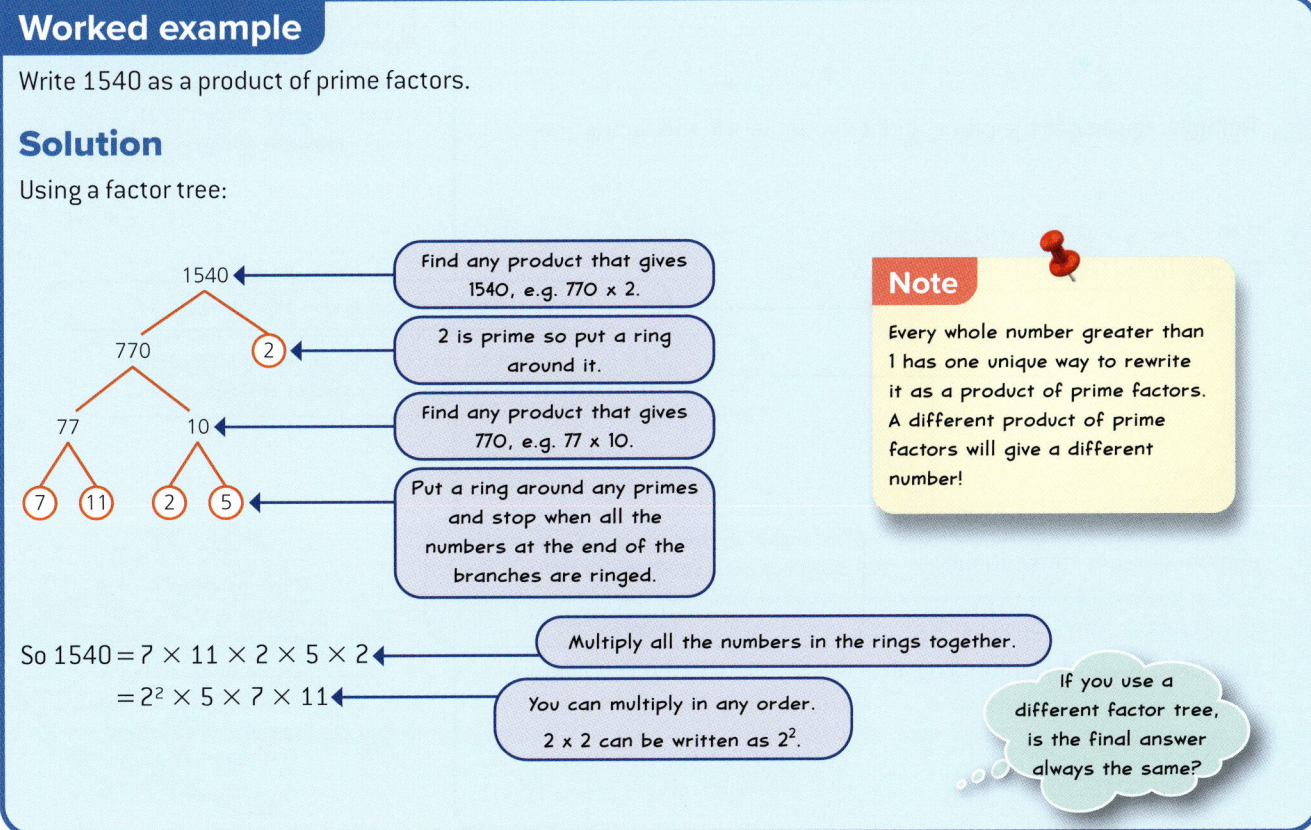

Find any product that gives 1540, e.g. 770 × 2.

2 is prime so put a ring around it.

Find any product that gives 770, e.g. 77 × 10.

Put a ring around any primes and stop when all the numbers at the end of the branches are ringed.

> **Note**
> Every whole number greater than 1 has one unique way to rewrite it as a product of prime factors. A different product of prime factors will give a different number!

So $1540 = 7 \times 11 \times 2 \times 5 \times 2$
$ = 2^2 \times 5 \times 7 \times 11$

Multiply all the numbers in the rings together.

You can multiply in any order. 2 × 2 can be written as 2^2.

If you use a different factor tree, is the final answer always the same?

▶ Finding the HCF and LCM of two numbers

The highest common factor (HCF) of two numbers is the largest factor that they share.
You can find the HCF of two numbers by listing their factors, as in this example for 20 and 30.

Factors of 20: 1 2 4 5 10 20
Factors of 30: 1 2 3 5 6 10 15 30

The highest number in both lists is 10 so this is the HCF of 20 and 30.

> The HCF is often quite a small number.

The lowest common multiple (LCM) of two numbers is the lowest multiple that they share.
You can find the LCM of two numbers by listing their multiples, as in this example, again using 20 and 30.

Multiples of 20: 20 40 60 80 100 …
Multiples of 30: 30 60 90 120 150 …

The lowest number in both lists is 60 so this is the LCM of 20 and 30.
The next example shows you how to use a Venn diagram to help you find the HCF and LCM of two numbers.

Worked example

a Show that 270 written as a product of prime factors is $2 \times 3^3 \times 5$.
b Write 315 as a product of prime factors.
c Find the HCF and LCM of 270 and 315.

Solution

a $2 \times 3^3 \times 5 = 2 \times 27 \times 5$
    $= 10 \times 27 = 270$

b
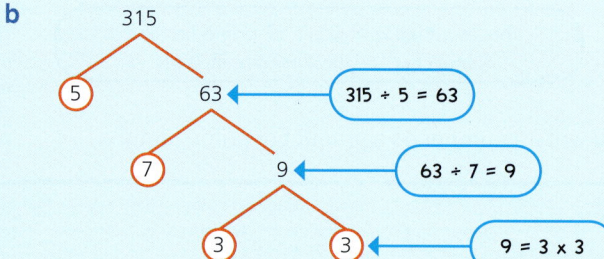

> 315 ÷ 5 = 63
> 63 ÷ 7 = 9
> 9 = 3 × 3

Multiply together all the prime (circled) numbers in the factor tree:
$315 = 5 \times 7 \times 3 \times 3$
$= 3^2 \times 5 \times 7$

c $270 = 2 \times 3 \times 3 \times 3 \times 5$ and $315 = 3 \times 3 \times 5 \times 7$
Placing these factors in a Venn diagram gives

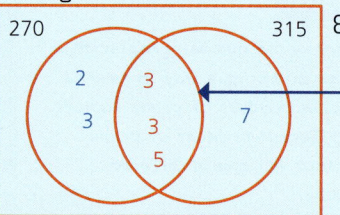

> The epsilon sign ε represents the universal set.

> 3, 3 and 5 are common factors so they go in the intersection. The intersection is the middle/crossover of the two circles.

The HCF is found by multiplying the numbers in the intersection:
$3 \times 3 \times 5 = 45$
So the HCF of 270 and 315 is 45.

> These are the common factors of both numbers so when you multiply them together you'll get the highest common factor.

The LCM is found by multiplying all of the numbers in the Venn diagram:
              315
$2 \times \underbrace{3 \times 3 \times 3 \times 5 \times 7}_{} = 1890$
    $\underbrace{}_{270}$

Remember

▶ Writing a number as a product of its prime factors means writing all of the prime factors as a multiplication.
▶ Use index notation to write a product of prime factors neatly.

Note

Venn diagrams are covered in full in chapter 15 (unit 15.2).

> Check: $1890 = 6 \times 315$ which means 1890 is a multiple of 315 and $1890 = 270 \times 7$ which means 1890 is also a multiple of 270. Can you see why this method works?

1.3 Now try these

Band 1 questions

1 Work out these.
  a $3 \times 5 \times 7$
  b $3 \times 7 \times 11$
  c $2 \times 3 \times 5$
  d $2 \times 3 \times 11$
  e 2^3
  f 2^4
  g $3^2 \times 5$
  h $2^2 \times 3$
  i $2^2 \times 5^2$

2 a What are the first ten multiples of 5?
  b What are the first ten multiples of 6?
  c What is the lowest common multiple (LCM) of 5 and 6?

3 a List the first eight multiples of 30.
  b List the first eight multiples of 24.
  c What is the lowest common multiple (LCM) of 30 and 24?

4 Find the lowest common multiple (LCM) of each pair of numbers.
  a 3 and 5
  b 4 and 6
  c 9 and 12

5 a What are the factors of 12?
  b What are the factors of 16?
  c What is the highest common factor (HCF) of 12 and 16?

6 Find the HCF of each pair of numbers.
  a 15 and 20
  b 18 and 24

Band 2 questions

7 a Copy and complete this diagram to find the prime factors of 18.
  b Write 18 as the product of prime factors.
  c Rewrite your answer using indices.

8 a Complete this list of the factors of 60.
       1, 2, 3, ..., 60
  b Write a list of the prime factors of 60.
  c Write 60 as a product of prime factors.

9 Draw factor trees to write these numbers as products of prime factors.
  a 12
  b 8
  c 15
  d 20
  e 30

10 Write each of these numbers as a product of its prime factors.
  Write your answers using indices.
  a 50
  b 140
  c 84
  d 36
  e 200

11 Find the **i** LCM and **ii** HCF of these pairs of numbers by first finding their prime factors and then using a Venn diagram.
  a 64 and 72
  b 20 and 35
  c 16 and 28

12 a Find the HCF of 90 and 360.
  b Find the LCM of 90 and 360.
  c What do you notice about your answers to parts **a** and **b**?
     Find another pair of numbers with this pattern.

13 a Find a number with prime factors of only 13, 17 and 19.
  b Find a number between 100 and 200 with prime factors that are all even.
  c Find a pair of numbers that have prime factors of only 2, 3 and 5.

Band 3 questions

14 a Show the prime factors of 24 and 90 on a Venn diagram.

b Use the Venn diagram to find the HCF and LCM of 24 and 90.

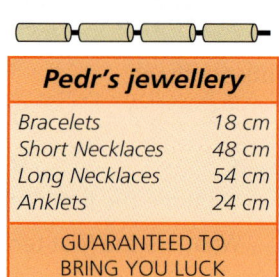

Pedr's jewellery

Bracelets	18 cm
Short Necklaces	48 cm
Long Necklaces	54 cm
Anklets	24 cm

GUARANTEED TO BRING YOU LUCK

15 Pedr makes rosewood jewellery.

Pieces of wood are joined together to make bracelets, necklaces and anklets.

All the pieces of wood are the same length.

Look at the poster.

What is the greatest possible length for one of the pieces of wood?

16 Find the LCM and HCF of 90, 75 and 60.

17 These dials are set at 0:

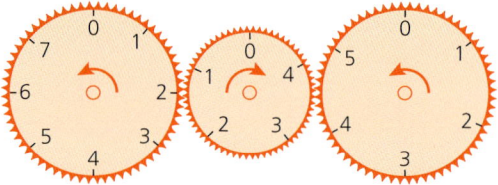

After the left dial has been turned through one complete turn they look like this:

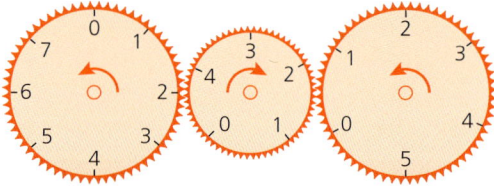

a Draw a diagram to show what they will look like after two complete turns of the left dial.

b How many complete turns of the left dial are needed before the first two dials are both set to 0?

c How many complete turns of the left dial are needed before all three dials are again set to 0?

18

BUSES AVAILABLE FROM THIS STOP

	1b	3a	4
Departing every	5 mins	8 mins	10 mins
First departure	9.00 a.m.	9.00 a.m.	9.00 a.m.

All three buses leave together at 9.00 a.m.

When is the next time that all three buses leave together?

Key words

Here is a list of the key words you met in this chapter.

Cube	Cube root	Factor	Highest common factor (HCF)
Indices	Lowest common multiple (LCM)	Multiple	Power
Prime	Product	Square	Square root
Standard form	Venn diagram		

Use the glossary at the back of this book to check any you are unsure about.

1 Powers and indices

Review exercise: powers and indices

Band 1 questions

1 Multiply out these.
  a $2^2 \times 3^2$
  b 3×5^2
  c 2×7^2
  d $2^3 \times 3^3$

2 Write 64 as a power of
  a 2
  b 4
  c 8
  d 64

3 Write the number in each statement as a power of 10.
  a There are 100 steps to the top of the tower.
  b The winner won by just 0.1 of a second.
  c The car costs £10 000.

4 In the number 4.12×10^6, the first digit has a value of 4 million or 4 000 000.
The numbers in the table are written in standard form.
Copy the tables and fill in the value of the first digit for each number.

a

Number	Value of first digit
6.1×10^4	
3.52×10^4	
2.9×10^7	
1.352×10^7	
4.5×10^9	
1.236×10^9	

b

Number	Value of first digit
2×10^{-2}	
1.46×10^{-2}	
3×10^{-4}	
6.2×10^{-4}	
5×10^{-6}	
3.21×10^{-6}	

5 Find the LCM and HCF of these pairs of numbers by first finding their prime factors.
  a 8 and 14
  b 30 and 35
  c 18 and 24

Band 2 questions

6 Find the values of these.
  a $(2^2)^3$
  b $2^2 \times 2^3$
  c $2^2 + 2^3$
  d $2^3 \times 2^3$
  e $2^3 \div 2^3$
  f $2^5 \div 2^3$
  g $2^{99} \div 2^{96}$
  h $\dfrac{2^5}{(2 \times 2 \times 2 \times 2 \times 2)}$

7 Find the LCM and HCF of these pairs of numbers by first finding their prime factors.
  a 80 and 100
  b 210 and 240

8 Work out the missing digits in these. There may be more than one answer.
  a $\square^2 = 25$
  b $\square\square^2 = \square\square 5$
  c $\square^3 = \square\square 6$
  d $\square^6 = \square^3$

Curriculum for Wales Mastering Mathematics: Book 3

Fluency

9 Write these numbers in standard form.
  a 2000
  b 32 000
  c 1450
  d 36 000 000

10 Write these numbers in standard form.
  a 0.067
  b 0.00341
  c 0.000 006
  d 0.23

11 Write these as ordinary numbers.
  a 2×10^3
  b 1.4×10^2
  c 4.56×10^4
  d 5.6×10^5
  e 3.576×10^{12}
  f 2.7×10^{-3}
  g 8.32×10^{-7}
  h 4.9×10^{-10}

Strategic competence

12 Write these numbers in standard form.
  a China has an estimated population of 1 393 000 000.
  b It takes 0.000 000 003 3 seconds for light to travel a distance of 1 metre.
  c The world's longest river, the Nile, is 6695 km long.
  d An amoeba is 0.0005 metres across.
  e The Star Wars films had a worldwide box office gross of £10 320 000 000.

Fluency

13 Write each expression as a single power of 3 or 5.
  a $3^4 \times 3^2$
  b $3^5 \div 3^2$
  c $(3^2)^4$
  d $5^2 \div 5^2$
  e $5^2 \div 5^4$
  f $5^2 \div 5^3$

Band 3 questions

Logical reasoning

14 Lowri has got all of her homework wrong.
The numbers are correct, but they should be written in standard form.
Correct Lowri's homework.

  a 100×10^5 ✗
  b 0.3×10^9 ✗
  c 0.5×10^{-3} ✗
  d 200×10^{-6} ✗

15 Write these sets of numbers in order of size, starting with the smallest.
  a 2.3×10^4 32 000 5.47×10^3 1.36×10^3 40 thousand
  b 4×10^{-5} 3.7×10^{-4} 1.8×10^{-4} 0.000 65   0.000 03

16 Nomsa thinks she has found a counter-example to the rule that any number has only one set of prime factors. She says,

> 'My number is 24 871. I can factorise it in two ways. It can be 209 × 119 or it can be 187 × 133. I have found an exception to the rule!'

Is Nomsa correct?

17 The prime factors of a cube number can be grouped in threes like this:
$216 = (2 \times 2 \times 2) \times (3 \times 3 \times 3)$ is the cube of (2×3).
  a $15 = 3 \times 5$. Write down the cube of 15.
  b What is the cube root of $3^3 \times 7^3$?

$64 = 2 \times 2 \times 2 \times 2 \times 2 \times 2 = 2^6$ is a square number and a cube number.
  c Show that its factors can be grouped in pairs and in threes.
  d Find two more numbers that are both square numbers and cube numbers.

In a motor race, the faster car completes the circuit in 5 minutes and the slower car takes 7 minutes.

  a How long is it before the faster car overtakes the slower car at the same time as they pass through the starting position?

  b How many laps has each car completed?

  c Does the faster car overtake the slower car anywhere else on the circuit?

  d If so, where does this happen and after how long?

  a You are given that $\dfrac{2^3 \times 2^{2n}}{2^5} = (2^2)^3$. *(Write each side as a single power of 2.)*

  Work out the value of n.

  b You are given that $3^m \times 3^{m+1} = (3^m)^3 \times 3^4$.

  Work out the value of m.

2 Fractions

Coming up...
- Understanding equivalent fractions
- Adding, subtracting, multiplying and dividing fractions
- Understanding mixed numbers
- Converting between mixed numbers and improper fractions

Fraction arithmagons

The number in each square is the SUM of the numbers in the circles adjacent to it.

Make a copy of each arithmagon and fill in the missing numbers.

①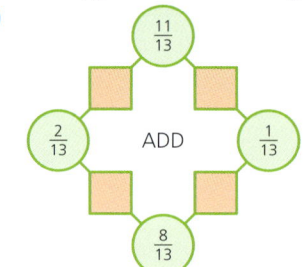

②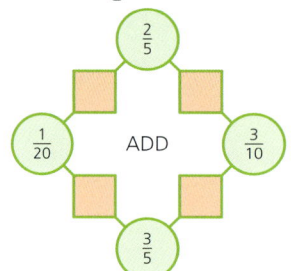

③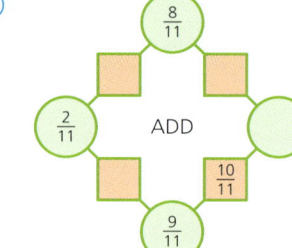

④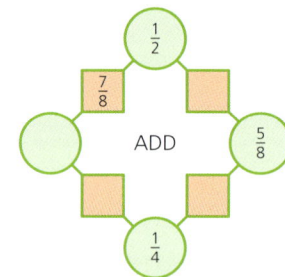

⑤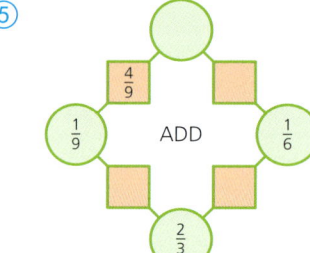

⑥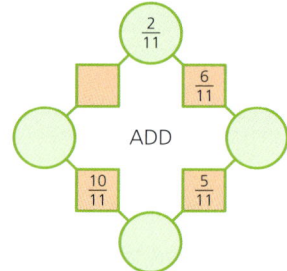

2.1 Fractions review

Skill checker

① Work out what fraction of this square is

   a   stripes

   b   orange

   c   white.

② Jamal has drawn this pie chart to show how he spends his pocket money.

Find what fraction of his money he spends on

a entertainment

b food

c clothes and travel.

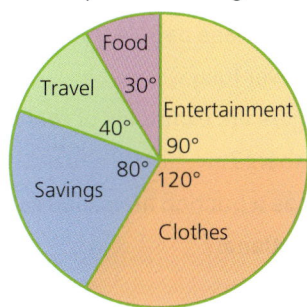

▶ Fractions to decimals

Remember you can change a fraction into a decimal by dividing the numerator (top) by the denominator (bottom).

For example, $\frac{3}{8} = 3 \div 8 = 0.375$

▶ Equivalent fractions

You can find **equivalent fractions** by multiplying or dividing **both the top and bottom** of a fraction by the same number.

So

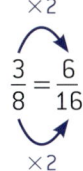

and

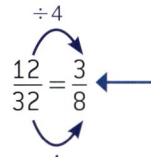

This is called simplifying or cancelling. You can't simplify $\frac{3}{8}$ any further, so you say $\frac{3}{8}$ is in its 'lowest terms' or 'simplest form'.

Note

Use your calculator to check that $6 \div 16 = 0.375$ and $12 \div 32 = 0.375$.

Equivalent fractions simplify to give the same fraction in its lowest terms.

For example

$$\overset{\div 5}{\underset{\div 5}{\frac{10}{15} = \frac{2}{3}}} \qquad \overset{\div 8}{\underset{\div 8}{\frac{16}{24} = \frac{2}{3}}}$$

$\frac{10}{15}$ and $\frac{16}{24}$ are equivalent as they both simplify to give $\frac{2}{3}$

You can use equivalent fractions to rewrite a decimal as a fraction.

For example, 0.375 means 375 thousandths which is $\frac{375}{1000}$

Simplifying gives $\frac{375}{1000} = \frac{75}{200} = \frac{15}{40} = \frac{3}{8}$

Activity

① What is a fraction? Look at what these people think.
Who is right? Explain your answer fully.

② Fractions, decimals and integers are examples of **real numbers**.
How many real numbers are there?

③ Numbers that can be written as a fraction are called **rational numbers**.
Which of these numbers are **rational**?

$$\pi \qquad 7 \qquad -\frac{1}{2} \qquad 1.25$$

Show how to write each number as a fraction.
How many rational numbers are there?

④ Berwyn says there are more positive integers than square numbers.
Is Berwyn correct?

> Amir: It is a type of number
>
> Carwyn: It is a part of a whole
>
> Hikaru: It is a number – all numbers are fractions
>
> Jaana: It is a decimal – every decimal is a fraction
>
> Lili: It is a proportion

▶ Adding and subtracting fractions

Worked example

Work out

a $\frac{3}{4} + \frac{1}{5}$

b $\frac{13}{18} - \frac{2}{3}$

Solution

a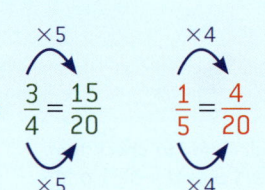

$$\frac{3}{4} + \frac{1}{5} = \frac{15}{20} + \frac{4}{20}$$

$$= \frac{19}{20}$$

b $\frac{13}{18}$

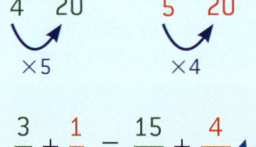

$$\frac{13}{18} - \frac{2}{3} = \frac{13}{18} - \frac{12}{18}$$

$$= \frac{1}{18}$$

> 20 is the lowest common denominator. It is the smallest number that is a multiple of both 4 and 5.

> 18 is a multiple of 3 so you don't need to rewrite both fractions to get the lowest common denominator. If you didn't spot this then you could rewrite both fractions with a common denominator of 3 × 18 = 54 and then simplify your answer.

Remember

You can only add or subtract fractions when they have the same denominator (bottom).

▶ Multiplying fractions

This number line shows that $6 \times \frac{2}{3} = 4$.

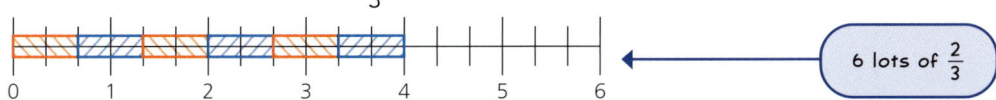

6 lots of $\frac{2}{3}$

Remember it doesn't matter which order you multiply so $\frac{2}{3} \times 6 = 4$.

The number line also shows that $\frac{2}{3}$ of $6 = 4$.

To shade $\frac{2}{3}$ of 6 you need to shade up to 4.

$6 \times \frac{2}{3}, \frac{2}{3} \times 6$ and $\frac{2}{3}$ of 6 are all different ways of writing the same thing.

You can also work out $\frac{2}{3}$ of 6 by finding $\frac{1}{3}$ of 6 and then doubling your answer.

$\frac{1}{3}$ of $6 = 2$

Remember: finding $\frac{1}{3}$ is the same as dividing by 3.

So $\frac{2}{3} \times 6 = 4$.

You can also multiply fractions together.

For example, $\frac{1}{4} \times \frac{3}{5}$ means $\frac{1}{4}$ of $\frac{3}{5}$.

In maths, 'of' means multiply.

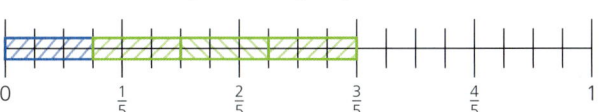

The number line shows that $\frac{1}{4}$ of $\frac{3}{5} = \frac{3}{20}$.

So $\frac{1}{4} \times \frac{3}{5} = \frac{3}{20}$.

Remember
To multiply fractions you multiply the numerators (tops) and then the denominators (bottoms).

Worked example

Find **a** $\frac{3}{7}$ of £84   **b** $\frac{3}{5} \times \frac{7}{9} \times \frac{10}{21}$

Solution

a $\frac{1}{7}$ of £84 = £84 ÷ 7 = £12

So $\frac{3}{7}$ of £84 = 3 × £12 = £36

This is the same as $\frac{3}{7} \times 84$ and $84 \times \frac{3}{7}$.

b **Method 1: Multiply then cancel**

$\frac{3}{5} \times \frac{7}{9} \times \frac{10}{21} = \frac{3 \times 7 \times 10}{5 \times 9 \times 21} = \frac{210}{945}$

You can simplify this fraction.

$\frac{210}{945} \xrightarrow{\div 5} \frac{42}{189} \xrightarrow{\div 3} \frac{14}{63} \xrightarrow{\div 7} \frac{2}{9}$

Method 2: Cancel then multiply

$$\frac{3}{5} \times \frac{7}{9} \times \frac{10}{21} = \frac{\cancel{3}^1}{\cancel{5}^1} \times \frac{\cancel{7}^1}{\cancel{9}^3} \times \frac{\cancel{10}^2}{\cancel{21}^3}$$

$$= \frac{1}{1} \times \frac{1}{3} \times \frac{2}{3}$$

$$= \frac{1 \times 1 \times 2}{1 \times 3 \times 3}$$

$$= \frac{2}{9}$$

Can you see how to cancel this in one step? Which method do you think is easier?

Multiplying by 3 and dividing by 9 is the same as dividing by 3.
Multiplying by 7 and then dividing by 21 is the same as dividing by 3.
Multiplying by 10 and then dividing by 5 is the same as multiplying by 2.

▶ Dividing fractions

Remember: $1 \div \frac{1}{5}$ means 'how many fifths go into 1?'

So $1 \div \frac{1}{5} = 5$ ◀ *Dividing by $\frac{1}{5}$ is the same as multiplying by 5.*

and so $6 \div \frac{1}{5} = 30$. ◀ *6 is six times bigger than 1, so the answer is 6 times bigger!*

$6 \div \frac{3}{5} = 10$ ◀ *$\frac{3}{5}$ is three times bigger than $\frac{1}{5}$, so the answer is one third the size.*

When you divide by $\frac{3}{5}$ you multiply by 5 (to divide by $\frac{1}{5}$) and then divide by 3.

So dividing by $\frac{3}{5}$ is the same as multiplying by $\frac{5}{3}$. ◀ $6 \times \frac{5}{3} = \frac{6 \times 5}{3} = \frac{30}{3} = 10$

$\frac{5}{3}$ is called the **reciprocal** of $\frac{3}{5}$.

Note

The reciprocal of a number is '1 ÷ the number' or '1 over the number'. When you have a fraction you just turn it upside down.

Worked example

Work out

a $8 \div \frac{2}{3}$

b $\frac{2}{5} \div \frac{6}{7}$

2   Fractions

Solution

a $8 \div \dfrac{2}{3} = 8 \times \dfrac{3}{2}$ ← Turn the 2nd fraction 'upside down' and multiply.

$= \dfrac{8 \times 3}{2}$

$= \dfrac{24}{2}$

$= 12$

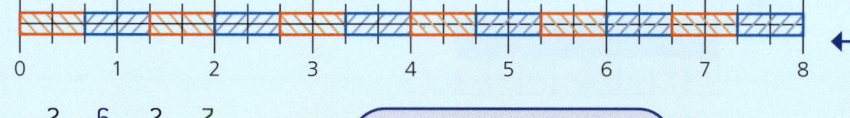

The diagram shows that $\dfrac{2}{3}$ 'goes into' 8 twelve times.

b $\dfrac{2}{5} \div \dfrac{6}{7} = \dfrac{2}{5} \times \dfrac{7}{6}$ ← Turn the 2nd fraction 'upside down' and multiply.

$= \dfrac{14}{30}$ ← You can simplify this: divide 'top' and 'bottom' by 2.

$= \dfrac{7}{15}$

Cross-curricular activity

In your Welsh lessons, find out how to say different fractions in Welsh. For example, 'half' is 'hanner' in Welsh.

How would you say a quarter?

How would you say a more complicated fraction, for example, seven elevenths $\left(\dfrac{7}{11}\right)$?

2.1 Now try these

Band 1 questions

1. Simplify these fractions.

   a $\dfrac{80}{90}$ b $\dfrac{5}{10}$ c $\dfrac{25}{100}$ d $\dfrac{8}{12}$ e $\dfrac{15}{20}$ f $\dfrac{7}{21}$

2. Write these fractions as decimals.

   The first one has been done for you.

   a $\dfrac{2}{5} = \dfrac{4}{10} = 0.4$ b $\dfrac{1}{5}$ c $\dfrac{1}{2}$ d $\dfrac{1}{4}$ e $\dfrac{3}{4}$ f $\dfrac{7}{8}$

3. Work out the following.

   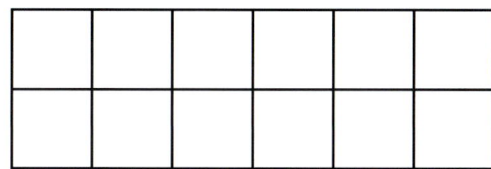

   Use the grid to help you.

   a $\dfrac{9}{12} - \dfrac{2}{12}$ b $\dfrac{1}{4} + \dfrac{1}{2}$ c $\dfrac{1}{6} + \dfrac{2}{3}$ d $\dfrac{1}{3} + \dfrac{1}{2}$ e $\dfrac{2}{3} - \dfrac{1}{2}$ f $\dfrac{11}{12} - \dfrac{1}{2}$

27

4 Work out these.

a $\dfrac{1}{5}$ of 20   b $\dfrac{1}{3} \times 12$ c $60 \times \dfrac{1}{10}$ d $\dfrac{1}{6} \times 30$ e $\dfrac{1}{7}$ of 63   f $72 \times \dfrac{1}{12}$

5 This bar model diagram illustrates $4 \div \dfrac{1}{2} = 8$

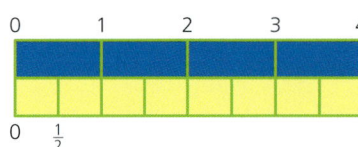

State the division represented by each of the following bar model diagrams.

a 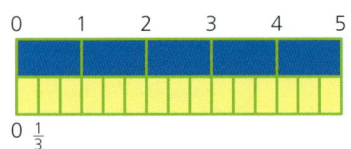   b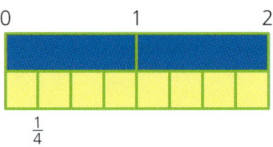

c    d

Band 2 questions

6 Work out these.

a $4 \div \dfrac{1}{2}$ b $3 \div \dfrac{1}{10}$ c $2 \div \dfrac{1}{6}$ d $3 \div \dfrac{1}{7}$ e $4 \div \dfrac{1}{5}$ f $10 \div \dfrac{1}{9}$

7 Write down the reciprocal of each of these.

a 8   b $\dfrac{1}{9}$ c $\dfrac{4}{7}$

8 Write these fractions in order of size, starting with the smallest.

$\dfrac{2}{5}$ $\dfrac{3}{10}$ $\dfrac{1}{2}$ $\dfrac{7}{25}$

9 Work out these.
Give each answer as a fraction in its lowest terms.

a $\dfrac{3}{7} + \dfrac{4}{9}$ b $\dfrac{5}{6} - \dfrac{2}{3}$ c $\dfrac{7}{9} - \dfrac{5}{18}$ d $\dfrac{1}{6} + \dfrac{1}{4}$ e $\dfrac{3}{8} + \dfrac{5}{12}$ f $\dfrac{2}{3} - \dfrac{2}{7}$

10 Work out these.
Give each answer as a fraction in its lowest terms.

a $\dfrac{1}{2} \times \dfrac{1}{5}$ b $\dfrac{1}{3} \times \dfrac{2}{7}$ c $\dfrac{2}{3} \times \dfrac{3}{8}$ d $\dfrac{5}{6} \times \dfrac{3}{4}$ e $\dfrac{3}{11} \times \dfrac{5}{6}$ f $\dfrac{4}{9} \times \dfrac{7}{12}$

11 Use a bar model diagram to help you work out these.

a $4 \div \dfrac{1}{3}$ b $6 \div \dfrac{2}{3}$ c $\dfrac{1}{2} \div \dfrac{1}{4}$ d $\dfrac{1}{6} \div 3$ e $\dfrac{1}{4} \div 5$ f $\dfrac{2}{5} \div 5$

12. Write each of these decimals as a fraction in its lowest terms.
    a 0.4
    b 0.95
    c 0.875

13. Jada, Sabra and Meena share £60 between them.
    Jada gets $\frac{1}{2}$ and Sabra gets $\frac{1}{3}$.
    a What fraction does Meena get?
    b How much money does Meena get?

Band 3 questions

14. Work out these.
    Give each answer as a fraction in its lowest terms.
    a $\frac{1}{5} \div \frac{1}{3}$
    b $\frac{1}{7} \div \frac{1}{2}$
    c $\frac{1}{6} \div \frac{2}{3}$
    d $\frac{2}{5} \div 4$
    e $\frac{4}{9} \div 3$
    f $\frac{7}{12} \div \frac{2}{3}$

15. There are 240 pupils in Year 9 at Hendy High.
    The students are surveyed to find out how they get to school.
    160 students walk to school.
    45 students catch the bus.
    $\frac{1}{12}$ of the students cycle.
    The rest of the pupils travel to school by car.
    Find what fraction of Year 9 pupils:
    a walk to school
    b catch the bus
    c travel by car.
    Give each of your answers as a fraction in its lowest terms.

16. Work out these.
    Give each answer as a fraction in its lowest terms.
    a $4 \times \frac{2}{5} - 5 \times \frac{1}{6}$
    b $\left(\frac{3}{5} - \frac{1}{4}\right)^2$
    c $\frac{8}{9} - \frac{2}{3} \times \frac{5}{6}$

17. Ioan is a fisherman.
    One day he lands 600 kg of fish.
    $\frac{2}{3}$ of this is flat fish.
    $\frac{4}{5}$ of the rest is codling.
    The remainder is a variety of other species.
    What weight of Ioan's fish were neither flatfish nor codling?

18. Nerys has read $\frac{5}{7}$ of her book.
    She has 100 pages left to read.
    How many pages has she read so far?

2.2 Mixed numbers

Skill checker

Complete each of these puzzles.

▶ The number in each circle is the sum of the numbers in the rectangles.
▶ The number in the triangle is the product of the numbers in the rectangles.

Remember

Product means multiply.

a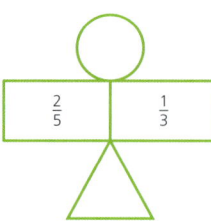
$\frac{2}{5}$ $\frac{1}{3}$

b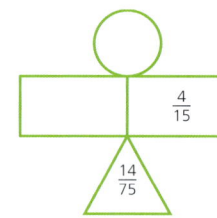
$\frac{19}{24}$ ... $\frac{5}{12}$

c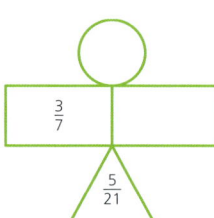
$\frac{3}{7}$ ... $\frac{5}{21}$

d
$\frac{4}{15}$ ... $\frac{14}{75}$

▶ Converting between mixed numbers and top-heavy fractions

Fractions can be used to represent numbers which are greater than 1.

This number line shows $\frac{5}{2}$.

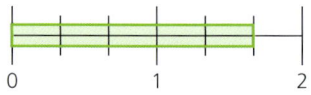

There are 2 halves in one whole.

So there are 4 halves in 2 wholes. 5 halves make 2 and a half.

This number line shows $\frac{5}{3}$.

There are 3 thirds in one whole. 5 thirds make 1 and 2 thirds.

Fractions like $\frac{5}{2}$ and $\frac{5}{3}$ are called **top heavy** or **improper fractions**.

They can also be written as **mixed numbers**.

A mixed number is a mix of a whole number and a fraction.

For example, $\frac{5}{2} = 2\frac{1}{2}$ and $\frac{5}{3} = 1\frac{2}{3}$

Every improper fraction can be written as a mixed number.

2 Fractions

Worked example

a Express $\dfrac{19}{5}$ as a mixed number.

b Express $2\dfrac{7}{8}$ as an improper fraction.

Solution

a You change $\dfrac{19}{5}$ to a mixed number by working out $19 \div 5$.

$19 \div 5 = 3$ remainder 4

So $\dfrac{19}{5} = 3\dfrac{4}{5}$.

> You can also split the improper fraction into two fractions: $\dfrac{19}{5} = \dfrac{15}{5} + \dfrac{4}{5} = 3 + \dfrac{4}{5} = 3\dfrac{4}{5}$.

b There are 16 eighths in 2 so $2 = \dfrac{16}{8}$

So $2\dfrac{7}{8} = \dfrac{16}{8} + \dfrac{7}{8}$

$= \dfrac{23}{8}$.

You can add and subtract mixed numbers by dealing with the whole numbers first or converting them to improper fractions.

Worked example

Work out:

a $2\dfrac{1}{3} + 1\dfrac{4}{5}$

b $2\dfrac{1}{3} - 1\dfrac{4}{5}$

Solution

Method 1 Working with the whole numbers and the fractions separately

First write the fraction parts with a common denominator

$\dfrac{1}{3} = \dfrac{5}{15}$ (×5) $\dfrac{4}{5} = \dfrac{12}{15}$ (×3)

a $2\dfrac{1}{3} + 1\dfrac{4}{5} = 2\dfrac{5}{15} + 1\dfrac{12}{15}$

$= 2 + 1 + \dfrac{5}{15} + \dfrac{12}{15}$

$= 3 + \dfrac{17}{15}$

> Write $\dfrac{17}{15}$ as a mixed number.

$= 3 + 1 + \dfrac{2}{15}$

$= 4\dfrac{2}{15}$

31

b $2\frac{1}{3} - 1\frac{4}{5} = 2\frac{5}{15} - 1\frac{12}{15}$

$= 2 - 1 + \frac{5}{15} - \frac{12}{15}$

$= 1 - \frac{7}{15}$

$= \frac{8}{15}$

Method 2: Converting to improper fractions

Start by writing $2\frac{1}{3}$ and $1\frac{4}{5}$ as top-heavy fractions.

$2\frac{1}{3} = \frac{6}{3} + \frac{1}{3} = \frac{7}{3}$ and $1\frac{4}{5} = \frac{5}{5} + \frac{4}{5} = \frac{9}{5}$

a $2\frac{1}{3} + 1\frac{4}{5} = \frac{7}{3} + \frac{9}{5}$

$= \frac{35}{15} + \frac{27}{15}$

$= \frac{62}{15}$ ← $62 \div 15 = 4$ remainder 2

$= \frac{60}{15} + \frac{2}{15}$

$= 4\frac{2}{15}$

Note
Rewrite the top-heavy fractions to equivalent fractions with the same denominator.

×5: $\frac{7}{3} = \frac{35}{15}$ ×5

×3: $\frac{9}{5} = \frac{27}{15}$ ×3

b $2\frac{1}{3} - 1\frac{4}{5} = \frac{7}{3} - \frac{9}{5}$

$= \frac{35}{15} - \frac{27}{15}$

$= \frac{8}{15}$

Which method do you prefer?

2.2 Now try these

Band 1 questions

1 The diagram shows the mixed number $1\frac{5}{6}$.

    a How many sixths are shaded in the diagram?

    b Write $1\frac{5}{6}$ as a top-heavy fraction.

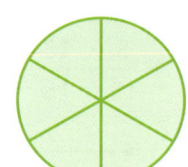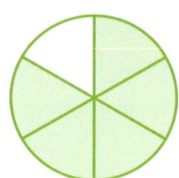

2   Fractions

2
a Colour a copy of this diagram to show the mixed number $1\frac{3}{4}$.
b Write $1\frac{3}{4}$ as a top-heavy fraction.

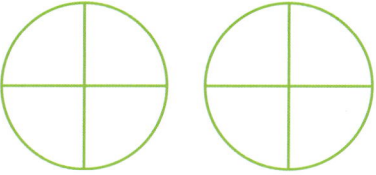

3 Colour a copy of these diagrams to show these fractions.

a $\frac{11}{8}$ b $\frac{16}{6}$

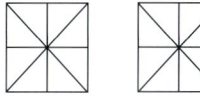   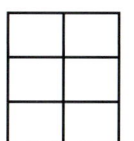   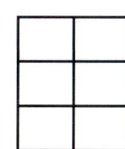   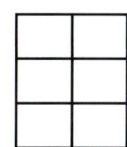

4 Find the missing number in each statement.

a $1 = \frac{\square}{3}$ b $1 = \frac{\square}{4}$ c $1 = \frac{\square}{5}$ d $2 = \frac{\square}{3}$ e $2 = \frac{\square}{4}$

f $2 = \frac{\square}{5}$ g $3 = \frac{\square}{3}$ h $3 = \frac{\square}{4}$ i $3 = \frac{\square}{5}$

5 Copy and complete these to change the mixed numbers to top-heavy fractions.

a $1\frac{1}{2} = \frac{\square}{2} + \frac{1}{2} = \frac{\square}{2}$ b $1\frac{1}{4} = \frac{\square}{4} + \frac{1}{4} = \frac{\square}{4}$

c $1\frac{2}{3} = \frac{\square}{3} + \frac{2}{3} = \frac{\square}{3}$ d $1\frac{5}{6} = \frac{\square}{6} + \frac{5}{6} = \frac{\square}{6}$

Band 2 questions

6 Write these improper fractions as mixed numbers.

a $\frac{8}{5} = 1\frac{\square}{5}$ b $\frac{7}{2} = \square\frac{1}{2}$ c $\frac{11}{9} = \square\frac{2}{\square}$ d $\frac{13}{8} = \square\frac{\square}{8}$ e $\frac{11}{3} = 3\frac{\square}{\square}$ f $\frac{17}{6} = \square\frac{5}{\square}$

7 Write these mixed numbers as improper fractions.

a $2\frac{3}{4} = \frac{\square}{4}$ b $6\frac{1}{2} = \frac{\square}{2}$ c $2\frac{3}{5} = \frac{\square}{5}$ d $2\frac{4}{7} = \frac{18}{\square}$ e $3\frac{1}{9} = \frac{\square}{9}$ f $3\frac{1}{8} = \frac{\square}{\square}$

8 Write these as mixed numbers.

a 80 minutes in hours   b 2325 grams in kilograms   c 24 days in weeks
d 420 centimetres in metres   e 500 seconds in minutes

9 Work out these.

Give your answers as mixed numbers.

a $\frac{2}{5} + \frac{4}{5}$ b $\frac{4}{5} + \frac{7}{10}$ c $\frac{1}{2} + \frac{2}{3}$ d $\frac{13}{4} + \frac{15}{7}$ e $\frac{15}{8} + \frac{5}{3}$ f $\frac{17}{6} + \frac{7}{4}$

10 Work out these.

Give your answers as mixed numbers where appropriate.

a $\frac{7}{4} - \frac{7}{8}$ b $\frac{5}{3} - \frac{5}{6}$ c $\frac{13}{8} - \frac{9}{16}$ d $\frac{5}{2} - \frac{5}{3}$ e $\frac{10}{3} - \frac{15}{8}$ f $\frac{13}{4} - \frac{19}{10}$

33

Fluency

11 Work out these.

Give your answers as mixed numbers where appropriate.

a $1\frac{1}{2} + 1\frac{7}{10}$ b $6\frac{2}{3} - 2\frac{7}{9}$ c $2\frac{1}{3} - \frac{7}{8}$ d $1\frac{1}{2} - \frac{9}{10}$ e $5\frac{1}{4} + 2\frac{7}{8}$ f $2\frac{1}{3} + 1\frac{7}{9}$

Strategic competence

12 Tomos needs $3\frac{1}{2}$ cups of dried fruit for a cake recipe.

He only has $\frac{3}{4}$ cup of sultanas and $1\frac{1}{3}$ cups of raisins.

He has plenty of currants.

How many cups of currants does he need to use?

Cross-curricular activity

Find a recipe for Welsh Cakes (or ask your Food Technology teacher to give you one).

If you were going to make, for example, 500 g of Welsh cake mixture, work out what fraction of each ingredient you would need.

Band 3 questions

Logical reasoning

13 Put these fractions into size order, smallest first.

$\frac{13}{6}$ $2\frac{1}{5}$ $1\frac{3}{4}$ $\frac{9}{5}$ $2\frac{1}{3}$

Fluency

14 Work out these.

Give your answers as mixed numbers.

a $1\frac{3}{4} + 3\frac{1}{2} - \frac{1}{4}$ b $5\frac{1}{3} - 1\frac{1}{5} + \frac{13}{15}$ c $3\frac{5}{6} + 4\frac{7}{8} - 1\frac{2}{3}$

d $4\frac{4}{5} - 1\frac{8}{9} - 1\frac{1}{9}$ e $3\frac{2}{3} + 1\frac{4}{5} + 2\frac{1}{2}$ f $2\frac{5}{8} + 3\frac{6}{7} - 4\frac{1}{2}$

Strategic competence

15 A children's play area needs to be fenced off from the local park.

The rectangle shows the dimensions of the play area.

Calculate the amount of fencing needed.

Give your answer as a mixed number.

$12\frac{3}{5}$ m

$5\frac{5}{6}$ m

16 A cross country race circuit is 2500 metres.

   a On Monday Ailsa runs 7000 metres.

      How many circuits is this? (Give your answer as a mixed number.)

   b On Tuesday Ailsa runs 9000 metres. How many circuits is this?

   c Add your answers to parts **a** and **b**.

   d How many metres does Ailsa run on Monday and Tuesday together? Convert your answer to a number of circuits.

Cross-curricular activity

Ask your PE teacher what the length of your school's cross country race circuit is.

What would the answers to Question 16 be if Ailsa ran the same distances on your school's cross country race circuit?

2.3 Multiplying and dividing mixed numbers

Skill checker

Here is Arwyn's maths homework.

1. $\dfrac{45}{7} = 6\dfrac{3}{7}$
2. $2\dfrac{3}{4} = \dfrac{9}{4}$
3. $\dfrac{5}{7} - \dfrac{2}{3} = \dfrac{3}{4}$
4. $\dfrac{\cancel{6}^2}{7} \times \dfrac{2}{\cancel{3}^3} = \dfrac{4}{21}$
5. $\dfrac{25}{9} = 2\dfrac{7}{9}$
6. $3\dfrac{9}{10} = \dfrac{39}{10}$
7. $\dfrac{44}{15} = 3\dfrac{1}{15}$
8. $\dfrac{2}{3} \div \dfrac{1}{2} = \dfrac{3}{2} \times \dfrac{1}{2} = \dfrac{3}{4}$
9. $\dfrac{33}{8} - \dfrac{9}{4} = 1\dfrac{7}{8}$
10. $3\dfrac{5}{11} = \dfrac{37}{11}$

Which questions has he got wrong?
What should the correct answers be?

▶ Calculations with mixed numbers

To multiply or divide mixed numbers, you need to start by rewriting each number as an improper fraction.

Worked example

Work out:

a. $2\dfrac{2}{5} \times 1\dfrac{1}{9}$

b. $2\dfrac{2}{5} \div 1\dfrac{1}{9}$

Solution

$2\dfrac{2}{5} = \dfrac{10}{5} + \dfrac{2}{5} = \dfrac{12}{5}$ and $1\dfrac{1}{9} = \dfrac{9}{9} + \dfrac{1}{9} = \dfrac{10}{9}$ ← Write $2\dfrac{2}{5}$ and $1\dfrac{1}{9}$ as top-heavy fractions.

a. **Method 1: Cancel and then multiply**

So $2\dfrac{2}{5} \times 1\dfrac{1}{9} = \dfrac{12}{5} \times \dfrac{10}{9}$

$= \dfrac{\cancel{12}^4}{\cancel{5}^1} \times \dfrac{\cancel{10}^2}{\cancel{9}^3}$ ← Cancel the common factors.

$= \dfrac{4}{1} \times \dfrac{2}{3} = \dfrac{8}{3}$

$= 2\dfrac{2}{3}$

Method 2: Multiply and then cancel

So $2\frac{2}{5} \times 1\frac{1}{9} = \frac{12}{5} \times \frac{10}{9}$

$= \frac{120}{45}$ ← Divide top and bottom by 15.

$= \frac{8}{3}$

$= 2\frac{2}{3}$

b $2\frac{2}{5} \div 1\frac{1}{9} = \frac{12}{5} \div \frac{10}{9}$ ← Flip the 2nd fraction and then multiply. $\frac{9}{10}$ is the reciprocal of $\frac{10}{9}$.

$= \frac{\cancel{12}^{6}}{5} \times \frac{9}{\cancel{10}_{5}}$ ← 2 is a factor of 12 and 10, so you can cancel.

$= \frac{6}{5} \times \frac{9}{5}$

$= \frac{54}{25}$

$= 2\frac{4}{25}$

2.3 Now try these

Band 1 questions

1 Find the reciprocal of each of these fractions.

   a $\frac{5}{2}$ **b** $\frac{7}{3}$ **c** $\frac{9}{4}$

2 Work out these.

   a **i** $\frac{1}{2} \times 6$ **ii** $\frac{3}{2} \times 6$ **iii** $6 \times \frac{3}{2}$

   b **i** $\frac{1}{5}$ of 10   **ii** $\frac{6}{5}$ of 10   **iii** $\frac{6}{5} \times 10$

   c **i** $24 \times \frac{1}{8}$ **ii** $24 \times \frac{11}{8}$ **iii** $\frac{11}{8} \times 24$

What do you notice about your answers?

3 Work out these.

   a **i** $1 \div \frac{1}{2}$ **ii** $5 \div \frac{1}{2}$ **iii** $5 \div \frac{3}{2}$

   b **i** $1 \div \frac{1}{3}$ **ii** $4 \div \frac{1}{3}$ **iii** $4 \div \frac{5}{3}$

   c **i** $1 \div \frac{1}{4}$ **ii** $2 \div \frac{1}{4}$ **iii** $2 \div \frac{5}{4}$

4
i Write each of these fractions as a top-heavy fraction.
ii Write down the reciprocal of each fraction.

a $2\frac{1}{2}$ b $1\frac{3}{4}$ c $2\frac{1}{4}$ d $4\frac{2}{3}$ e $3\frac{2}{5}$ f $2\frac{7}{8}$

5 Four people share $2\frac{1}{2}$ pizzas equally.

What fraction of a whole pizza do they get each?

Band 2 questions

6 Work out these.

Give your answers as mixed numbers where appropriate.

a $2\frac{1}{5} \times \frac{1}{3}$ b $1\frac{1}{4} \times \frac{4}{7}$ c $1\frac{4}{5} \times 1\frac{2}{3}$ d $2\frac{1}{7} \times 1\frac{3}{8}$ e $3\frac{7}{10} \times 2\frac{1}{2}$ f $1\frac{7}{8} \times 3\frac{4}{5}$

7 Work out these.

a $1\frac{3}{4} \div 2$ b $2\frac{3}{8} \div 4$ c $4\frac{2}{3} \div 5$

8 Work out these.

a i $1\frac{1}{4} \div 1\frac{1}{2}$ ii $1\frac{1}{2} \div 1\frac{1}{4}$ b i $2\frac{3}{5} \div 1\frac{1}{3}$ ii $1\frac{1}{3} \div 2\frac{3}{5}$

c i $4\frac{3}{4} \div 2\frac{1}{6}$ ii $2\frac{1}{6} \div 4\frac{3}{4}$

What do you notice about your answers?

9 Calculate the area of each of these shapes.

a 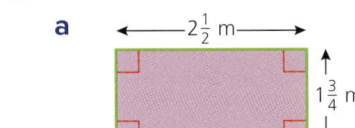 b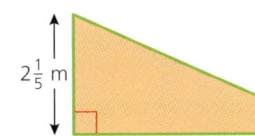

10 George paints $\frac{2}{5}$ of his garage in $1\frac{1}{2}$ hours.

How long would he take to paint:

a $\frac{1}{5}$ b $\frac{3}{5}$ c all of the garage?

Band 3 questions

11 Work out these.

Give your answers as mixed numbers where appropriate.

a $2\frac{1}{4} \times 3\frac{2}{5} \times \frac{2}{17}$ b $2\frac{3}{5} \times 2\frac{2}{3} \times 1\frac{2}{13}$ c $1\frac{3}{4} + 2\frac{1}{5} \times 3\frac{2}{3}$

d $8\frac{1}{2} - 4\frac{1}{3} \times 1\frac{7}{8}$ e $2\frac{1}{4} \times 1\frac{1}{5} + 2\frac{3}{10}$ f $2\frac{2}{3} \times 1\frac{3}{4} \times 2\frac{1}{2}$

g $1\frac{7}{8} \div 3\frac{1}{2} + 2\frac{1}{2}$ h $2\frac{3}{4} - 1\frac{5}{6} \div 2\frac{1}{3}$ i $3\frac{3}{4} \div 2\frac{2}{3} \div 1\frac{1}{2}$

Strategic competence

12 Sandor goes cycling on Friday, Saturday and Sunday.

He cycles $32\frac{3}{4}$ km altogether.

On Friday he cycles $5\frac{4}{5}$ km.

On Saturday he cycles three times as far as he did on Friday.

How far did Sandor cycle on Sunday?

13 Krisztina is a long-distance runner.

She runs at a steady pace of $7\frac{1}{2}$ miles each hour.

How far does she run in:

a $2\frac{1}{2}$ hours

b 3 hours 20 minutes?

Logical reasoning

14 Dai says that the reciprocal of 2 is $\frac{1}{2}$ and the reciprocal of $\frac{1}{3}$ is 3.

So the reciprocal of $2\frac{1}{3}$ is $3\frac{1}{2}$.

Is Dai correct? Explain your answer fully.

Strategic competence

15 The area of a rectangle is $10\frac{4}{5}$ m².

The width of the rectangle is $6\frac{1}{4}$ m.

Calculate the perimeter of the rectangle.

Key words

Here is a list of the key words you met in this chapter.

Denominator	Equivalent	Fraction	Improper fraction
Lowest terms	Mixed number	Numerator	Reciprocal

Use the glossary at the back of this book to check any you are unsure about.

Review exercise: fractions

Band 1 questions

1 Look at this table.

$\frac{2}{4}$	$\frac{8}{12}$	$\frac{9}{27}$	$\frac{5}{10}$
$\frac{2}{6}$	$\frac{15}{20}$	$\frac{6}{8}$	$\frac{3}{12}$
$\frac{6}{10}$	$\frac{9}{12}$	$\frac{12}{16}$	$\frac{3}{9}$
$\frac{7}{14}$	$\frac{11}{33}$	$\frac{4}{16}$	$\frac{14}{28}$

    a Write down all the fractions that are equivalent to:

        i $\frac{1}{2}$ **ii** $\frac{1}{3}$ **iii** $\frac{3}{4}$

    b Write all the other fractions in the table in their simplest form.

2 Say whether these statements are true or false.

    If false, change it to make a true statement.

    a 20 minutes is $\frac{1}{3}$ of an hour.   **b** 10 grams are $\frac{1}{10}$ of a kilogram.

    c 20 pence are $\frac{1}{5}$ of £1.   **d** 3 millimetres are $\frac{3}{10}$ of a centimetre.

    e 5 weeks are $\frac{1}{10}$ of a year.

3 Change these mixed numbers to top-heavy fractions.

    a $1\frac{4}{5}$ **b** $1\frac{2}{7}$ **c** $1\frac{4}{9}$

4 Mali is trying to limit her screen time to 3 hours a day.

    One day she uses her screen time like this.

Computer games	$\frac{1}{2}$ hour
Watching videos on the internet	$\frac{3}{4}$ hour
TV	$\frac{1}{2}$ hour
Social media	$\frac{5}{6}$ hour

    a How much screen time has she used up?

    b How much screen time does she have left? Give your answer in minutes.

Curriculum for Wales Mastering Mathematics: Book 3

Band 2 questions

5 Write each of these sets of fractions in order of size, starting with the smallest.

a $\dfrac{3}{4} \quad \dfrac{5}{6} \quad \dfrac{3}{7} \quad \dfrac{7}{10} \quad \dfrac{4}{5} \quad \dfrac{9}{10}$

b $\dfrac{9}{21} \quad \dfrac{40}{50} \quad \dfrac{1}{3} \quad \dfrac{2}{4} \quad \dfrac{15}{18}$

c $\dfrac{4}{8} \quad \dfrac{6}{10} \quad \dfrac{50}{60} \quad \dfrac{13}{39} \quad \dfrac{16}{20}$

6 Alun, Bryn and Cerys share £72 between them.

Alun gets £32, Bryn gets £18 and Cerys gets the rest.

What fraction does each get?

Write the fractions in their simplest form.

7 Work out these. Give your answers as mixed numbers where appropriate.

a $\dfrac{39}{10} + \dfrac{13}{8}$

b $\dfrac{9}{2} - \dfrac{16}{9}$

c $\dfrac{1}{2} + \dfrac{2}{3}$

8 Work out these. Give your answers as mixed numbers where appropriate.

a $5 \times 1\dfrac{1}{2}$

b $6 \times 2\dfrac{2}{3}$

c $5\dfrac{1}{4} \times 1\dfrac{1}{3}$

9 Jac can buy two different tins of sweets. He likes toffees best.

a In Tin A, he finds that $\dfrac{1}{5}$ are soft toffees and $\dfrac{1}{4}$ are hard toffees.

What fraction of Tin A are toffees?

b In Tin B, $\dfrac{3}{10}$ are truffles and $\dfrac{1}{4}$ are plain chocolate, the rest are toffees.

What fraction of Tin B are toffees?

c Tin A costs £5 and holds approximately 20 sweets. Tin B costs £3 and holds approximately 15 sweets. Which tin is the better buy?

10 Work these out. Give your answers as mixed numbers where appropriate.

a $1\dfrac{9}{10} + 2\dfrac{5}{6}$

b $8 - 5\dfrac{1}{3}$

c $6\dfrac{6}{7} + 3\dfrac{7}{10}$

d $5\dfrac{2}{7} - 2\dfrac{6}{11}$

e $2\dfrac{1}{3} - 1\dfrac{4}{5}$

f $2\dfrac{1}{4} - 1\dfrac{5}{6}$

g $3\dfrac{2}{5} - 2\dfrac{7}{8}$

h $4\dfrac{1}{2} - 1\dfrac{9}{10}$

i $3\dfrac{5}{8} + 2\dfrac{3}{4}$

Band 3 questions

11 Give your answers to these questions as mixed numbers.

a A glass contains 200 ml. How many glasses amount to 750 ml?

b Mair earns £9 per hour. How many hours would she need to work to make £100?

c Abdul takes 4 minutes to read one page. How many pages does he read in 15 minutes?

d A bag of sugar weighs 250 grams. How many bags are needed to get 1300 grams?

12 The area of a rectangle is $\frac{3}{10}$ m².

The width of the rectangle is $\frac{3}{4}$ m.

Work out the perimeter of the rectangle.

13 The formula for speed is

speed = distance ÷ time taken

Work out the speed of an object that travels

  a $\frac{4}{5}$ of a kilometre in $\frac{3}{4}$ of an hour   **b** $\frac{3}{10}$ of a kilometre in $\frac{4}{5}$ of an hour

  c $\frac{5}{8}$ of a kilometre in $\frac{1}{3}$ of an hour   **d**   6 kilometres in $\frac{1}{10}$ of an hour.

14   **a**   Catrin says that the square of $2\frac{3}{4}$ is $4\frac{9}{16}$.

       Is Catrin correct? Explain your answer fully.

   b   Work out these.

       i $\left(1\frac{2}{3}\right)^2$ **ii** $\left(2\frac{1}{4}\right)^2 - \left(1\frac{1}{2}\right)^2$ **iii** $\left(2\frac{1}{4} - 1\frac{1}{2}\right)^2$

   c   Show that $\sqrt{\frac{9}{16}} = \frac{3}{4}$.

   d   Work out these.

       i $\sqrt{\frac{1}{25}}$ **ii** $\sqrt{\frac{81}{100}} - \sqrt{\frac{16}{25}}$ **iii** $\sqrt{1\frac{11}{25}}$

15 A square has an area of $1\frac{9}{16}$ m².

Work out the perimeter of the square.

3 Accuracy

Coming up...
- Rounding using significant figures
- Approximating the answer to calculations
- Finding the limits of accuracy

Estimators

Play this estimation game with a partner.

The winner is the person who gave the better estimate in more rounds.

Round 1: How long?

Estimate the length of this line.

Round 2: Just a minute!

You'll need one stopwatch each – or take it in turns.

Start a stopwatch and shut your eyes.

Open your eyes and stop the clock when you think 1 minute has passed.

The person closer to a minute wins.

Round 3: Spider senses!

You have 10 seconds to estimate the number of 🕷 in this rectangle.

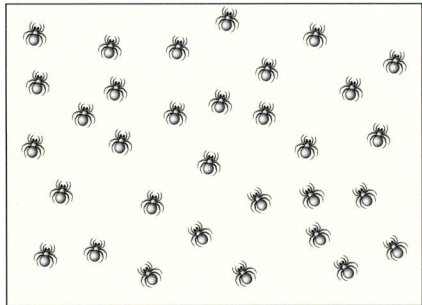

Count the spiders to see who was closer.

Round 4: How big?

Estimate the area of this triangle.

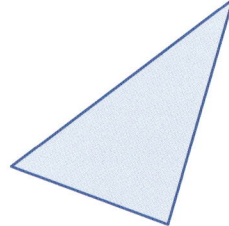

Measure the area to see who is closer.

Who won?
What strategies did you use?

3 Accuracy

3.1 Significant figures

Skill checker

① Round the numbers on these calculator displays to the nearest integer (whole number).
  a 345.67299
  b 127.94535
  c 4367.8437

② Round the numbers on these calculator displays to 1 decimal place.
  a 6.48279
  b 19.873245
  c 7.014255

③ Round the numbers on these calculator displays to 2 decimal places.
  a 1.8745597
  b 0.476589324
  c 0.8758758T

④ Write the numbers on these calculator displays to 3 decimal places.
  a 37.964375
  b 8.6749523
  c 90.090909
  d 0.6666666

▶ Rounding using significant figures

Look at this number.

 3 1 9 062

The digit that tells you the most about the size of the number is the 3 — it tells you the number is around 300 000. You say that 3 is the 1st significant figure. ← 1st s.f. or 1st sig. fig.

The next most significant digit is the '1' — that is the 2nd significant figure.

- The **1st significant figure** of a number is the **first non-zero digit** in that number (reading from left to right).
- The **2nd significant figure** is the digit (**including zero**) immediately to the right of the 1st significant figure and so on.

Significant figures are used in rounding. You are usually asked to round to 1, 2 or 3 significant figures.

Here is how you round to 3 significant figures.

- Underline the first 3 significant figures.
- Look at the next significant figure.
  ▶ If it is 4 or less, leave the 3rd significant figure as it is.
  ▶ If it is 5 or more, round the 3rd significant figure UP.
- Replace any other digits BEFORE the decimal point with a 0.

When you round you must still keep the size of the number. So in numbers greater than 10 it is important to fill in the zeros to the right of the significant figures so you can still see how big the unrounded number was.

319 062 = 300 000 to 1 s.f. ✓

319 062 = 3 to 1 s.f. ✗

Curriculum for Wales Mastering Mathematics: Book 3

In numbers less than 1, you need to keep the place-holding zeros but you don't need to replace the other digits with 0.

0.000 704 91 = 0.000 70 to 2 s.f. ✓

0.000 704 91 = 0.70 to 2 s.f. ✗

0.000 704 91 = 0.0007 to 2 s.f. ✗

When you round to 2 or more significant figures, then you must give 2 significant figures even if one of them is 0.

It's wrong because it only has 1 significant figure!

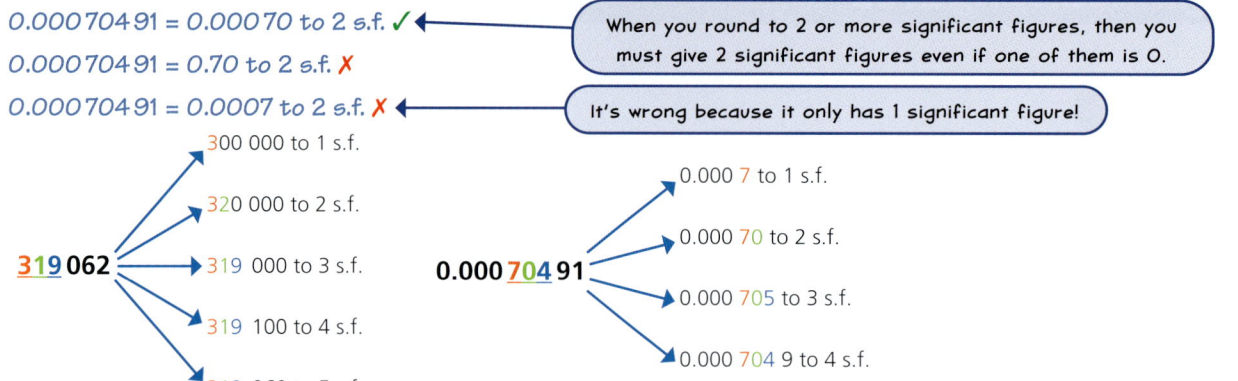

Discussion activity

① How should you round negative numbers?
Jamal and Chiara are rounding −15 to one significant figure.
Jamal says, '−15 should round up to −10, as the 5 means you round up'.
Chiara says, '15 rounds to 20, so −15 should round to −20 to mirror that.'
Who do you think is right? Why?

② Always rounding '0.5' up can cause a problem.
Estimate the answer to 4.5 + 6.5 + 7.5 + 9.5.
Jamal says, 'The answer is roughly 5 + 7 + 8 + 10 which is 30.'
Chiara says, 'I am going to use banker's rounding to get 4 + 6 + 8 + 10 which gives 28.'
What is the actual answer to 4.5 + 6.5 + 7.5 + 9.5?
What do you think the rule is for 'banker's rounding'?
Why did Chiara round some of the numbers down and others up?
What are the advantages and disadvantages of rounding in this way?

③ Investigate other methods for rounding.

3.1 Now try these

Band 1 questions

Fluency

① Write these numbers correct to 1 significant figure.
    a 43.34   **b** 372.12   **c** 8934   **d** 21 456   **e** 1.765

② **a** Copy these numbers and underline the first 2 significant figures in each of them.
    i 45.3   **ii** 12 706   **iii** 587 210   **iv** 408   **v** 301 999
   b Use your answers to part **a** to write each number correct to 2 significant figures.

③ **a** Copy these numbers and underline the first 3 significant figures in each of them.
    i 2745.3   **ii** 127 453   **iii** 50 087 210   **iv** 3002   **v** 7009
   b Write each of the numbers correct to 3 significant figures.

Logical reasoning

④ The first significant figure is underlined in red in each of these numbers.

    4136   12.8   0.523   0.000 063 2   4 200 000

Is the first significant figure always the first digit? Explain your answer.

44

3 Accuracy

Band 2 questions

5 Round these numbers to
  i the nearest 10   **ii** 2 significant figures.
  a 254   **b** 1785   **c** 21.35   **d** 103.67   **e** 458 361
  Are your answers the same? Why/why not?

6 Write these numbers correct to
  i 1 decimal place   **ii** 2 significant figures.
  a 34.43   **b** 12.372   **c** 1.567   **d** 0.432   **e** 0.2561
  Are your answers the same? Why/why not?

7 Write the number 324 860 correct to
  a 1 s.f.   **b** 2 s.f.   **c** 3 s.f.   **d** 4 s.f.   **e** 5 s.f.

8 Write the numbers on these calculator displays to 3 significant figures.
  a 65.9083461   **b** 108.275621   **c** 0.94860321

9 Write these numbers correct to the number of significant figures shown in brackets.
  a 0.003 45 **(1)**   **b** 0.000 456 8 **(2)**   **c** 0.100 04 **(3)**   **d** 0.000 000 899 **(1)**
  e 0.070 99 **(2)**   **f** 0.700 515 **(3)**

10 Use your calculator to do these calculations.
  Give your answers correct to 3 significant figures.
  a $56 \times 112 \div 37$ **b** $5.6 \times 11.2 \div 3.7$ **c** $0.56 \times 1.12 \div 37$

Band 3 questions

11 **a** Carys says that 3.98 rounded to 2 significant figures is 4.
        Explain why Carys is incorrect.
   b Round each of these numbers to the number of significant figures shown in brackets.
        i 19.9 **(2)**   **ii** 797 123 **(2)**   **iii** 0.8595 **(3)**   **iv** 0.003 598 **(3)**

12 Each of the statements **a** to **e** contains a mistake.
   Rewrite the statements correctly and explain the mistakes.
   a 354 = 35 correct to 2 significant figures.
   b The digit 4 is the second significant figure in the number 20 453.
   c 2135 = 2130 correct to 3 significant figures.
   d 0.4196 = 0.42 correct to 3 significant figures.
   e 0.23 is exactly the same as 0.230.

13 Here are five answers from calculator displays.
   3.4457982   3.4561207   3.5471035   3.5568156   3.4672331
   i Match each of them to three of the answers below.
   ii Then write each of the calculator answers to 3 significant figures.
   a 3.45 to 2 d.p.   **b** 3.6 to 1 d.p.   **c** 3.456 to 3 d.p.
   d 3.5 to 1 d.p.   **e** 3.557 to 3 d.p. **f** 3.47 to 2 d.p.
   g 3.46 to 2 d.p.   **h** 3.547 to 3 d.p. **i** 3.4 to 1 d.p.
   j 3.446 to 3 d.p. **k** 3.56 to 2 d.p.   **l** 3.467 to 3 d.p.
   m 3.55 to 2 d.p.

   Remember
   d.p. means decimal places

14 Write the number 0.999 999 correct to
   a 1 s.f.   **b** 2 s.f.   **c** 3 s.f.   **d** 4 s.f.   **e** 5 s.f.

45

Curriculum for Wales Mastering Mathematics: Book 3

3.2 Approximating

Skill checker

Aki asks you to check his homework for him.

Which questions has he got right?

Find and correct any mistakes that he has made.

1. $4 + 2 \times 5 = 30$
2. $2 \times 5^2 = 100$
3. $\dfrac{10-4}{8} \times 12 = 9$
4. $3^2 - 2^2 + 1^2 = 1$
5. $\sqrt{3^2 + 4^2} = 7$
6. $(2 + 5)(8 - 2 \times 3) = 14$
7. $\dfrac{1 - 2 + 3 - 4 + 5}{1 + 2 - 3 + 4 - 5} = 3$
8. $\dfrac{\sqrt{144}}{\sqrt[3]{27}} = 4$
9. $1 + \dfrac{3 \times 4 + 3}{1 + 2^2} = 3$
10. $\dfrac{10 - 4 + 3^2}{5} = 3$

▶ Using rounding

You can use rounding to 1 significant figure to get an estimate to a calculation without having to work out the exact answer. It is also a useful way to check that an answer is about right.

Worked example

Eliza and Albie work out the answer to

$$\dfrac{64.9 \times (43.3 + 7.7)}{30.4 - 19.1}$$

Eliza says that the answer is 298 to 3 significant figures.

Albie says that the answer is 91.7 to 3 significant figures.

Estimate the answer to the calculation.

Who is definitely wrong? Is the other person definitely right?

Solution

$$\dfrac{64.9 \times (43.3 + 7.7)}{30.4 - 19.1} \approx \dfrac{60 \times (40 + 10)}{30 - 20}$$

≈ means approximately equal to

$$\dfrac{60 \times (40 + 10)}{30 - 20} = \dfrac{60 \times 50}{10}$$

$$= 6 \times 50$$

$$= 300$$

So $\dfrac{64.9 \times (43.3 + 7.7)}{30.4 - 19.1} \approx 300$

Albie is definitely wrong.

Eliza's answer is about right, but the only way to tell for sure is to work out the calculation exactly.

3 Accuracy

Approximating is about
- rounding numbers for a calculation so that you can do it in your head
- making a rough calculation to anticipate what sort of answer to expect
- recognising when an error has been made
- giving a number to the level of accuracy that suits the context.

3.2 Now try these

Band 1 questions

1. Is each of these true or false? Explain your answer.
   a   76.3 is about 76.
   b   457.8 is about 48.
   c   141.2 is about 141.
   d   9607 is about 9600.
   e   47 132 is about 4713.
   f   191 is about 200.

2. A greengrocer has 203 kg of potatoes. Estimate how many 5 kg bags she can fill.

3. Haydn and three friends go out for a meal. They share the cost of the meal equally.
   Haydn has £20. Use estimation to check that he has enough money.
   Write down the calculations you used.

4 Seasons Pizzeria	
4 pizzas	£38
2 garlic bread	£12.50
2 side salads	£7
4 colas	£5.60

4. Here is a problem and some calculations.

   > There are 1230 pupils in the school. There are 47 tutor groups. Approximately how many pupils are in each tutor group?

   1230 × 47   1230 ÷ 47   1200 × 50   1200 ÷ 50   1215 ÷ 45

   a   Which two calculations will not help to solve the problem?
   b   Which two calculations give approximate solutions to the problem?
   c   Which calculation would you use?

5. Caron is reading a book that has 217 pages. She reads about 40 pages each evening.
   Roughly how many days will it take Caron to read the book?

Band 2 questions

6. a   There are 36 eggs in a tray. A box of eggs contains 12 trays of eggs.
      Approximately how many eggs are in a box?
   b   Approximately how many 62-seater coaches are needed to take a school of 1796 students on a trip?
   c   A bottle of cola contains 1950 ml. Approximately how many millilitres are there in 11 bottles?
   d   A bottle of cola contains 1950 ml. 205 ml are needed to fill a cup.
      Approximately how many cups can be filled from one bottle?
   e   A job pays £214 per week. Approximately how much is this in one year (52 weeks)?
   f   One pint of milk is sufficient for 22 cups of tea.
      Approximately how many pints are needed for 485 cups of tea?

7 One of the four numbers below is the correct answer for 59×81.

    4779   4803   4239   4885

An approximate answer is $60 \times 80 = 4800$.

Without using a calculator, choose the correct answer and explain your choice.

8 For each of these calculations, there is a choice of answers.
Use an approximation to help you select the correct answer.

  a 3.2×8.9
      i 284.8   **ii** 28.48   **iii** 54.8

  b $72 \div 12.5$
      i 9.45   **ii** 3.76   **iii** 5.76

  c $\dfrac{110 \times 94}{55}$
      i 188   **ii** 1880   **iii** 18.8

9 For each of the calculations below
  i estimate the answer
  ii work out the correct answer on your calculator
  iii compare your estimate with the correct answer.

  a $134 \div 8.9$ **b** 79×8.31 **c** $2.3 + 45.9 + 123.7$

  d $897 \div 19.25$ **e** 21.3×47.2 **f** $56 \times 72 \times 12$

  g $634 \div 3.9$ **h** $\dfrac{47.6 + 16.87}{6.9}$ **i** $\dfrac{87.6 - 11.25}{18.9}$

Band 3 questions

10 Alun works five days every week.
In a typical day he spends about 1 hour 40 minutes travelling to and from work.

  a Estimate the number of hours Alun spends travelling in a full working year of 48 weeks.

  b Approximately how many days a year does he spend travelling?

11 Efa has a part-time holiday job.
The table shows how long Efa worked at different rates of pay last week.

	Hours worked	Rate of pay
Normal pay	25 hours 20 minutes	£8.15
Overtime	4 hours 45 minutes	£9.60
Sunday	3 hours 10 minutes	£13.20

Which four of these calculations help to estimate Efa's earnings for the week?

  a $3 \times 13 = 39$ **b** $25 \times 8 = 200$ **c** $10 \times 13 = 130$ **d** $5 \times 10 = 50$

  e $8 \times 10 = 80$ **f** $26 + 5 + 3 = 33$ **g** $40 + 200 + 50 = 290$

12 In these calculations, each box represents an operation (+, −, × or ÷).
   Use estimation to find which operations to use to complete the calculations.
   a (37 ☐ 21) ☐ 223 = 1000
   b (756 ☐ 18) ☐ 29 = 1218
   c 27 ☐ (36 ☐ 18) = 675
   d 31 ☐ (87 ☐ 19) = 2108
   e 476 ☐ (2040 ☐ 24) = 391
   f (3461 ☐ 276) ☐ 101 = 37
   g (967 ☐ 34) ☐ (1023 ☐ 654) = 369 369
   h (29 ☐ 8²) ☐ 9 = 84
   i (619 ☐ 316) ☐ 425 ☐ 196 = 924
   j 6975 ☐ (36 ☐ 39) = 93

13 Look how Teleri estimates the answer to $\sqrt{37.35} \times \sqrt{78}$.

   $\sqrt{37.35} \times \sqrt{78} \approx \sqrt{36} \times \sqrt{81}$
   $\approx 6 \times 9$
   ≈ 54

   a Why did Teleri round the numbers to 36 and 81?
   b Use your calculator to work out $\sqrt{37.35} \times \sqrt{78}$ correct to 2 decimal places.
      Was Teleri's method accurate?
   c Use Teleri's method to estimate the answers to these calculations.
      i $\sqrt{15.6}$
      ii $\sqrt{8.9} \times \sqrt{4.3}$
      iii $\dfrac{\sqrt{103.5}}{\sqrt{38.9} + \sqrt{17.3}}$

3.3 Accuracy

Skill checker

Play a game of rounding '4 in a line'.

You will need
- four 10-sided dice
- 2 coloured pens.

Throw the dice and use the scores to give you a number between 1000 and 9999. If a number is less than 1000, throw again.

These dice show 5854

Round your number to one significant figure and place a counter in the matching box on the grid.

The winner is first the player to get a line of 4 counters.

The line can be horizontal, vertical or diagonal.

3000	2000	5000	6000	2000	4000
6000	0	7000	1000	9000	8000
8000	2000	9000	6000	3000	4000
3000	10 000	1000	9000	10 000	6000
4000	9000	5000	9000	2000	1000
10 000	1000	8000	2000	7000	5000

▶ Upper and lower bounds

The number of people at a concert was 24 000 correct to the nearest thousand.
How many people could have attended the concert?
You can use a number line to help you work this out

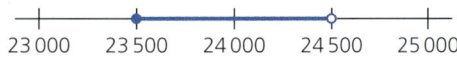

23 000   23 500   24 000   24 500   25 000

The smallest number that rounds up to 24 000 to the nearest thousand is 23 500. ← *23 499 rounds to 23 000 to the nearest thousand.*
24 500 rounds up to 25 000 to the nearest thousand.
So the maximum number of people at the concert is 24 499.
When you measure something, it is important to say how accurate your measurement was.

> **Discussion activity**
> What is the maximum capacity of the Principality Stadium in Cardiff? What would this be to the nearest 1000?

Kamari says the length of a line is 7 cm to the nearest centimetre.

The line could be as short as 6.5 cm as 6.5 cm rounds up to 7 cm to the nearest centimetre.
6.5 cm is called the **lower bound**. ← *6.4999 cm would round down to 6 cm.*

You can write this as an inequality using l for the length of the line:

$6.5 \leq l$ ← *l is greater than or equal to 6.5.*

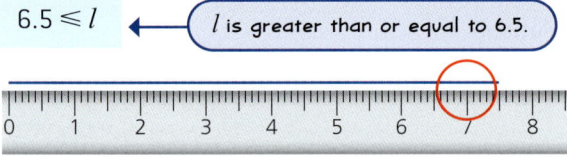

The line can't quite be as long as 7.5 cm but can get very, very close to it. ← *The line could be 7.499 999... cm long.*
So 7.5 cm is called the **upper bound**.
As an inequality this is written

$l < 7.5$ ← *l is less than 7.5.*

You can write the inequalities together like this:

$6.5 \leq l < 7.5$ ← *l is greater than or equal to 6.5 but is less than 7.5.*

Worked example

Seren draws a square with a side length of 21 cm to the nearest centimetre.

a What are the lower and upper bounds of the side length, s, of the square?

b What are the lower and upper bounds of the perimeter, P, of the square?

Give each of your answers as an inequality.

Solution

a The lower bound for s is 20.5 cm ← *20.5 is $\frac{1}{2}$ cm less than 21 cm.*
 The upper bound for s is 21.5 cm ← *21.5 is $\frac{1}{2}$ cm more than 21 cm.*

```
         20.5        21.5
    ├─────┼──────┼─────┤
    20          21          22
```

$20.5 \leq s < 21.5$

b Lower bound of perimeter = $4 \times 20.5 = 82$ cm
 Upper bound of perimeter = $4 \times 21.5 = 86$ cm
 $82 \leq P < 86$

3.3 Now try these

Band 1 questions

1. There are 80 people in a cinema to the nearest 10 people.
   What is the
   a smallest possible number of people at the cinema?
   b largest possible number of people at the cinema?

2. Cai buys some sweets for his younger sister's birthday party.
   He says, 'I have 400 sweets to the nearest 100 sweets.'
   What is the
   a smallest number of sweets that Cai could have?
   b largest number of sweets that Cai could have?

3. Write down the
   i lower bounds   ii upper bounds
   for the number that
   a rounds to 30 to the nearest 10
   b rounds to 1200 to the nearest 100
   c rounds to 3000 to the nearest 1000
   d rounds to 15 to the nearest integer (whole number).

4. 350 is accurate to 2 significant figures.
   a Explain why Dewi rounded 349 to 350 to 2 s.f.
   b Write down two other whole numbers smaller than 349 which will round to 350 to 2 s.f.

Band 2 questions

5. 2.56 is correct to 2 decimal places.
   a Explain why Dilys wrote 2.563 as 2.56 to 2 decimal places.
   b Write down two more numbers bigger than 2.563 that Dilys would still round to 2.56 to 2 decimal places.
   c Write down two numbers smaller than 2.56 which will round up to 2.56 to 2 decimal places.
   d What are the biggest and the smallest numbers that can round to 2.56 to 2 decimal places?

6. a Write these numbers correct to 2 significant figures.
      i   1.24   ii   1.237   iii 1.1652
      iv   1.151   v   1.249   vi   1.249 99
      What do you notice?
   b Kate measures the length of a line as 1.2 m correct to 2 significant figures.
      i   Write down the lower and upper bounds for the length of the line l.
          Lower bound = __ metres
          Upper bound = __ metres
      ii Complete the inequality for l.
          $\square \leq l < \square$

7. Bryn weighs 67 kg to the nearest kilogram.
   a Round these weights to the nearest kilogram.
      67.4 kg, 66.97 kg, 66.52 kg, 67.499 9 kg
   b Write down the smallest possible value for Bryn's weight.
   c Write down the largest possible value for Bryn's weight.
   d Write Bryn's weight, w, as an inequality.

8. A box of soap powder is labelled 4 kg.
   a The label is correct to the nearest kilogram.
      What are the minimum and the maximum amounts of soap powder in the box?
   b Write this as an inequality.

Curriculum for Wales Mastering Mathematics: Book 3

Fluency

9. A bottle of apple juice holds 75 cl to the nearest centilitre.
   a. What is the largest possible error when measuring to the nearest centilitre?
   b. Write down the smallest possible value for the volume of apple juice.
   c. Write down the largest possible value for the volume of apple juice.
   d. Write the volume of apple juice as an inequality.

10. 10 Swiss francs are worth £4.30 to the nearest 10 pence.
    a. What is the largest possible error when measuring to the nearest 10 pence?
    b. Write down the smallest possible value for 10 Swiss francs.
    c. Write down the largest possible value for 10 Swiss francs.
    d. Write the value of 10 Swiss francs in pounds as an inequality.

Logical reasoning

11. Explain why it might not be safe for these people to travel together in the lift.
    - Dafydd 65 kg
    - Bryn 92 kg
    - Bronwen 74 kg
    - Pat 54 kg
    - Pedr 86 kg
    - Elwyn 95 kg
    - Ahmed 89 kg
    - Marc 93 kg

    Lift
    Maximum safe load
    8 persons or
    650 kg

Band 3 questions

Strategic competence

12. The angles a and b are 100° and 140° (to the nearest degree).

    a. Find the upper and lower bounds of a and b.
    b. What is the smallest angle c can be?

13. The measurements of this photograph are accurate to the nearest millimetre.

    6.3 cm
    8.7 cm

    a. Write the upper and lower bounds of each measurement.
    b. Work out the smallest possible value for the perimeter of the photograph.
    c. Work out the largest possible value for the perimeter.
    d. Work out the smallest and largest values for the area of the photograph.

14. Carlos is designing a central heating flue duct to take waste fumes out through the wall.
    The hole for the duct has to be sealed.
    There is a 16.5 cm square plate with a 4.5 cm radius circle removed from it.
    The measurements are given accurate to 1 decimal place.
    a. Calculate the maximum and minimum areas of the square sheet of metal needed.
    b. Calculate the maximum and minimum areas of the circle to be removed.
    c. Calculate the maximum area of the metal seal (the metal remaining after the circle has been cut out).
    d. Write an inequality to show the upper and lower bounds of the area of the metal seal.

Key words

Here is a list of the key words you met in this chapter.

| Accuracy | Approximate | Decimal | Estimate |
| Lower bound | Rounding | Significant figure | Upper bound |

Use the glossary at the back of this book to check any you are unsure about.

Review exercise: accuracy

Band 1 questions

1. Zac does several calculations.
   For each one, he estimates the answer then uses his calculator to do the calculation accurately.
   Here are his results.
   a Rough 6, calculator 7.162
   b Rough 0.1, calculator 10.29
   c Rough 450, calculator 489.2
   d Rough 24, calculator 2.414
   e Rough 0.048, calculator 0.0511
   f Rough 17.5, calculator 84.23
   Which estimates suggest that an error has been made in using the calculator?

2. a Copy these numbers and underline the second significant figure in each of them.
   i 23.7   ii 1056   iii 0.437   iv 0.00257   v 759 218
   b Write each of the numbers in part **a** to 2 significant figures.

3. Write the number 17 032.9 correct to
   a 1 significant figure
   b 2 significant figures
   c 3 significant figures
   d 4 significant figures.

4. Ioan orders three portions of fish and chips, two pie and chips and a jumbo sausage.

Afonffordd Fisheries	
Fish and chips	£5.95
Pie and chips	£3.90
Jumbo sausage	£2.25

   a Round each price to the nearest pound.
   b Use your answers to estimate the cost of Ioan's order.
   c Ioan has £30. Is this enough?

5. 750 g is a measurement accurate to 2 significant figures.
   a Explain why 746 is 750 correct to 2 significant figures.
   b What is the smallest number that will give 750 when rounded to 2 significant figures?
   c Write down the largest number that will give 750 when rounded to 2 significant figures.

6. Glyn's height is being measured. He is 178 cm tall to the nearest centimetre.
   a What is the largest possible error when measuring to the nearest centimetre?
   b Write down the smallest possible value for Glyn's height.
   c Write down the largest possible value for Glyn's height.
   d Write Glyn's height as an inequality:
      $\square \leqslant$ Glyn's height $< \square$

Band 2 questions

7. Write these numbers correct to the number of significant figures shown in brackets.
   a 18.34 **(2)**   b 41.359 **(3)**   c 1246 **(3)**   d 0.015 **(1)**
   e 460 **(1)**   f 4986 **(2)**   g 0.0204 **(2)** h 0.1096 **(3)**

8 Ari saved regularly every week so that he had £189.50 in a year.
Approximately how much did he save each week?

9 Write each of these measurements as an inequality.

$\square \leq x < \square$

a 45 cm to the nearest cm
b 124 litres to the nearest litre
c 58 grams to the nearest gram
d 4.1 cm to the nearest mm
e 2.5 km to the nearest metre

10 At a Christmas market, 1794 entrance tickets were sold. The total amount of money made was £16 146.

a Estimate the average amount each person spent on entrance tickets.
b At last year's Christmas market, they sold 2057 tickets and made £14 399.
Estimate how much more a ticket to the Christmas market costs this year.

11 Quick checks by rounding the numbers will tell you that three of these answers are wrong.
Which three?

a $9.2 \times 17.5 = 161$
b $45.8 \div 17 = 2.694...$
c $28.9 \times 11.2 = 40.1$
d $3.2 \times 2.9 \times 19.1 = 177.2...$
e $\dfrac{29}{14.7} = 14.3$
f $\dfrac{13.2 \times 21.6}{49} = 58.18...$

Band 3 questions

12 Use your calculator to do these calculations.
Give your answers to the degree of accuracy shown in brackets.

a $(11.47 - 3.56)^2$ **(3 s.f.)**
b 3.7^2 **(3 s.f.)**
c $\pi \times 20.8^2$ **(2 s.f.)**
d 0.025^2 **(2 s.f.)**
e $(9.56 + 3.37)^2 \times (9.56 + 3.37)^2 \div 0.0102$ **(2 s.f.)**

13 This question is about the numbers 123.5678 and 0.123 567 8.

a Round these numbers to
    i   3 decimal places
    ii 3 significant figures.
b In part **a** which of the numbers gives the same answer?
c Write down another number which is the same when rounded to 3 decimal places and 3 significant figures.

14 Here are six calculations:

a $(49.7 \times 43) \div (12.6 - 7.4)$
b $\sqrt{8.7^2 + 7.8^2}$
c $(675.6 - 123.2) \div (17.8 + 12.3)$
d $\sqrt[3]{1.2^2 + 2.8^2 + 3.9^2}$
e $\pi \times 5.2^2$
f $\sqrt{\dfrac{314.16}{\pi}}$

The correct answers, to 1 decimal place, are given below but they have been mixed up.

| 2.9 | 18.4 | 411.0 | 84.9 | 10.0 | 4.1 |

Use estimation to match each answer to the correct calculation.

15 A paperback book is 2.6 cm thick, measured to the nearest millimetre, and contains 378 numbered pages.

a What is the maximum thickness of the book?
b What is the maximum thickness of one sheet of paper in millimetres?
Give your answer to 2 significant figures.

Take care! Sheets are not the same as pages.

Consolidation 1: Chapters 1–3

Band 1 questions

1. Write these fractions in their simplest form.

   a $\dfrac{40}{80}$ b $\dfrac{12}{16}$ c $\dfrac{25}{30}$ d $\dfrac{12}{18}$

   e $\dfrac{21}{35}$ f $\dfrac{24}{32}$ g $\dfrac{12}{20}$ h $\dfrac{240}{360}$

2. Write the correct inequality sign (< or >) between each pair of fractions.

   a $\dfrac{3}{4} \square \dfrac{7}{8}$ b $\dfrac{2}{3} \square \dfrac{5}{6}$ c $\dfrac{9}{20} \square \dfrac{3}{7}$ d $\dfrac{8}{9} \square \dfrac{4}{5}$

3. Copy and complete this cross-number puzzle.

   Across
   1 2^4
   4 7^2
   5 The square root of 25
   8 The cube root of 8
   9 11^3

   Down
   2 4^3
   3 3^1
   6 11^2
   7 The square root of 169

4. A bus arrives at a bus stop. It is already $\dfrac{3}{4}$ full.

   The number of people standing at the stop could fill $\dfrac{1}{3}$ of the bus.

   What fraction of a bus load are left at the bus stop?

5. Here are Eira's exam marks.

   a Copy and complete this statement.

   $\dfrac{10}{25} = \dfrac{\square}{100}$

   b Write $\dfrac{10}{25}$ as a percentage.

   c Write the rest of Eira's marks as percentages.

   d Which is Eira's
      i   best subject
      ii worst subject?

Subject	Mark
French (Ffrangeg)	10 out of 25
English (Saesneg)	15 out of 20
Science (Gwyddoniaeth)	21 out of 50
Maths (Mathemateg)	18 out of 40

Band 2 questions

6 Work out these.

Give each answer as a fraction in its lowest terms.

a $\frac{1}{4} + \frac{1}{5} + \frac{1}{10}$
b $\frac{1}{2} + \frac{1}{4} - \frac{5}{8}$
c $\frac{1}{4} - \frac{9}{16} + \frac{3}{8}$

7 Estimate the volume of this room.

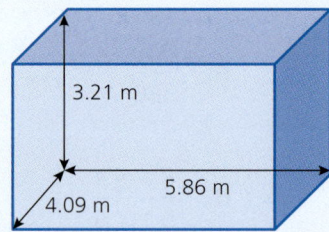

8 Estimate the number of minutes in a year. Give your answer to 1 significant figure.

9 Write each expression as a single power of 2.

a $2^5 \times 2^2$
b $2^8 \div 2^3$
c $(2^3)^2$
d $2^4 \div 2^4$
e $2^4 \div 2^6$
f $\frac{2^3 \times 2^5}{2^2}$

10 Work out these.

Give your answers as mixed numbers where appropriate.

a $\frac{17}{8} - \frac{11}{6}$
b $\frac{8}{3} + \frac{19}{5}$
c $\frac{27}{5} - \frac{19}{7}$

11 The answers below were found using a calculator. Write each of the answers

i correct to 2 significant figures
ii correct to 3 significant figures.

a 49.737 cm²
b £283 721
c 7.8241 cm
d 0.067 36 m
e 0.000 484 2
f 8.937 kg
g 10.785 m
h £37 694
i 40.038 cm³
j 0.706 83 m²
k 40.96 kg
l 20.81 litres
m 0.9008 km
n 5.942 m
o 10.94 cm²

12 To get to work, Irwen walks $\frac{2}{3}$ km to the bus stop.

She catches the bus to the train station, a distance of $3\frac{3}{4}$ km.

Her train journey is $14\frac{4}{5}$ km.

Finally, she walks $\frac{1}{2}$ km to her office.

What is the total length of Irwen's journey to work?

Give your answer as a mixed number.

13 Write the numbers in these statements in standard form.

a There are 1500 pupils in the school.
b This week's lottery jackpot is worth £3.6 million.
c The distance from Cardiff to New York is about 3000 miles.

14 Write these as ordinary numbers.

a A red blood cell has a diameter of 7×10^{-3} mm.
b The Gobi desert covers an area of 1.04×10^6 km².
c The total weight of fish in the world's oceans is estimated at 7.6×10^8 tonnes.
d The radius of a uranium atom is 8.68×10^{-15} metres.
e The Nou Camp stadium in Barcelona can hold 1.15×10^5 people.

Consolidation 1

Band 3 questions

15 Write down a number that lies between each of these pairs of numbers.
  a 2.3×10^4 and 2.3×10^6
  b 1.2×10^3 and 2.0×10^3

16 Calculate the area of each of these shapes.
  a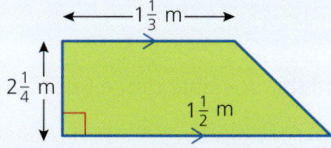
  b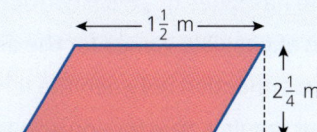

17 Sometimes the mixed number $3\frac{1}{7}$ is used as an approximation to π.
Use this value to work out the circumference and area of these circles without a calculator.
  a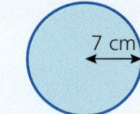
  b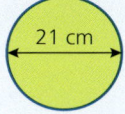
  c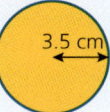

18 Some fractions and their equivalent decimal values are given below.
Write each decimal correct to 3 significant figures.

  a $\frac{2}{3} = 0.\dot{6}$
  b $\frac{5}{6} = 0.8\dot{3}$ — A dot over a digit shows that the digit recurs, e.g. $0.8\dot{3} = 0.833\,33...$
  c $\frac{5}{11} = 0.\dot{4}\dot{5}$
  d $\frac{5}{37} = 0.\dot{1}3\dot{5}$ — Two dots show that the digits between the dots recur, e.g. $0.\dot{1}3\dot{5} = 0.135\,135\,135...$
  e $\frac{5}{12} = 0.416$
  f $\frac{5}{7} = 0.\dot{7}1428\dot{5}$
  g $\frac{11}{12} = 0.916$
  h $\frac{503}{999} = 0.\dot{5}0\dot{3}$
  i $\frac{19}{27} = 0.\dot{7}0\dot{9}$

19 You are told that $T = \dfrac{7.8^2 + 5.79^2}{17.67}$
  a Estimate an approximate value for T by rounding all the numbers to 1 significant figure.
  b Use your calculator to find the true value of T correct to 3 significant figures.

20 A megabyte is approximately one million bytes.
A gigabyte is approximately 1000 megabytes.
A terabyte is approximately 1000 gigabytes.
Write down approximately how many bytes are in a terabyte.
Give your answer in standard form.

21 Simplify these algebraic expressions.
The first one has been done for you.
  a $\frac{x}{3} + \frac{x}{2} = \frac{2x}{6} + \frac{3x}{6} = \frac{5x}{6}$
  b $\frac{m}{4} + \frac{m}{5}$
  c $\frac{d}{2} + \frac{2d}{7}$
  d $\frac{x}{5} + \frac{2x}{9}$
  e $\frac{e}{3} - \frac{e}{5}$
  f $\frac{5h}{6} - \frac{h}{4}$

22 Each side of a piece of square card is 21 cm correct to the nearest centimetre.

This means that each side could be anywhere between 20.5 cm and 21.5 cm.

The area of the square could be:
- largest possible area = 21.5 × 21.5 = 462.25 cm²
- smallest possible area = 20.5 × 20.5 = 420.25 cm²

So the range of area for this square is 462.25 − 420.25 = 42 cm².

a Another square has sides of 16 cm correct to the nearest cm.

Find the range of possible values for the area of this square.

b Can you find a connection between the length of the side of a square and the possible range for its area?

c Prove that your result is true for any square.

4 Percentages

Coming up...
- Percentages review
- Using multipliers
- Reverse percentages
- Appreciation and depreciation

Find the magic number

See if you can find the magic number in the red square.
After following all of the clues you will get back to the magic number.
Be careful: there is only one number that works!

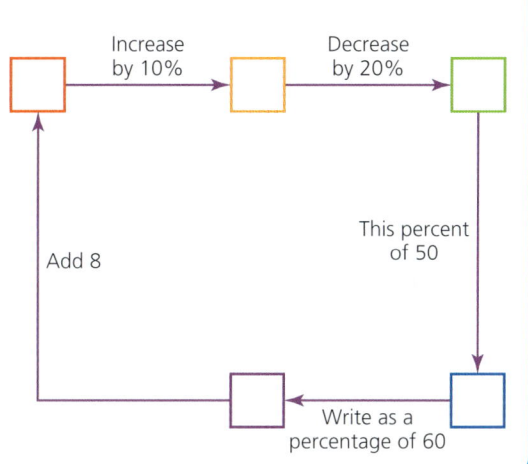

4.1 Percentages review

Skill checker

Look at the grid.
- a What percentage of the squares are coloured blue (the plain squares)?
- b What percentage of the numbers are prime numbers?
- c What percentage of the blue (plain) squares are square numbers?
- d What percentage of the numbers in the first two rows are even numbers?
- e What percentage of the squares containing square numbers are coloured in the yellow and black diagonal stripes?
- f Can you think up some of your own questions?

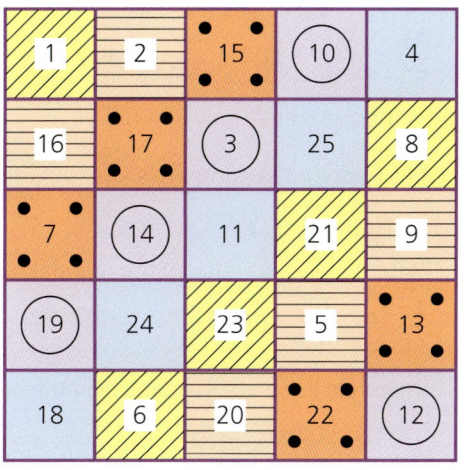

▶ Percentages recap

Recall the important skills you have learnt relating to percentages:
- Percentages add up to 100%.
- Finding a percentage of a quantity, for example 20% of £300.
- Writing one number as a percentage of another, for example writing 6 cm as a percentage of 10 cm.
- Converting between fractions, percentages and decimals.
- How to calculate percentage increases and decreases.
- How to calculate percentage change.

Worked example

There are fifty seats in the school canteen. At 12:15 the canteen is full.
At 12:30, 20% of pupils leave the canteen.

a What percentage of the pupils remain in the canteen?
b How many pupils left the canteen?
c How many pupils remain in the canteen?
d At 12:45 five more pupils enter the canteen. How full is the canteen now? Give your answer as a percentage.
e Write your answer to part **d** as a fraction and as a decimal.

At 1:00 the canteen is completely full.
Mr Griffiths, who is on duty, finds an extra 15 seats.

f What is the percentage increase in the number of seats available?
g Write the new number of seats as a percentage of the original number.

Solution

a If 20% leave, 80% remain in the canteen.
b 20% of 50 is $\frac{1}{5}$ of 50, which is 10 pupils who left the canteen.
c 80% of 50 is $\frac{4}{5}$ of 50, which is 40 pupils who remain.
d $40 + 5 = 45$
   The canteen is now $\frac{45}{50}$ full.
   $\frac{45}{50} = \frac{90}{100} = 90\%$
   So the canteen is now 90% full.
e As a fraction the canteen is $\frac{45}{50}$ full, which simplifies to $\frac{9}{10}$ full. As a decimal this is 0.9.
f The number of seats increases by 15.
   Write this as a fraction of the original number and convert to a percentage:
   $\frac{15}{50} = \frac{30}{100} = 30\%$
   There has been a 30% increase in the number of seats.
g There are now 65 seats.
   Write this as a fraction of the original number:
   $\frac{65}{50} = \frac{130}{100} = 130\%$

4 Percentages

Activity

This is a game for two players.

Player 1 starts with two pots of marbles, each containing 10 marbles.

Player 2 does not watch what Player 1 is doing.

Player 1 makes four moves and describes them, making at least two moves that involve percentages.

For example:

a   I'm transferring 10% of the marbles from pot 1 to pot 2.

b   I'm transferring three marbles from pot 2 to pot 1.

c   I'm transferring 25% of the marbles from pot 1 to pot 2.

d   I'm removing one third of the marbles from pot 1.

Player 2 must try to keep track of how many marbles are in each pot.

After these four moves, if Player 2 gives the correct number of marbles in each pot, they win the game. Otherwise Player 1 wins.

Swap positions and play again.

4.1 Now try these

You can use a calculator in this exercise, except where you see the non-calculator symbol.

Band 1 questions

1. Find these:
   a $22\frac{1}{2}$% of 80 m²
   b   5% of 56 mm
   c $7\frac{1}{2}$% of 80 km
   d $37\frac{1}{2}$% of 48 cm
   e $62\frac{1}{2}$% of 8 mm
   f   3% of 64 cm²
   g   74% of 200 kg
   h   85% of 16 g

2. Betsan's season ticket for the train costs £342 a quarter.
   Fares are to increase by 5.5%.
   A quarter means a quarter of a year, which is 3 months.
   a   How much more will Betsan's season ticket cost?
   b   How much will Betsan pay per quarter after the increase?

3. Last year the amount Teilo paid for his car insurance was £420.
   This year Teilo's insurance premium will increase by 12%.
   How much will Teilo pay this year?

4. A mathematics book has 260 pages, of which 35% are on algebra, 25% are on geometry and the remainder are on arithmetic.
   How many pages are there on each topic?

Curriculum for Wales Mastering Mathematics: Book 3

Band 2 questions

5 Copy and complete this table.
Give all fractions in their simplest form.
The first row has been done for you.

Fraction	Decimal	Percentage
$\frac{19}{20}$	0.95	95%
$\frac{9}{10}$		
	0.65	
		24%
$\frac{31}{40}$		
	0.675	
		42.5%
$\frac{6}{5}$		
	2.3	
		180%

6 Alfredo runs a pizza restaurant.
Customers leave feedback about the restaurant on a website.
The scores are shown below.

> **Restaurant Feedback**
> Quality of food 19 out of 20
> Service 8 out of 10
> Cleanliness 17 out of 20

   a Write all three scores as percentages.
       i Quality of food   **ii** Service   **iii** Cleanliness
   b What is the restaurant's total score out of 50?
   c Write the total score as a percentage.

7 **a** Copy and complete the following to work out 15% of £3900.
      100% is £3900
      1% is £3900 ÷ 100 = £ _____
      15% is 15 × £ _____ = £ _____

   b Now copy and complete these:
       i An increase of 15% gives £3900 + £ _____ = £ _____
      ii A decrease of 15% gives £3900 − £ _____ = £ _____
   c Steffan thinks it would be easier to use a different method.
      Copy and complete Steffan's working.
      10% is £3900 ÷ 10 = £ _____
      5% is £ _____ ÷ 2 = £ _____
      15% is £ _____ + £ _____ = £ _____

4 Percentages

8 a Copy and complete this table to find 10%, 5% and 1% of £270.

Percentage	100%	10%	5%	1%
Amount	£270			

b Now copy and complete the following:
  i 15% of £270 is £___ + £___ = £___
  ii 40% of £270 is 4 × £___ = £___
  iii 16% of £270 is £___ + £___ + £___ = £___
  iv 35% of £270 is 3 × £_____ + £_____ = £_____
  v 27% of £270 is _____
  vi 110% of £270 is _____

9 Alok's phone bill will be increased by 8% from the first quarter of the year to the second quarter.

His phone bill was £64.50 for the first quarter.

How much will Alok pay for his phone bill in the second quarter?

A quarter means a quarter of a year, which is 3 months.

10 Twenty four percent of a field is sold to build houses.

If the field was originally 15 acres in size, how large is it after the sale?

An acre is a measure of area, roughly equal to 4000 m².

Band 3 questions

11 What percentage of each shape is coloured?

a 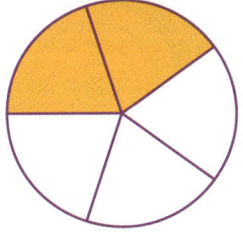 b 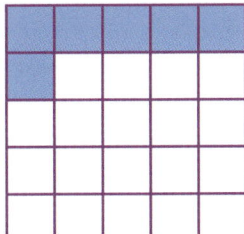 c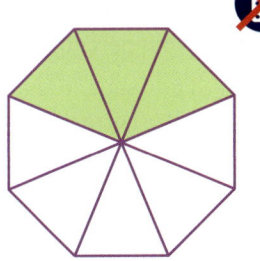

d 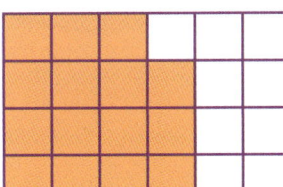 e  f

g 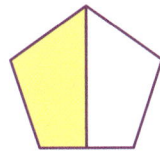 h  i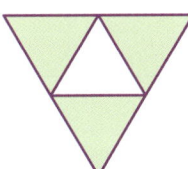

12 The interest rate on Sahar's savings account is 2.5%.

If she invests £300, how much simple interest will she receive

a in one year
b in five years?

You learnt about simple interest in Book 2 Chapter 12

63

13 Jadir and Carys did a survey about which British cheese people preferred.

The results are shown in the pie chart below, but one of the numbers has been lost.

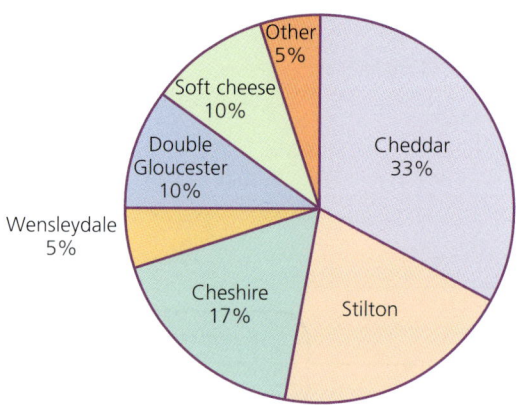

a   What percentage said they preferred Stilton?

b   Jadir and Carys asked 180 people altogether.

How many of them said they preferred

  i   Double Gloucester?   ii   Wensleydale?

14 Carlo is practising his archery.

He fires five arrows, labelled A to E on the diagram.

a   Look at where the arrows landed and calculate his total score.

b   What is the maximum score with five arrows?

c   What percentage of the maximum did Carlo get?

d   Carlo says, 'If I hadn't fired arrow C, my percentage would have been 70%!'
Is he correct? Explain your answer.

15 The bar chart shows the number of hours Ffion worked each day one week.

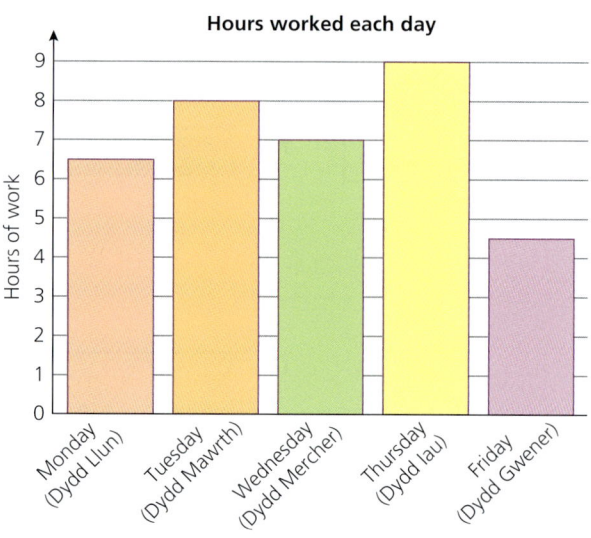

a   How many hours did Ffion work in total this week?

b   What percentage of those hours were on Wednesday?

c   Ffion worked fewer hours on Friday than she did on Thursday.
What was the percentage decrease from Thursday to Friday?

d   What was the percentage increase in Ffion's hours from Wednesday to Thursday?

e   Ffion had an important deadline to meet this week.
She had to write a report for her manager by _____ morning.
On which morning do you think she had to deliver her report?

4 Percentages

4.2 Using multipliers

Skill checker

Find the new price for these items.
Copy and complete the table.

Item	Old price	Percentage change	New price
Bag of flour	£1	15% increase	
Scarf	£8.50	20% decrease	
Jacket	£49.50	10% off in the sale	
Blueberries	£2.90	Reduced by 30%	
Skirt	£36	Price going up by 25%	
Jeans	£45	Price cut by 40%	
Tin foil	£1.80	5% added to price	

▶ Percentage increase using a multiplier

Worked example

There are three ways to work out percentage increases and decreases.
All three methods are shown below for this question:
Find the amount when £120 is

a increased by 15%
b decreased by 15%.

Solution

Method 1. Finding the increase or decrease
You have used this method in Book 2.
100% is £120
1% is £120 ÷ 100 = £1.20
15% is 15 × £1.20 = £18

a An increase of 15% gives £120 + £18 = £138
b A decrease of 15% gives £120 − £18 = £102

Method 2. Using a table
You have used this method in Book 2.

Percentage	100%	10%	5%	15%
Amount	£120	£120 ÷ 10 = £12	£12 ÷ 2 = £6	£12 + £6 = £18

a An increase of 15% gives £120 + £18 = £138
b A decrease of 15% gives £120 − £18 = £102

Curriculum for Wales Mastering Mathematics: Book 3

Conceptual understanding

Method 3. Using a multiplier

This is a calculator method that you may not have learnt before.

It is usually quicker than the other two methods.

a To increase by 15% multiply by 1.15

£120 × 1.15 = £138

> Add 15 to 100 and divide by 100
> $$\frac{100 + 15}{100} = 1.15$$

b To decrease by 15% multiply by 0.85

£120 × 0.85 = £102

> Why does this work?
> Increasing a quantity by 15% is the same as finding 115% of the quantity.
> To find 115%, multiply by $\frac{115}{100}$ or 1.15.

> Why does this work?
> Decreasing a quantity by 15% is the same as finding 85% of the quantity.
> To find 85%, multiply by $\frac{85}{100}$ or 0.85.

> Subtract 15 from 100 and divide by 100
> $$\frac{100 - 15}{100} = 0.85$$

Worked example

Using a multiplier

a increase £48 by 25%
b decrease £48 by 25%
c increase 150 kg by 5%
d decrease 150 kg by 5%
e increase 15 cm by $33\frac{1}{3}$%
f decrease 15 cm by $33\frac{1}{3}$%
g increase 6 km by 100%
h increase 150 sandwiches by 200%
i increase 500 litres by 250%

Solution

a To increase £48 by 25%, multiply by 1.25
   £48 × 1.25 = £60

b To decrease £48 by 25%, multiply by 0.75
   £48 × 0.75 = £36

c To increase 150 kg by 5%, multiply by 1.05
   150 × 1.05 = 157.5 kg

d To decrease 150 kg by 5%, multiply by 0.95
   150 × 0.95 = 142.5 kg

e To increase 15 cm by $33\frac{1}{3}$%
   $33\frac{1}{3}$% = 33.333...%
   So the multiplier is 1.33333... or $1\frac{1}{3}$.
   $15 \times 1.\dot{3} = 20$

   > Some calculators have a button for recurring decimals.

   or
   $15 \times 1\frac{1}{3} = 20$

   The answer is 20 cm.

f Decreasing by $33\frac{1}{3}$% is the same as finding $66\frac{2}{3}$% of the number.
   So the multiplier is 0.6666... or $\frac{2}{3}$.
   $15 \times 0.\dot{6} = 10$

   or
   $15 \times \frac{2}{3} = 10$

   The answer is 10 cm.

   > Be careful with multipliers that are recurring decimals.
   > Using an incorrect multiplier of 0.6 gives an answer of 9 cm.
   > Using an incorrect multiplier of 0.66 gives an answer of 9.9 cm.
   > Although these are close to the correct answer, they are not exact!

g To increase by 100% use a multiplier of 2.
   6 × 2 = 12 km

   To find the multiplier, add 100 to 100 and divide by 100:
   $$\frac{100 + 100}{2} = 2$$

h To increase by 200% use a multiplier of 3.
   150 × 3 = 450 sandwiches

   To find the multiplier, add 200 to 100 and divide by 100:
   $$\frac{100 + 200}{100} = 3$$

i To increase by 250% use a multiplier of 3.5.
   500 × 3.5 = 1750 litres

   To find the multiplier, add 250 to 100 and divide by 100:
   $$\frac{100 + 250}{100} = 3.5$$

Worked example

Anneka earns £21 000 per year. She is given a 3% increase.
Using a multiplier, calculate how much she now earns in a year.

Solution

To increase by 3%, use a multiplier of 1.03.
   £21 000 × 1.03 = £21 630

4.2 Now try these

Band 1 questions

1 What multiplier would you use to
   a increase by 12%   b decrease by 12%   c increase by 20%
   d decrease by 20%   e increase by 7%   f decrease by 7%
   g increase by 80%   h decrease by 80%   i increase by 100%
   j decrease by 50%   k increase by 150%   l increase by 300%

2 Use a multiplier to find
   a 20 cm increased by 25%   b £400 increased by 30%   c £7500 increased by 22%
   d 620 metres decreased by 15%   e 160 kg decreased by 40%

3 Jac is selling a computer games console. He originally asks for £80.
   After a few days the console has not been sold. Jac reduces the price by 20%.
   a Using a multiplier, find the reduced price of the console.
   b Calculate the reduction in price.

4 The price of each of these items goes up by 20%.
   Find the new prices using a multiplier.
   a Textbook £15   b Jeans £45   c Smart speaker £75   d Football ticket £25

5 These items go down in value by the percentage shown. Find the new value of each item using a multiplier.
   a Bicycle £300, down 20%   b Guitar £440, down 25%   c Model car £30, down 50%

6 Using a multiplier, find the sale price of each of these items.

SALE 10% off

a  £40

b  £65

c 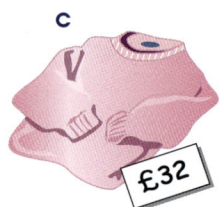 £32

7 Sara gets a 3% pay rise. Her salary before the rise was £29 500.
   a Calculate her new salary using a multiplier.
   b What is her salary increase?

Band 2 questions

8 Wyn bought a new bicycle three years ago, costing £300.
   Now it has lost 30% of its value.
   Using a multiplier, find its new value.

9 The area of sand on a beach is 160 m².
   When the tide goes out, the area of sand increases by 400%.
   Using a multiplier, find the area of sand when the tide is out.

10 a Increase 25 by 100%.
   b Increase 48 by 100%.
   c Increase 1200 by 100%.
   d Copy this sentence and complete the missing word:
      Increasing by 100% is the same as d_____g.
   e Increase 250 by 200%.
   f Increase 2 by 200%.
   g Increase 60 by 200%.
   h Copy this sentence and complete the missing word:
      Increasing by 200% is the same as _____ing.

Band 3 questions

11 Ciaran is thinking about putting some money into a savings account.
   Here is an advert for the account.

> **Mynydd Building Society**
> *Young Savers Savings Account*
> Balances below £1000 receive 2.5% simple interest per annum
> Balances £1000 or above receive 3.5% simple interest per annum

Ciaran is trying to decide whether to put £900 or £1000 into the account.
   a If Ciaran put £900 into the savings account, how much simple interest would he receive each year?
   b If Ciaran put £1000 into the savings account, how much simple interest would he receive each year?
Ciaran plans to keep his money in the account for 5 years.
   c Find the amount of interest he would receive over 5 years on both £900 and £1000.
   d What is the difference in the total amount of interest he would receive over 5 years?

12 A shopkeeper bought a box of 200 chocolate bars for £50.
    a How much did she pay for each bar?
    b She made a profit of 20% on each bar. How much did she sell each bar for?
    c How much money did she receive from the sale of all the bars?
    d How much profit did she make from selling all the bars of chocolate?
    e The shopkeeper paid £50 for the bars of chocolate and made a profit of 20% on each bar.
       Find 20% of £50.
       What do you notice about your answer to part d and this value? Explain your answer.

13 A shop has this offer on all its office chairs.

    Prices reduced by £5 or 10% whichever is greater!

    a What would be the sale price for a chair which normally costs
        i   £20   ii   £35   iii   £60   iv   £74?
    b Elin buys an office chair. The sale price is the same whichever way the discount is given.
        i   What is the normal price of the chair Elin buys?
        ii What is the sale price?

14 A youth club has some special offers as shown in the diagrams.
    a Using multipliers, work out the offer prices for the six activities.
    b If you wanted to spend exactly £10, how could you do it?

    Cinema
    Usually £4.52
    25% off

    Salsa Dancing
    Normally £4.50
    Save 18%

    TEN PIN BOWLING
    Reduced from £5.07 by $33\frac{1}{3}$%

    DIVING CLASSES
    12% off £4.00 usual charge

    Clay pigeon shooting
    Standard price £3.80
    Get 15% off

    Abseiling
    Normal price £3.90
    Now 10% off

15 When water is frozen, its volume increases by 4%.
    a 5 litres of water is frozen. How many cm³ of ice will be made? ← 1 litre = 1000 cm³
    b A machine makes identical ice cubes, each with a volume of 10 cm³.
       2.5 litres of water are used to make these ice cubes.
       How many ice cubes are made?

4.3 Appreciation and depreciation

Appreciation is the **increase** in value of **an item** over time.

Depreciation is the **decrease** in value of **an item** over time.

We can use the idea of a multiplier from Section 4.2 (Using multipliers) to work out calculations of appreciation and depreciation.

Skill checker

Copy this multiplier trail.

Fill in the shaded squares with the correct multipliers to get from one number to the next. The first one has been done for you.

Give your answers as fractions or decimals.

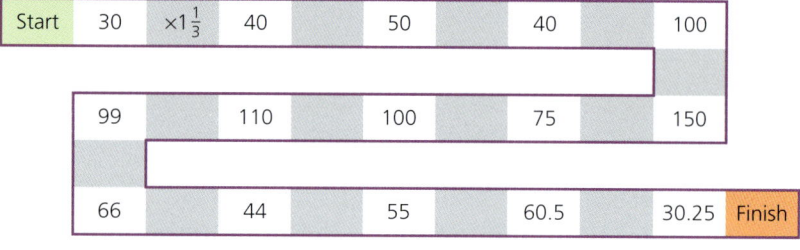

Worked example

Nerys buys an antique painting for £1000.
It appreciates in value by 15% every year.
What will the painting be worth after 3 years?

Solution

At the end of each year 15% will be added by multiplying by 1.15.

Value after 1 year: £1000 × 1.15 = £1150

Value after 2 years: £1150 × 1.15 = £1322.50

Value after 3 years: £1322.50 × 1.15 = £1520.88

The complete calculation is: £1000 × 1.15 × 1.15 × 1.15

This means that the final amount could have been worked out in one calculation:

£1000 × 1.15^3 = £1520.88

Notice the power represents the number of times the original value is increased (or decreased) by the multiplier.

Worked example

Glyn buys a car for £20 000.
It depreciates by 21% each year.
What will it be worth after 5 years?

Solution

Consider the original value of the car to be 100%.

At the end of each year 21% will be subtracted from the original value.

This means there will be 100% − 21% = 79% remaining, which is the same as multiplying by 0.79.

So, the calculation is: £20 000 × 0.79^5 = £6154.11

4 Percentages

4.3 Now try these

Band 1 questions

1. An amount appreciates by 15% each year.
   Write down the multiplier that calculates the appreciated amount after 3 years.
   ×0.45   ×0.15^3 ×0.03^{15} ×1.15^3 ×0.3^{15}

2. An amount depreciates by 7% each year.
   Write down the multiplier that calculates the depreciated amount after 5 years.
   ×0.35   ×0.07^5 ×0.93^5 ×0.05^7 ×0.07^{93}

3. Write down the multiplier for the following appreciations:
   a 12% increase each year for two years.
   b 27% increase each month for 4 months.
   c An appreciation of 7% each day for 10 days.
   d 3.7% appreciation each week for 12 weeks.

4. Write down the multiplier for the following depreciations:
   a 11% decrease each year for 3 years.
   b 21% decrease each month for 7 months.
   c A depreciation of 3% each day for 6 days.
   d 2.3% depreciation each week for 17 weeks.

Band 2 questions

5. Heulwen bought a necklace in an antique shop for £570.
   The necklace appreciates in value by 15% each year.
   What will her necklace be worth in 2 years' time?

6. A new mobile phone cost £850.
   It depreciates in value by 20% each year.
   What will it be worth in 3 years' time?

7. A painting by the Welsh artist Augustus John was bought for £1 million.
   It is known to appreciate in value by about 37% each year.
   What will be its estimated worth after 10 years?
   Give your answer to the nearest hundred thousand. (What would this be in standard form?)

   Cross-curricular activity
   Find out more about the life and work of Augustus John.

8. Zoltan buys a new car for £50 000 which depreciates in value by 23.5% each year.
   He sells it 3 years later.
   How much does he sell it for? (Give your answer to the nearest £10.)

Band 3 questions

9. The population of a town in Carmarthenshire is 130 000 and it is growing at 3.5% per annum.
   a What will the population be in 4 years' time? (Give your answer to the nearest 1000.)
   b In how many years' time will the population be more than 170 000?

10. The population of an island is reducing by a quarter every year. If the current population is 1 million, what will it be after 5 years? (Give your answer to the nearest 100.)

11. There are 500 oak trees in a forest. Every year there are 10% fewer of these oak trees due to a disease.
    After how many years will there be half the number of oak trees in the forest?

4.4 Reverse percentages

Skill checker

Work these out without using a calculator:

① $51.3 \div 0.2$
② $71.4 \div 0.8$
③ $2.68 \div 0.04$
④ $0.362 \div 0.25$
⑤ $7.2 \div 0.75$

▶ Finding the original price or size

Sometimes you know the cost or size of something and the percentage change but not the original cost or size. Reverse percentages can be used to find the original cost or size of a quantity.

Worked example

A coat costs £45 in a sale. It has been reduced by 10%. What was its original price?

Solution

This percentage bar shows what you know and what you need to find out.

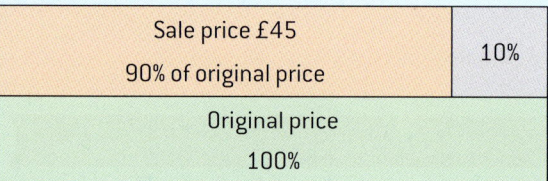

90% of the original price is £45

Turn this into an equation:

$0.9 \times x = 45$

$x = \dfrac{45}{0.9} = £50$

90% is 0.9. Replace the word 'of' with a multiplication sign. Use x for the original price.

The original price of the coat was £50.

Worked example

A multi-storey car park has a total area of $14\,400\,m^2$.
It is 12.5% bigger than last year after the addition of an extra storey.
a What was the area of the car park last year in m^2?
b What is the size in m^2 of the new storey?

Solution

a The car park is 12.5% bigger than last year, so it is now 112.5% of its original size.
112.5% of the original size is 14 400
$1.125x = 14\,400$

Use x for the original size.

$x = \dfrac{14\,400}{1.125} = 12\,800$

Last year the car park had an area of $12\,800\,m^2$.

b The area of the extra storey is the difference between the new size and the original size.
$14\,400 - 12\,800 = 1600\,m^2$
The extra storey has an area of $1600\,m^2$.

Worked example

The sale price of a pair of designer sunglasses is £96.

a   The reduction in the sale is 20%. What percentage of the original price is the sale price?

b   Find the original price.

Solution

a $100\% - 20\% = 80\%$

The sale price is 80% of the original price.

b   80% of the original price is £96. As an equation, using x for the original price:

$0.8x = 96$

$x = \dfrac{96}{0.8} = £120$

The original price of the sunglasses was £120.

▶ VAT for shoppers

VAT is a tax added to the price of some goods in shops.

In 2022, VAT is charged at 20%.

Usually the prices in shops already have VAT added.

Worked example

A school bag is priced at £24.

How much VAT was included in the price?

Solution

Look at the diagram below.

Total price including VAT	
£24	
120%	
Price without VAT	20%
100%	

The price including VAT is 120% of the price without VAT.

Using x for the price without VAT:

$1.2x = 24$

$x = \dfrac{24}{1.2} = £20$

The price before VAT was £20.

The VAT added was £24 − £20 = £4.

▶ Percentage profit and loss

> **Worked example**
>
> A violin is sold in a music shop for £55 at a percentage profit of 10%.
> What did the shop buy it for?
>
> **Solution**
>
> 110% of the original price is £55.
>
> $1.1x = 55$ (where x is the original price).
>
> $x = \dfrac{55}{1.1} = £50$
>
> The violin was bought by the shop for £50.

Communication using symbols

> **Worked example**
>
> Mr Jones sells his house at a 15% loss.
> The sale price is £425 000.
> How much did Mr Jones buy the house for?
>
> **Solution**
>
> £425 000 is 85% of the original price.
>
> $0.85x = 425\,000$
>
> $x = \dfrac{425\,000}{0.85}$
>
> $x = 500\,000$
>
> Mr Jones bought the house for £500 000.

Conceptual understanding

> **Remember**
>
> The original amount is always represented by 100%.

Activity

VAT for businesses

① Not all businesses charge VAT, but if a business or self-employed person earns more than a threshold amount (in 2020 this was £85 000 per year), they have to register for VAT. This means that

   i   they have to charge VAT on all their bills

   ii they can claim back any VAT they pay when they buy anything for the business.

The second point is important, as it is money the business can reclaim. They have to do the calculations first though and work out how much they can claim. Many invoices just give the total. You have to carry out a reverse percentage calculation to find the VAT paid.

Your friend Gwilym is running his own printing business, but he does not understand VAT very well.

Gwilym buys a new printer for the business for £240 including VAT.

Help him to calculate how much VAT he could claim back.

Fill in the blank spaces below. If x is the price before VAT was added, then

$1.2x = £\_\_\_$

$x = \dfrac{£\_\_\_}{1.2} = £\_\_\_$

To work out the VAT:

VAT = £240 − £___ = £___

The price of the printer before VAT was £___.

The VAT added was £___, so this is the amount Gwilym can claim back.

② A few years ago, VAT was charged at 17.5%. How would Gwilym calculate what to claim back if this happened again?

To find the price before VAT, the equation would be

$1.175x = £\_\_\_$

$x = \dfrac{£\_\_\_}{1.175} = £\_\_\_$

③ What about if VAT goes up to 22.5%?

$1.225x = £\_\_\_$

$x = \dfrac{£\_\_\_}{1.225} = £\_\_\_$

④ Mali has her own landscape gardening business. She is earning about £35 000 each year and is wondering whether to register for VAT. She does not have to as she is earning way under the government threshold.

She asks you whether she should register anyway and claim back the VAT on the things she has to buy. If she registers though, she will have to charge her customers VAT. Would you advise her to register or not? What sort of information might you need from her to help make the decision?

4.4 Now try these

Band 1 questions

① Find 100% if
- a 20% is 13
- b 45% is 675 metres
- c 15% is 24 party hats
- d 40% is 388
- e 25% is 1925 cockroaches
- f 60% is 534 litres
- g 75% is 4425
- h 95% is 399 cm
- i 75% is 4875 miles
- j 80% is 2000 kg

② Sale prices in a department store are 80% of the original price.

Find the original price for each of these items.
- a Towels: sale price £12
- b Table lamp: sale price £35
- c Table and chairs: sale price £90
- d Cutlery set: sale price £450

③ A bottle of shampoo says it contains 30% extra free.

The bottle contains 780 ml of shampoo.

What volume of shampoo is in a bottle that does not have the extra 30%?

④ House prices in Cardiff have risen by 3% in the last month.
- a Taking last month's value as 100%, what percentage is the current value?
- b A house in Cardiff is sold for £785 000.

   What was the value of the house one month ago?

   (Round your answer to the nearest pound.)

5 Sale prices in a shoe shop are 70% of the original price.

Find the original price for each of these items.

a Trainers: £35
b Boots: £42
c Shoes: £45.50

Band 2 questions

6 These prices include VAT at 20%, so they are 120% of the original prices.

Find the price before VAT.

a £72
b £240
c £2160
d £64.80

7 VAT (value added tax) is 20%. It is added to the original price (100%) of an item.

The prices of the items shown include VAT. For each of these items, find

i the original price
ii the amount of the VAT.

a £300
b £168
c £204

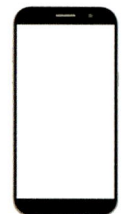
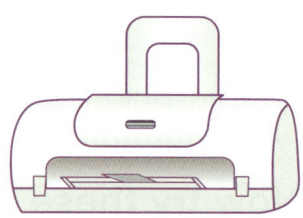

8 Miguel bought a car for £6800 and later sold it for £6120.

What percentage loss did Miguel make?

9 Ranjeet pays tax at 24%. Find how much money he earns when the amount left after tax is

a £18 240
b £6399.20
c £91.20

10 These prices include VAT at 20%. Find the price before VAT is added.

a Printer cartridge: £22.50
b Telephone bill: £103.20
c Burger and chips: £9.12
d Hooded top: £43.20

11 Mabon has some business expenses this month.

On each of these items, VAT has been paid at 20%.

For each item, work out

i the cost before VAT was added
ii the amount of VAT that was added.

a He bought some building materials for £180.
b He received a telephone bill of £135.

12 A market stallholder sells fruit and vegetables at a profit of 40%.

Find the price the stallholder paid for these items.

a Cauliflowers sold for 98p each
b Apples sold for £1.82 per kg
c New potatoes sold for £2.80 per bag
d Bananas sold for £1.05 per kg
e Mushrooms sold for £3.50 per kg
f Carrots sold for 56p per kg

4 Percentages

Band 3 questions

13 Tabitha stays at three different hotels on Friday, Saturday and Sunday.

Her hotel bills are

Grand Salisbury Hotel: Friday £79.20

Best Northern Hotel: Saturday £64.80

Trafalgar Rooms: Sunday £84

   a The hotel bills for Friday and Sunday nights include VAT at 20%.

   Find the cost at these two hotels before VAT is added.

   b The hotel bill for Saturday night did not include VAT.

   Which was the most expensive hotel Tabitha stayed at?

   Which was the cheapest?

14 There is roughly 29.94 million km^3 of ice in the ice sheet of Greenland.

The ice, however, is melting. It is thought that the ice sheet has lost roughly 0.2% of its volume in the last 30 years.

How much ice was there in the Greenland ice sheet 30 years ago?

15 The VAT rate is 20%.

   a Find a rule for working out the VAT paid on a cost that includes VAT.

   b A restaurant buys a new cooker costing £450 before VAT.

   Check your rule on this item by doing the following:

   i Work out the VAT that is paid.

   ii What is the total cost including VAT?

   iii Now use your rule from part **a**.

Does your answer agree with your answer to **b** part **i**?

> **Key words**
>
> Here is a list of the key words you met in this chapter.
>
> | Appreciation | Depreciation | Percentage | Percentage change | Percentage increase |
> | Percentage loss | Percentage profit | Reverse percentage | VAT | |
>
> Use the glossary at the back of this book to check any you are unsure about.

Review exercise: percentages

Band 1 questions

1. Wyoming and Colorado are the only rectangular states of the 50 states in the USA.
   What percentage of the states are rectangular?

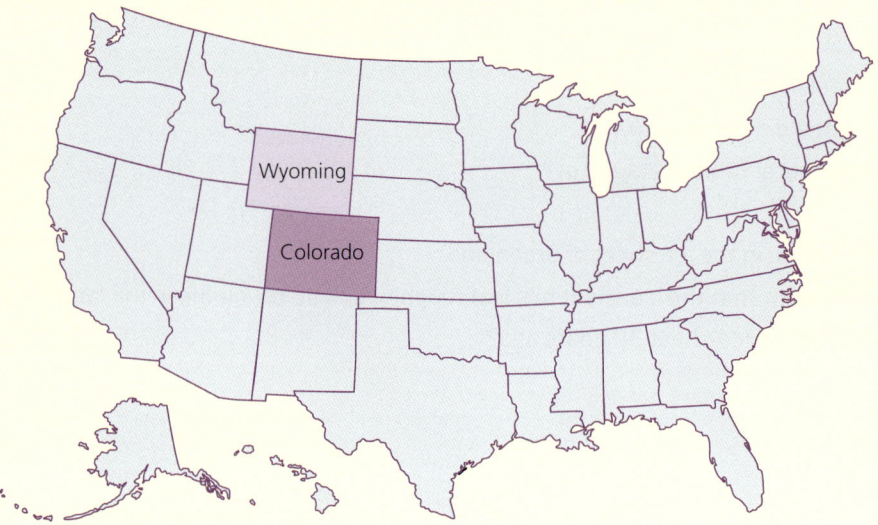

2. Dylan's phone is 85% charged. Write this
   a as a fraction in its simplest form
   b as a decimal.

3. A shop is selling goods at 10% off.
   Using a multiplier, find the sale price of each of these items.
   a Shoes £30
   b Jeans £55
   c Jumper £42

4. Find 100% if
   a 50% is 16
   b 25% is 1350
   c 10% is 64
   d 70% is 3780 litres
   e 75% is 4.875 cm
   f 60% is 252 m
   g 90% is 3060
   h 45% is 1575 tonnes
   i 85% is 7055 years
   j 15% is 27

Band 2 questions

5. A valuable painting is bought at auction for £192 000.
   It is later sold at a loss for £120 000.
   What is the percentage loss?

6. The government awards a 4% pay increase to all public sector workers.
   Work out the new salary for
   a Beatrice, who earned £15 000
   b Jemaine, who earned £17 000
   c Brett, who earned £24 000

7. Cars depreciate in value as they get older.
   a By using a multiplier, find the value of each of these cars after one year's depreciation.
      i   Peugeot 806: originally £17 500, depreciation 9%
      ii Chrysler Neon: originally £13 496, depreciation 25%
      iii Vauxhall Astra: originally £16 140, depreciation 15%
   b What would the value of these cars be after 3 years' depreciation?

Remember
Appreciate in value means go up in value. Depreciate in value means go down in value.

4 Percentages

8 The value of these items has increased. Find their new values using a multiplier.
  a An old postage stamp: £5, increased by 20%
  b Antique pottery: £150, increased by 65%
  c A diamond ring: £1500, increased by 124%
  d A vintage car: £1800, increased by 26%
  e For parts **a** to **d**, calculate the new values after 4 years of appreciation.

9 a The cost price of a bedroom rug is £120. A carpet shop sells it at a profit of 12%.
      What is the selling price?
  b A book shop buys a maths textbook for £18. A more recent edition has been written by the same author, so the book shop sells the book at a loss of 25%.
      What is the selling price?
  c The cost of some gym equipment is £140. It is sold at a profit of 40%.
      What is the selling price?
  d A pair of trainers is bought for £60. They are sold at a profit of 70%.
      What is the selling price?

Band 3 questions

10 These items are marked with the sale prices.

SALE 20% off

A  £40

B  £65

C 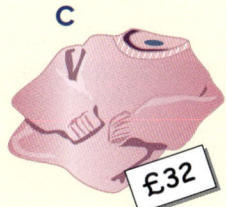 £32

  a What percentage of the original prices are the sale prices?
  b For each of the sale items find
      i   1% of the original price
      ii the original price.

11 Find the original price of these items.
  a Designer label jeans selling at £51.20 after a 20% reduction
  b Designer label shirt selling at £31.50 after a 30% reduction

12 Betsan has just bought an office for her business.
She needs to make a lot of purchases.
The prices below include VAT at 20%.
Find the price of each of these items before VAT was added.
  a A carpet £336
  b A wall painting £19.20
  c A computer printer £132
  d Table and chairs £462

13 a Ollie's weight increases from 55 kg to 63.25 kg between his 15th and 16th birthdays.
      What is the percentage increase in his weight during this year?
  b During the summer, Ollie's mum and dad both try to lose weight. Look at this table.

	Ollie's mum	Ollie's dad
Weight (start of summer)	76 kg	
Weight (end of summer)		76 kg

Ollie's mum's weight went down by 5%.
What is her weight at the **end** of the summer?
Ollie's dad's weight also went down by 5%.
What was his weight at the **start** of the summer?

Logical reasoning

14. Amol carried out a survey. He asked twenty pupils in his class about their favourite type of film.

    The pie chart shows the results of the survey.

    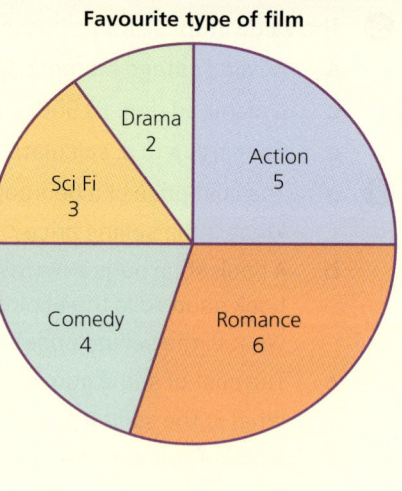
    Favourite type of film

    a   Find the percentage of pupils who said they preferred each type of film.

        Copy and complete this table.

        Put the types of film in order. Romance is top as it has the largest number of pupils.

Type of film	Number of pupils (out of 20)	Percentage (of 20 pupils)
Romance		

    b   The following day five more pupils asked Amol if they could be included in the survey.

        One of these five said they preferred drama films, one said action, one said romance, one said comedy and one said sci fi.

        Copy and complete this table, again listing the types of film in order, from largest number to smallest.

Type of film	Number of pupils (out of 25)	Percentage (of 25 pupils)

    c   After adding one extra pupil in each category, some of the percentages go up and some go down.

        Can you explain why this happens?

Strategic competence

15. The concentration of carbon dioxide (CO_2) in the earth's atmosphere is measured in parts per million (ppm).

    The concentration of CO_2 in the atmosphere in 1965 was roughly 320 ppm.

    By 2015 this had increased by about 25%.

    a   Using a multiplier, find the concentration of CO_2 in the atmosphere in 2015.

    b   Scientists warn that the concentration of CO_2 in the atmosphere may increase further.

        If the concentration of CO_2 in the atmosphere reaches 500 ppm, what will the percentage increase be, compared with

        i   the 2015 concentration   ii   the 1965 concentration?

    c   The 1965 level of 320 ppm was roughly 25% higher than what scientists call **pre-industrial levels**.

        Roughly what were pre-industrial levels of CO_2 in the atmosphere?

5 Using measures

Coming up...
- Using time in calculations
- Speed as a compound measure
- Conversion between metric units of area and volume
- Checking dimensions in a calculation

Guess the formula

① The units for speed are metres per second (m/s).
Which one of the following is the correct formula for speed?

$$\text{speed} = \text{distance} \times \text{time} \qquad \text{speed} = \frac{\text{distance}}{\text{time}} \qquad \text{speed} = \frac{\text{time}}{\text{distance}}$$

② The units for density are grams per cubic centimetre (g/cm³).
Which one of the following is the correct formula for density?

$$\text{density} = \frac{\text{mass}}{\text{volume}} \qquad \text{density} = \text{mass} \times \text{volume} \qquad \text{density} = \frac{\text{volume}}{\text{mass}}$$

③ The units for pressure are newtons per square metre (N/m²). *(A newton is a unit of force.)*
Which one of the following is the correct formula for pressure?

$$\text{pressure} = \text{force} \times \text{area} \qquad \text{pressure} = \frac{\text{area}}{\text{force}} \qquad \text{pressure} = \frac{\text{force}}{\text{area}}$$

5.1 Calculations involving time

Skill checker

① Convert the times on the clocks below to 24-hour digital format.

a	b	c	d

In the morning	In the evening	In the morning	In the evening

② How would you represent the following times on a clock face?

a 04:30   b 05:15   c 23:45   d 20:20

③ Write the following amounts of time, given in minutes, as a fraction of an hour.
   a 15 minutes   b 30 minutes   c 45 minutes   d 20 minutes   e 40 minutes

④ Write the following as hours, using decimal notation. *(Think of the minutes as a fraction of an hour and express this as a decimal.)*
   a 2 hours 15 minutes   b 3 hours 30 minutes
   c 4 hours 45 minutes   d 1 hour 20 minutes

▶ Calculations involving time

Remember, it is important to know how different units of time are related to each other.

1 day	24 hours
1 hour	60 minutes
1 minute	60 seconds

Worked example

How many minutes are there between these times?

a 7:23 a.m. and 7:51 a.m.

b `05:45` `06:30`

c 14:40 and 16:38

d 20:35 and 03:18

Solution

a As both times are within the same hour, find the difference between the minutes.

   51 − 23 = 28 minutes (or you could count on from 23 to 51).

b The left clock reads 5:45 a.m. (quarter to 6).

   The right clock reads 6:30 a.m. (half past 6).

   From 5:45 to 6 o'clock is 15 minutes.

   From 6 o'clock to 6:30 is 30 minutes.

   There are 45 minutes between 5:45 a.m. and 6:30 a.m.

c It is simpler to count on by 2 hours and adjust, because the minutes of the times are close.

   From 14:40 to 16:40 is 2 hours.

   15:38 is 2 minutes before 16:40.

   There is 1 hour 58 minutes (or 118 minutes) between 14:40 and 16:38.

d Using a timeline is a good way to visualise this type of question:

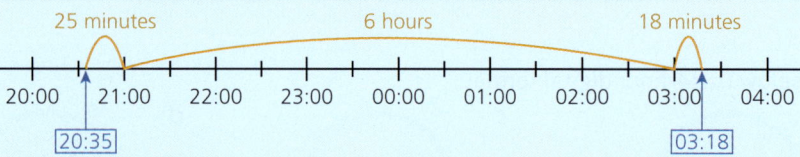

25 minutes + 6 hours + 18 minutes = 6 hours 43 minutes = 6 × 60 + 43 = 403 minutes

> **Activity**
>
> Can you show the answer to question part **c** on a timeline?

▶ Converting amounts in minutes into hours and minutes

As there are 60 minutes in an hour, work out how many lots of 60 are in the number of minutes to find the number of hours.

The remainder is the number of minutes added.

5 Using measures

Worked example

a Convert 180 minutes into hours and minutes.

b Convert 290 minutes into hours and minutes.
Now write your answer as a mixed number.

c Arwyn and his family go on holiday to Chicago in the USA.
The trip involves two plane journeys.
The first journey takes 6 hours and 45 minutes.
The second journey takes 2 hours and 20 minutes.
How long do the plane journeys take in total?

Solution

a $180 \div 60 = 3$
Answer is 3 hours.

b $290 \div 60 = 4$ remainder 50
Answer is 4 hours and 50 minutes.
This is $4\frac{50}{60} = 4\frac{5}{6}$ hours.

c You could work this out in two ways:

Method 1:

Add hours and minutes separately:

$6 + 2 = 8$ hours

$45 + 20 = 65$ minutes. This is 1 hour and 5 minutes.

Now add 8 hours and 1 hour and 5 minutes.

The answer is 9 hours and 5 minutes.

Method 2:

Convert times to minutes first and convert back at the end:

6 hours and 45 minutes is $6 \times 60 + 45 = 405$ minutes.

2 hours and 20 minutes is $2 \times 60 + 20 = 140$ minutes.

$405 + 140 = 545$ minutes.

$545 \div 60 = 9$ remainder 5.

The answer is 9 hours and 5 minutes.

Cross-curricular activity

Greenwich is in the south of England.

What is Greenwich Mean Time? How else is it known?

Why has the world map been split into time zones? Your History teacher could help you answer this.

How does a time zone impact on the time in different countries that you study in Geography or other subjects?

Think about what time it is now in different countries.

How were the time zones created?

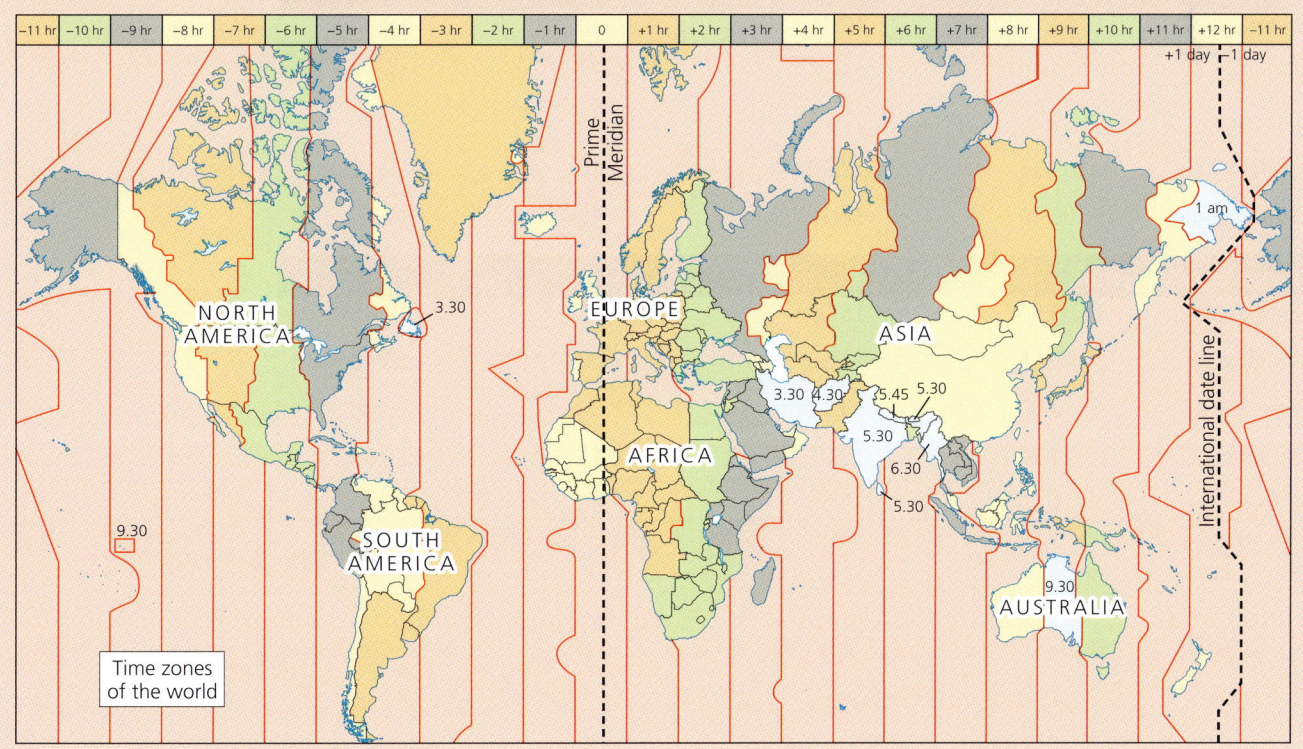

Time zones of the world

5.1 Now try these

Band 1 questions

For the following questions, calculate how long is between each pair of times. Give each answer in minutes and, if possible, hours and minutes.

1. 02:10 to 02:37
2. 14:23 to 14:55
3. 15:50 to 16:15
4. 11:10 to 12:30
5. 10:25 a.m. to 1:30 p.m.
6. 22:54 to 01:31
7. Lunchtime at school starts at 1:35 p.m. and ends at 2:25 p.m.
   How long does lunchtime last?
8. A hockey match started at 09:45 and the final whistle was blown at 11:30.
   How long did the match last?

Band 2 questions

9. At what times do the following films finish at the local cinema in Caerfyrddin?
   a. Starts at 14:04 and lasts 47 minutes.
   b. Starts at 15:35 and lasts 40 minutes.
   c. Starts at quarter past 11 and lasts 50 minutes.
   d. Starts at quarter to 3 and lasts 1 hour and 45 minutes.
   e. Starts at 13:35 and lasts 100 minutes.

Use the following train timetable, between Abernant and Eglwysgymin, to answer Questions 10, 11 and 12.

Abernant – Eglwysgymin				
From Abernant				
Abernant	09:15	12:10	16:20	18:50
Brynfach	09:20	12:15	16:25	18:55
Caerafon	09:30	12:25	16:35	19:05
Drefawr	09:35	12:30	16:40	19:10
Eglwysgymin	09:40	12:35	16:45	19:15
From Eglwysgymin				
Eglwysgymin	09:50	13:05	16:55	19:35
Drefawr	09:55	13:10	17:00	19:40
Caerafon	10:00	13:15	17:05	19:45
Brynfach	10:10	13:25	17:15	19:55
Abernant	10:15	13:30	17:20	20:00

10. a How long does the second train from Abernant take to get to Drefawr?
    b Dafydd lives in Brynfach. What is the last train he can catch to arrive in Caerafon by 17:00?
    c Elin catches the third train from Eglwysgymin to meet her friend in Abernant at 17:00.
       How many minutes late is she?
    d Ami takes the first train from Brynfach to Drefawr. She must return to Brynfach by 18:00 at the latest. How long can she spend in Drefawr?

11. Carys lives in Abernant and goes to meet her friend in Drefawr at 17:00.
    a At what time must she catch the train to Drefawr?
    b How long does the journey take?
    c How long must she wait in Drefawr before meeting her friend?
    d Carys must be back home by 20:30 at the latest.
       What time must she catch the train home?
    e How much time does Carys have to spend with her friend?

12. Dilys lives in Brynfach. She wants to watch a film which starts at 13:00 in the cinema in Caerafon. The film is 2 hours 15 minutes long.
    a At what time must she catch the train?
    b How long does the journey take?
    c After she arrives in Caerafon, how long is there before the film starts?
    d At what time does she catch the train home to Brynfach?
    e How long does she have to wait for the train?

Band 3 questions

13. Dai records the following four journey times of his favourite walks in Wales.

1 hour 20 minutes	10 hours 17 minutes	5 hours 25 minutes	3 hours 52 minutes

    What is the mean of the four time periods?

14. Gizella is flying from Cardiff to Budapest.
    At Cardiff Airport, the clocks show the current time in both cities:

    Cardiff   Budapest

    Cross-curricular activity
    Budapest is the capital city of which country?
    How many capital cities do you know in Europe (and the world)?

    a Her plane departs at 14:45 and takes 2 hours 25 minutes.
       At what time should she arrive in Budapest, in local Budapest time?
    b She returns to Cardiff the following week.
       Her plane from Budapest departs at 15:30.
       It normally takes 20 minutes longer to return to Cardiff.
       At what time does she arrive in Cardiff?

15. Hanako is flying from Tokyo to London.
    When the plane departs at 02:45 on Tuesday, she knows it is 17:45 on Monday in London.
    The flight takes 11 hours 20 minutes.
    At what time, and on what day, should she arrive in London?

    Cross-curricular activity
    It is usually the case that flying East is faster than flying West.
    Why could this be?

5.2 Speed

Skill checker

Substitute the numbers into the formulas.

a $A = BC$
   Find A if $B = 3$ and $C = 6$.

b $P = QR$
   Find P if $Q = -7$ and $R = 6$.

c $Z = \dfrac{X}{Y}$
   Find Z if $X = 12$ and $Y = 4$.

d $m = np$
   Find m if $n = 5$ and $p = 3.5$.

e $s = \dfrac{r}{t}$
   Find s if $r = 20$ and $t = 2.5$.

f $a = fg$
   Find a if $f = 1.5$ and $g = 10.5$.

g $k = \dfrac{F}{r}$
   Find k if $r = 16$ and $F = 2$.

h $c = pe$
   Find c if $p = \dfrac{1}{5}$ and $e = 7$.

i $d = \dfrac{q}{n}$
   Find d if $q = \dfrac{1}{8}$ and $n = \dfrac{1}{4}$.

▶ Speed as a compound measure

A compound measure is a measure involving two or more quantities.

An example is speed, which involves distance and time.

$$\text{speed} = \dfrac{\text{distance}}{\text{time}}$$

If distance is measured in kilometres and the time is in hours, then the unit for speed will be kilometres per hour (km/h).

If distance is measured in metres and the time is in seconds, then the unit for speed will be metres per second (m/s).

Other possible units for speed are miles per hour, centimetres per second, kilometres per minute, etc.

Worked example

Converting units of time

a Convert 6.5 minutes to minutes and seconds.

b Convert 3 hours 15 minutes to hours.

Solution

a 6.5 minutes = $6\dfrac{1}{2}$ minutes, or 6 minutes 30 seconds.

b 3 hours 15 minutes is $3\dfrac{1}{4}$ hours, or 3.25 hours.

Remember

3.25 hours **does not mean** 3 hours 25 minutes!

Worked example

Working with the formula for speed

Janos leaves Cardiff on a motorbike at 06:00 and travels to his parents' house 432 km away.

He arrives at 10:00.

a Find his average speed

   i in kilometres per hour

   ii in kilometres per minute

   iii in metres per second.

b Write the unit 'metres per second' another way.

Solution

a His journey takes from 6 o'clock to 10 o'clock so it lasts 4 hours.

  i average speed = $\dfrac{\text{distance}}{\text{time}} = \dfrac{432 \text{ kilometres}}{4 \text{ hours}}$ = 108 kilometres per hour

  ii For an answer in kilometres per minute, use the distance in kilometres and the time in minutes.

  $\dfrac{\text{distance}}{\text{time}} = \dfrac{432 \text{ kilometres}}{240 \text{ minutes}}$

  = 1.8 kilometres per minute

  (4 hours = 4 × 60 = 240 minutes)

  iii For an answer in metres per second, use the distance in metres and the time in seconds.

  $\dfrac{\text{distance}}{\text{time}} = \dfrac{432\,000 \text{ metres}}{14\,400 \text{ seconds}}$

  = 30 metres per second

  (432 kilometres = 432 × 1000 = 432 000 metres)

  (4 hours = 4 × 60 × 60 = 14 400 seconds)

b Another way of writing 'metres per second' is m/s.

▶ Using a formula triangle

For all compound measures, you may find it helpful to use a formula triangle.

Using the formula

speed = $\dfrac{\text{distance}}{\text{time}}$

begin labelling the triangle from the bottom left, then the top, then the bottom right.

To use the triangle:

- To find the speed, cover up s with your finger.

  The formula is $s = \dfrac{d}{t}$

- To find the distance, cover up d with your finger.
  The formula is $d = s \times t$

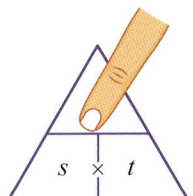

- To find the time, cover up t with your finger.
  The formula is $t = \dfrac{d}{s}$

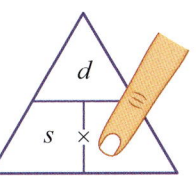

If you are not sure of the formula for a compound measure, you may be able to construct it using the units.

For example, speed is measured in metres per second (or miles per hour, etc.), so its formula is distance over time.

Curriculum for Wales **Mastering Mathematics: Book 3**

> **Worked example**
>
> a A truck travels 126 miles in 2 hours 15 minutes on a motorway.
>   What is the average speed of the truck while it is on the motorway?
> b The truck leaves the motorway and travels a further 100 miles at a speed of 50 miles per hour.
>   How long does it take for this part of the journey?
> c For the final part of its journey, the truck travels at an average speed of 40 miles per hour. It takes $1\frac{1}{2}$ hours to reach its final destination.
>   What distance did the truck travel during this third part of the journey?
>
> **Solution**
>
> a Convert 2 hours 15 minutes to a time in hours.
>   15 minutes is $\frac{1}{4}$ hour or 0.25 hours, so 2 hours 15 minutes = 2.25 hours.
>   Using the triangle and covering s you find that $s = \frac{d}{t}$
> $s = \frac{126}{2.25} = 56$ miles per hour
>
> b In this part of the question you are finding the time, so cover t in the triangle.
> $t = \frac{d}{s} = \frac{100}{50} = 2$ hours
>
> c Find the distance d for the third part of the journey. Covering d in the triangle gives $d = s \times t$.
> $d = 40 \times 1.5 = 60$ miles

5.2 Now try these

Band 1 questions

1. Change the units for these amounts of time.

   a Convert 2 hours to minutes.
   b Convert $2\frac{1}{4}$ hours to minutes.
   c Convert 2.75 hours to minutes.
   d Convert 2 hours 25 minutes to minutes.
   e Convert 2 minutes to seconds.
   f Convert $2\frac{1}{2}$ minutes to seconds.
   g Convert 2.5 minutes to seconds.
   h Convert 2 minutes 20 seconds to seconds.
   i Convert 3 days to hours.
   j Convert $2\frac{1}{3}$ days to hours.
   k Convert 1.25 days to hours.
   l Convert 2 days and 6 hours to hours.

2. Calculate the average speed for each of these objects, giving your answer with appropriate units.

   a A cricket ball travelling 60 metres to the edge of the cricket ground in 5 seconds.
   b A paper plane flying 4 metres in 3.2 seconds.
   c A leaf falling 10 metres from a tree, taking 16 seconds to reach the ground.
   d A jet plane travelling 800 miles from London to Madrid in Spain in 2 hours 30 minutes.
   e A rocket taking $8\frac{1}{2}$ minutes to travel 255 km into orbit.

3. A sprinter runs 200 m in 20 seconds.
   How fast, on average, was the sprinter running

   a in metres per second
   b in kilometres per hour?

5 Using measures

4. a Kamala drives a distance of 280 km in 4 hours.
      Work out Kamala's average speed.
   b A worm moves 8 metres in 32 minutes.
      Work out the worm's average speed in
      i   metres per minute
      ii centimetres per second.
   c Julia swims for 1 hour. She completes 58 lengths, each 25 metres.
      What is her average speed?

5. Work out the average speed of each of these ships and boats.
   a A speedboat travelling 12 km in $\frac{1}{4}$ hour.
   b A cargo ship travelling 19 200 km from Shanghai, China to the UK in 50 days.
   c A cruise ship covering 384 km in 12 hours.

Band 2 questions

6. Anna's train leaves Edinburgh at 10:26 in the morning and arrives in London at 2:56 that afternoon. The total distance for the journey is 337.5 miles.
   What is the train's average speed in miles per hour?

7. The national speed limit for cars on most roads is 60 miles per hour.
   A bird can fly at 45 metres per second.
   Which is faster, the car or the bird?
   Explain your answer.

8. The official world land speed record is 1223.7 kilometres per hour.
   It was set on 15 October 1997 by Andy Green in the jet-engine car Thrust SSC.
   What was Thrust SSC's average speed in metres per second?

9. A black mamba snake of Eastern Africa can move at up to 5 m/s.
   What is its average speed in miles per hour?
   (Use the fact that 8 kilometres is approximately 5 miles.)

10. Sound travels 1 kilometre in 3 seconds.
    Alana sees a fork of lightning.
    Five seconds later she hears the thunder.
    How many metres away is the lightning?

    Alana sees the lightning almost instantaneously because light travels very, very quickly.

Band 3 questions

11. A light year is the distance that light travels in one year.
    The speed of light is 300 million metres per second.
    How many kilometres are there in one light year?

12. Geraint is riding a motorbike at 35 miles per hour.
    Caron is driving her car at 12.5 metres per second.
    Convert both of these speeds into kilometres per hour, using the fact that 5 miles is approximately 8 kilometres.
    Who is travelling faster?

13. Arwel and Delyth live 2 km from school.
    This travel graph represents their journeys to school one day.

    a How far does Arwel travel in
       i   2 minutes
       ii 5 minutes?
    b What is Arwel's speed in
       i   metres per minute
       ii kilometres per hour?
    c What is Delyth's speed in
       i   metres per minute
       ii kilometres per hour?
    d One of them cycles and the other walks. Who cycles?

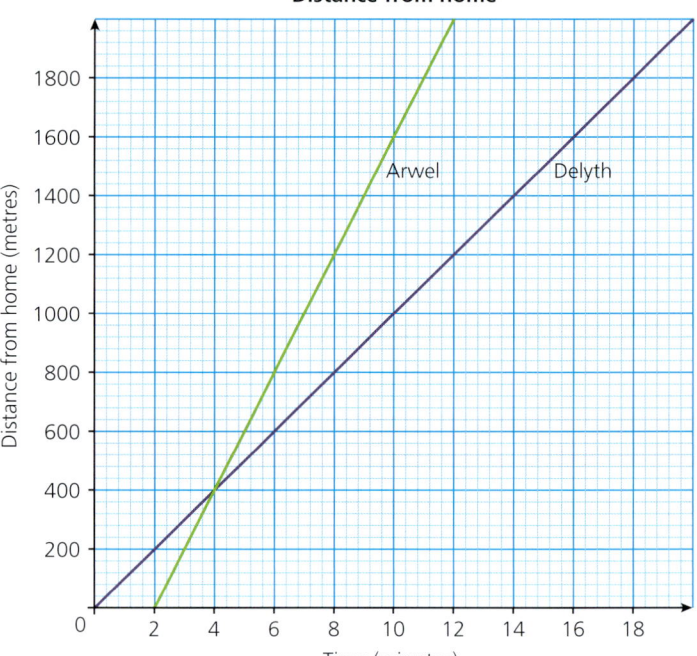

14. Rhodri is doing a research project on birds. He learns that most birds can fly, but others are described as flightless.
    Here is some information Rhodri finds about how fast some birds can travel.
    a A swallow can fly 100 m in 7.2 seconds.
    b A penguin can swim 2 km in $\frac{1}{4}$ hour.
    c An ostrich can run 200 m in 12 seconds.
    Rhodri works out the speed of each of these birds in
       i   metres per second
       ii kilometres per hour.
    Copy and complete the table.

	Speed (mps)	Speed (kph)
Swallow		
Penguin		
Ostrich		

15. Rowena has her own private aeroplane.
    She makes a five-leg tour of England, Wales and Scotland, starting in Penzance.

Leg	Start	End	Distance (km)	Time	Average speed (km/h)
1	Penzance	Swansea	200	20 minutes	
2	Swansea	Carlisle	370		370
3	Carlisle	Newcastle		30 minutes	
4	Newcastle	Aberdeen		45 minutes	324
5	Aberdeen	Penzance	814		407

    a Using the map on the following page, copy and complete the table.
       For one leg of the journey you will need to measure the distance with a ruler and convert to kilometres.
    b On which stretch of the journey does she have the highest average speed?
    c How far does Rowena fly in total and what is her total flying time?
    d What is her average speed for the entire tour?

5 Using measures

The map shows the first leg of her tour, from Penzance to Swansea.

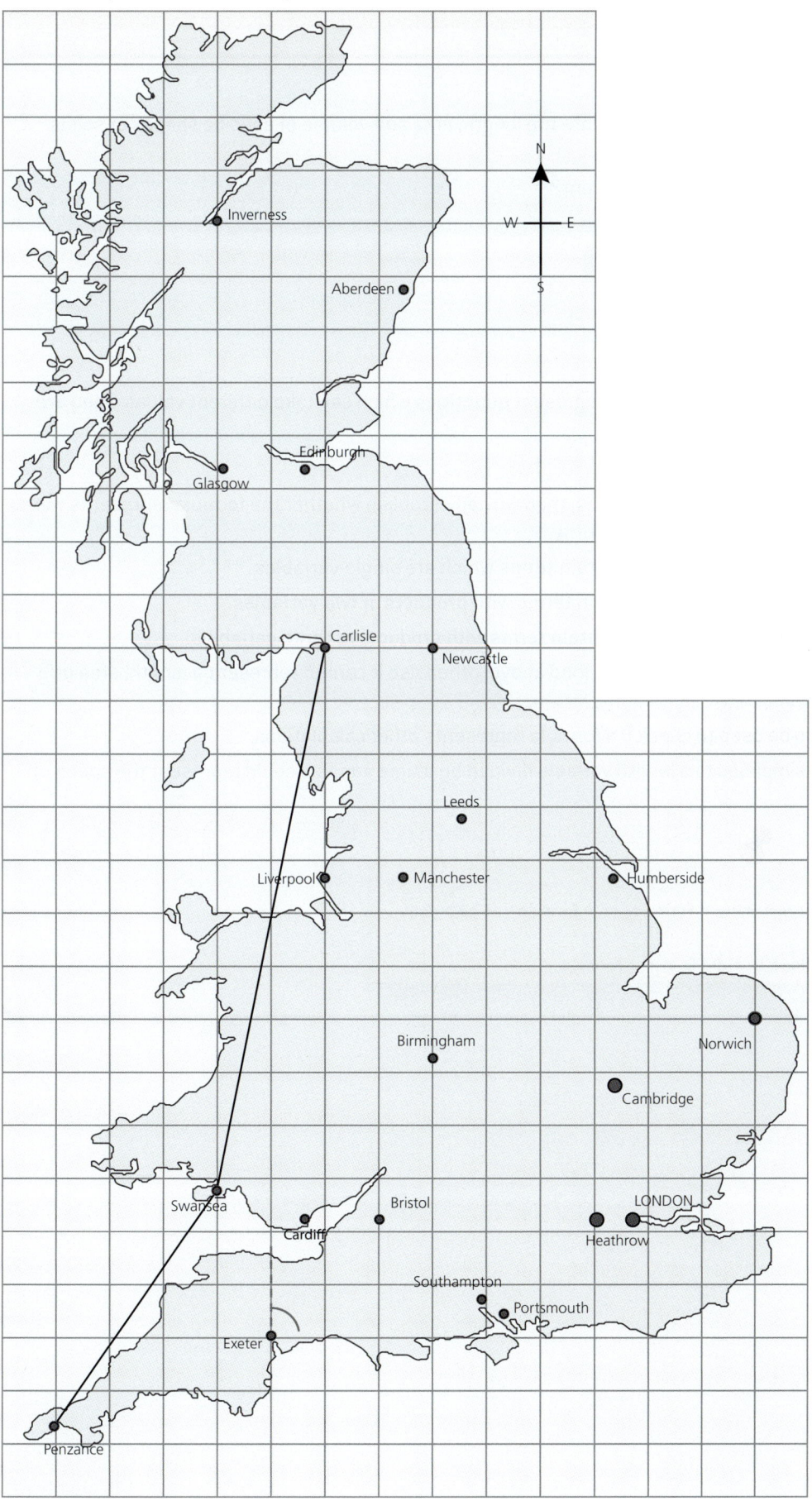

1 cm represents 50 km

5.3 Dimensions of a calculation

Skill checker

The following are examples of formulas used to calculate the length, area and volume of specific shapes or solids. You may have seen them before.

Do you remember which shapes or solids they represent?

Can you think of any other formulas? Write them down.

Length	Area	Volume
$2(l + w)$	$\frac{1}{2}bh$	lwh
$2\pi r$	πr^2	$\pi r^2 h$

In each of these formulas, the letters such as w or h represent quantities which can take different values. They are called **variables**.

Fixed numbers like 2, $\frac{1}{2}$, or π are called **constants**.

If all the single variables in a formula represent lengths, then we can establish whether the formula represents the calculation of a length (or perimeter), an area or a volume.

- A formula which calculates length can only contain terms which are single variables.
- A formula which calculates area can only contain terms with products of two variables.
- A formula which calculates volume can only contain terms with products of three variables.

A formula must have the **correct dimensions** as described above, otherwise it cannot represent a length, area or volume.

Dimensional analysis can also be used to check if a formula represents other calculations.

For example, a formula which simplifies to a length variable divided by a time variable could represent the speed of an object.

Note

The units used can also help you remember a formula, or a formula can help you remember which units to use.

'Per' contained in units means there is a divide in the formula.

5 Using measures

> **Worked example**
>
> a, b, r and h are single variables representing lengths.
>
> Decide whether each of these formulas could represent a length, an area, a volume or none of these.
>
> **1** $3a + 4b$
>
> This formula contains terms with single variables added together, so it could represent a length.
>
> **2** $\frac{1}{2}(a+b)h$
>
> (length + length) × length = another length × length = area
>
> This formula contains terms which simplify to 2 terms multiplied together, so the formula could represent an area.
>
> **3** $2\pi r + r^2$
>
> length + length × length = length + length² = length + area
>
> This formula contains terms with a single variable (2 and π are constants) and a variable squared.
>
> The terms are different, so the formula cannot represent length, area or volume.
>
> **4** $4\pi r^2$
>
> length² = area
>
> This formula contains a term with the variable squared (4 and π are constants).
>
> Therefore, this formula could represent an area.
>
> **5** This example shows how to check if the formula involves speed.
>
> From Section 5.2, we know that speed = distance ÷ time.
>
> In the following formula, t represents time, and a, b and h represent lengths:
>
> $\frac{a}{t} + \frac{br}{ht}$
>
> $\frac{\text{length}}{\text{time}} + \frac{\text{length} \times \text{length}}{\text{length} \times \text{time}} = \frac{\text{length}}{\text{time}} + \frac{\text{length}}{\text{time}} = $ speed + speed = speed

5.3 Now try these

Band 1 questions

1 In the following exercise, all the letters other than t are single variables that represent a length.

The variable t represents time.

Numbers and π are constants and have no dimensions.

Copy the table and place a tick in the correct box to indicate if the formula could represent a length, an area, a volume, a speed or none of these.

The first one has been done for you.

Formula	Length	Area	Volume	Speed	None of these
a^2		✓			
b					
c^3					
$2ab$					
$\frac{b}{t}$					
$a + b$					
$a + a^2$					

Fluency

Formula	Length	Area	Volume	Speed	None of these
abc					
πr^2					
$2\pi r$					
$a^2 + b^3$					

Band 2 questions

In the following questions, all the letters other than t are single variables that represent a length.

The variable t represents time.

Numbers and π are constants and have no dimensions.

Strategic competence

2 Which of the following expressions could be a length?
  a $4g$
  b xyt
  c $v + 2w$
  d $abct$

3 Which of the following expressions could be an area?
  a $h^2 - kt$
  b $7dg$
  c $2a^2 + \pi b^2$
  d ae^2

4 Which of the following expressions could be a volume?
  a $h^3 - c^3$
  b x^3
  c $a^2 + b^2 + c^2$
  d $4tvx^2$

5 Which of the following expressions could be a speed?
  a $\dfrac{d^3}{t}$
  b $\dfrac{x}{t^2}$
  c $\dfrac{2h}{t}$
  d $\dfrac{a+b+c}{\pi t}$

6 What does each of the following expressions represent: a length, an area, a volume or a speed?
  a $\pi r l$
  b $\dfrac{1}{3}\pi r^2 h$
  c $2l + 2w$

7 What does each of the following expressions represent: a length, an area, a volume, a speed or none of these?
  a $\dfrac{4}{3}\pi r^3$
  b $17d$
  c $4\pi r^2$
  d $\dfrac{3x - 2y^3}{t}$
  e $\dfrac{a+b}{5t}$

Band 3 questions

In the following questions, all the letters other than t are single variables that represent a length.

The variable t represents time.

State whether each of the following formulas could be for a length, an area, a volume, a speed or none of these.

Logical reasoning

8 $\dfrac{abc}{3}$

9 $3n + \pi b$

10 $6at$

11 $xy - 3ad$

12 $5(3e + f)$

13 $7(b + c + d)$

14 $ab(c + d - e)$

15 $\dfrac{wx + z^3}{a + t}$

16 $\dfrac{2a + 3b + 5c}{\pi t + 7t}$

5.4 Converting between metric units of area and volume

Skill checker

Copy and complete the following.

① There are 100 cm in 1 metre.
   To convert from metres to centimetres multiply by 100.
   To convert from centimetres to metres _____ by 100.
   Examples: 5.5 m = _____ cm
               659 cm = _____ m

② There are 10 mm in 1 cm.
   To convert from centimetres to millimetres _____ by 10.
   To convert from millimetres to centimetres divide by _____.
   Examples: 7.4 cm = _____ mm
               25.9 mm = _____ cm

③ There are 1000 m in 1 km.
   To convert from kilometres to metres _____ by 1000.
   To convert from metres to kilometres _____ by _____.
   Examples: 19 km = _____ m
               25 700 m = _____ km

▶ Units of area

Worked example

Converting 1 cm² to mm² (diagrams not-to-scale)

Look at this diagram. It shows there are 10 millimetres in 1 centimetre.

1 cm

Now look at this diagram showing 1 square centimetre (1 cm²).

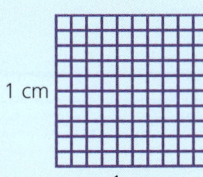
1 cm

1 cm

It has been divided into square millimetres.

How many square millimetres make up one square centimetre?

Solution

There are ten rows of ten.

So there are 100 square millimetres in one square centimetre.

$1 \text{ cm}^2 = 100 \text{ mm}^2$

To convert from square centimetres to square millimetres multiply by 10^2 or 100.

To convert from square millimetres to square centimetres divide by 10^2 or 100.

▶ Converting between metric units of volume

Worked example

Converting 1 cm³ to mm³

Now look at this picture of 1 cubic centimetre (1 cm³).

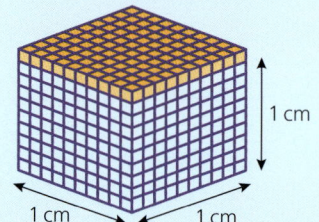

It has been divided into cubic millimetres.
How many cubic millimetres make up one cubic centimetre?

Solution

In the top layer (coloured orange) there are 100 cubic millimetres.

In the entire cube there are 10 layers of 100.

$10 \times 100 = 1000$

So there are 1000 mm³ in 1 cm³.

$1 \text{ cm}^3 = 1000 \text{ mm}^3$

To convert from cubic centimetres to cubic millimetres multiply by 10^3 or 1000.
To convert from cubic millimetres to cubic centimetres divide by 10^3 or 1000.

Worked example

a How many centimetres make up one metre?
b How many square centimetres make up one square metre?
c How many cubic centimetres make up one cubic metre?
d How many square centimetres make up five square metres?
e How many cubic centimetres make up five cubic metres?

Solution

a 1 m = 100 cm
b 1 m² = 10 000 cm² (ten thousand)   *Because $100^2 = 10000$*
c 1 m³ = 1 000 000 cm³ (one million)   *Because $100^3 = 1000000$*
d There are 10 000 cm² in one square metre.

   So, to convert from square metres to square centimetres multiply by 10 000.

   $5 \times 10\,000 = 50\,000$

   There are 50 000 cm² in 5 m².

e There are 1 000 000 cm³ in one cubic metre.

   So, to convert from cubic metres to cubic centimetres multiply by 1 000 000.

   $5 \times 1\,000\,000 = 5\,000\,000$

   So there are 5 000 000 cm³ in 5 m³.

Worked example

a Convert an area of 0.07 m² to cm².
b Convert a volume of 168 000 cm³ to m³.

Solution

a 1 m² = 10 000 cm²
   0.07 × 10 000 = 700
   So 0.07 m² = 700 cm²

b 1 m³ = 1 000 000 cm³
   168 000 ÷ 1 000 000 = 0.168
   So 168 000 cm³ = 0.168 m³

▶ Summary

	To convert
1 cm = 10 mm	From cm to mm: multiply by 10.
	From mm to cm: divide by 10.
$1\,cm^2 = 100\,mm^2$	From cm^2 to mm^2: multiply by 100. *(100 is 10^2)*
	From mm^2 to cm^2: divide by 100.
$1\,cm^3 = 1000\,mm^3$	From cm^3 to mm^3: multiply by 1000. *(1000 is 10^3)*
	From mm^3 to cm^3: divide by 1000.
1 m = 100 cm	From m to cm: multiply by 100.
	From cm to m: divide by 100.
$1\,m^2 = 10\,000\,cm^2$	From m^2 to cm^2: multiply by 10 000. *(10 000 is 100^2)*
	From cm^2 to m^2: divide by 10 000.
$1\,m^3 = 1\,000\,000\,cm^3$	From m^3 to cm^3: multiply by 1 000 000. *(1 000 000 is 100^3)*
	From cm^3 to m^3: divide by 1 000 000.
1 km = 1000 m	From km to m: multiply by 1000.
	From m to km: divide by 1000.
$1\,km^2 = 1\,000\,000\,m^2$	From km^2 to m^2: multiply by 1 000 000. *(1 000 000 is 1000^2)*
	From m^2 to km^2: divide by 1 000 000.
$1\,km^3 = 1\,000\,000\,000\,m^3$	From km^3 to m^3: multiply by 1 000 000 000. *(1 000 000 000 is 1000^3)*
	From m^3 to km^3: divide by 1 000 000 000.

5.4 Now try these

Band 1 questions

1 True or false?

Correct any false statements.

- **a** 1 cm = 10 mm
- **b** $1\,cm^2 = 10\,mm^2$
- **c** $1\,cm^3 = 10\,mm^3$
- **d** 1 m = 100 cm
- **e** $1\,m^2 = 100\,cm^2$
- **f** $1\,m^3 = 100\,cm^3$
- **g** 1 km = 1000 m
- **h** $1\,km^2 = 1000\,m^2$
- **i** $1\,km^3 = 1000\,m^3$

2 Copy and complete each statement. The first one has been done for you.

- **a** $4\,m^2 = 40\,000\,cm^2$
- **b** $60\,m^2 =$ _____ cm^2
- **c** _____ $cm^2 = 400\,mm^2$
- **d** $0.1\,km^2 =$ _____ m^2
- **e** $3\,m^3 =$ _____ cm^3
- **f** $55\,m^3 =$ _____ cm^3
- **g** _____ $cm^3 = 250\,mm^3$
- **h** $9.3\,km^3 =$ _____ m^3

3 The school playground is 500 square metres. How many square centimetres is this?

4 A shipping container can hold 50 cubic metres of cargo. How many cubic centimetres is this?

Band 2 questions

5 Convert each of these quantities into the units given.

- **a** $78\,mm^2$ into cm^2
- **b** $1650\,cm^2$ into m^2
- **c** $26\,000\,mm^2$ into cm^2
- **d** $1\,350\,000\,m^2$ into km^2
- **e** $6400\,mm^2$ into cm^2
- **f** $1900\,cm^2$ into m^2

6. Jac has been asked to paint three red squares on the school playground for a game.
   Each square should measure 1 metre by 1 metre.
   How many square centimetres of paint does Jac use?

7. a Find the volume of each of these cuboids in cubic centimetres.
   b Convert your answers to cubic millimetres.

   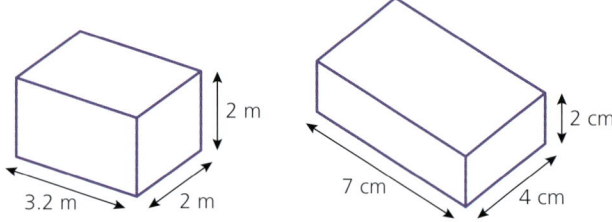

8. Priti wants to carpet two rectangular rooms.
   One of the rooms measures 3.5 m by 1.8 m.
   The other is 480 cm by 290 cm.
   How many square metres of carpet will she need?

9. A swimming pool is 2 m deep and has a base area of 435 m².
   a Find the volume of the pool in m³.
   b Find the volume of the pool in cm³.

10. A volcano erupts.
    3.2 cubic kilometres of the mountain is blown away by the blast.
    What is this in cubic metres?

Band 3 questions

11. The edge of one dice is 1.2 cm in length.
    a Find the volume of one dice in cubic centimetres.
    b Find the volume of one dice in cubic millimetres.
    c Find the volume of three dice in cubic millimetres.

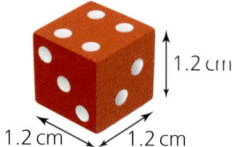

98

5 Using measures

12 Reuben's homework is below. There is one mistake in each of his answers.
Can you correct his work?

Question 1. Convert 5.8 m² to cm²

Step 1   1 m = 100 cm
         1 m² = 100 × 100 × 100 = 1 000 000 cm²
Step 2   5.8 m² = 5.8 × 1 000 000 = 5 800 000 cm²

Question 2. Convert 152 mm³ to cm³

Step 1   1 cm = 10 mm
         1 cm³ = 10 × 10 × 10 = 1000 mm³
Step 2   152 mm³ = 152 × 1000 = 152 000 cm³

13 Convert each of these measurements into the units given.
- a 1.2 m² into mm²
- b 0.001 km² into cm²
- c 50 million mm² into m²
- d 613 000 000 cm² into km²
- e 0.000 04 km² into mm²
- f 71 550 mm² into m²

14 Convert each of these measurements into the units given.
- a 4 million mm³ into km³
- b 0.0002 m³ into mm³
- c 875 cm³ into km³
- d 18 450 mm³ into m³
- e 27 cm³ into m³
- f 0.001 69 km³ into mm³

15 The table gives the volumes of three 3D shapes.
- a Which shape has the largest surface area? Show working to explain your answer.
- b Which shape has the largest volume? Show working to explain your answer.

Shape	Surface area	Volume
A	70 m²	1.15×10^7 cm³
B	750 000 cm²	8.96 m³
C	65 000 000 mm²	9.25×10^9 mm³

Key words

Here is a list of the key words you met in this chapter.

Centimetre	Compound measure	Dimension	Kilometre
Mass	Metre	Metric unit	Millimetre
Speed	Volume		

Use the glossary at the back of this book to check any you are unsure about.

Review exercise: using measures

Band 1 questions

1 A rugby match started at 11:20 and finished at half past one in the afternoon. How long did the match last?

2 a Convert $4\frac{1}{4}$ minutes to seconds.
  b Convert 3.25 hours to minutes.

3 a Pazir walks 16 km in 4 hours. What is his average speed?
  b Aoife travels by train a distance of 329 km in 3 hours 30 minutes. Work out the train's average speed in kilometres per hour.
  c Abdullah drives 120 miles in 1 hour 45 minutes. Find Abdullah's average speed.

4 The distance from London to Budapest is 1450 km.
A flight from London to Budapest takes 4 hours. At what average speed does the aeroplane fly?

5 A train travels 440 km in 2 hours 20 minutes.
Find the average speed of the train in km/h.

6 Convert each of these quantities into the units given.
  a 0.003 km³ into m³
  b 17 cm³ into mm³
  c 0.6 m³ into cm³
  d 0.236 km³ into m³
  e 0.000054 cm³ into mm³

7 The letters a, b and c are single variables representing length; t represents time.
What does each of the following expressions represent: a length, an area, a volume or a speed?
  a ab
  b $a + 3b$
  c $a - bt$
  d $\dfrac{b}{t}$
  e abc

Band 2 questions

8 Rhodri sits two exams on the same day.
The first one starts at 09:30 and lasts for 1 hour 45 minutes.
He then has a 50 minute lunch break before he sits his second exam which lasts for 2 hours 15 minutes.
At what time does he finish the second exam?

9 Ifan and Seren are travelling from their homes to the same meeting.
Ifan leaves his home at 08:00. He travels 35 miles at an average speed of 28 miles per hour.
Seren leaves her home at 07:10. She travels 48 miles in 2 hours.
  a How long does it take Ifan to reach the meeting?
  b At what speed does Seren travel?
  c Who travels faster?
  d Who reaches the meeting first?

5 Using measures

10 The time difference between Cardiff and Moscow is 3 hours.

Nihar flies from Moscow to Cardiff, and the flight time is 5 hours and 30 minutes.

If the plane departs at 14:00, what time will he arrive in Cardiff?

11 A supercar has a top speed of 270 kilometres per hour. What is this speed in metres per second?

12 The International Space Station travels at approximately 7700 metres per second.

How many kilometres does it travel in 1 hour?

> **Cross-curricular activity**
>
> Find out what the approximate circumference of the Earth is.
>
> Then work out how long it takes the International Space Station to travel around the Earth once.

13 An aeroplane flies at 1080 kilometres per hour.

   a How far would it travel, at this speed, in

   i 6 minutes   **ii** 1 minute   **iii** 15 seconds   **iv** 5 seconds?

   b Using the fact that 5 miles is approximately 8 kilometres, convert the aeroplane's speed into miles per hour.

14 a Find the area of each of these rectangles in square centimetres.

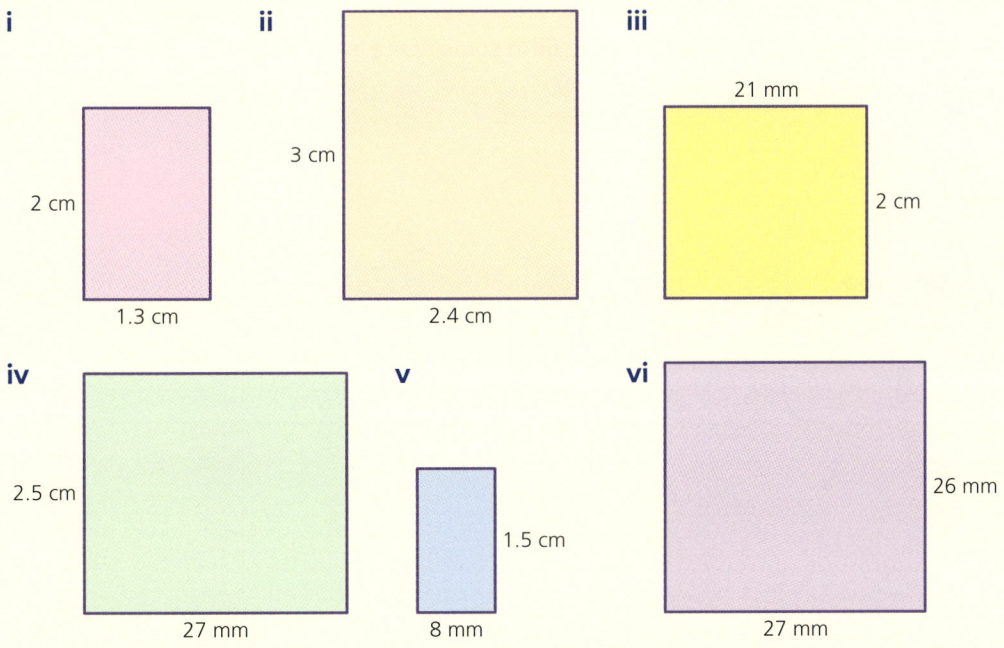

i — 2 cm by 1.3 cm

ii — 3 cm by 2.4 cm

iii — 21 mm by 2 cm

iv — 2.5 cm by 27 mm

v — 1.5 cm by 8 mm

vi — 26 mm by 27 mm

   b Convert your answers to square millimetres.

15 a Find the volume of each of these cuboids in cubic centimetres.

b Convert your answers to cubic millimetres.

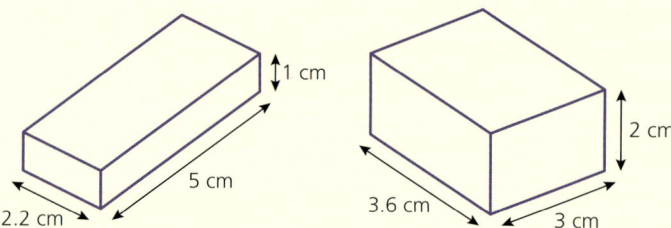

16 All the letters, except for *t*, are single variables that represent length.

The letter *t* represents a time.

What does each of the following expressions represent: a length, an area, a volume, a speed or none of these?

a $5a^2$

b ab^2

c $\dfrac{4\pi r^2}{t}$

d $\dfrac{3x - 2g}{t}$

e $5(w + 4f - x)$

17 Megan is in her mathematics exam.

In one question, she must work out the surface area of a cylinder.

She writes down two different formulas with the letters representing length.

Unfortunately, she cannot decide which one it is.

$2\pi r^2 + 2\pi r l$ OR $2\pi r^2 + 2\pi r^2 l$

Which one should she use? Explain your answer.

Band 3 questions

18 Gwyneth and Alun go on a cycling holiday to Bufi's Rock, shown on the map.

It is a small island and they cycle around it from Friday to Monday.

They keep a record of their journeys, but they forget to fill in some of the information.

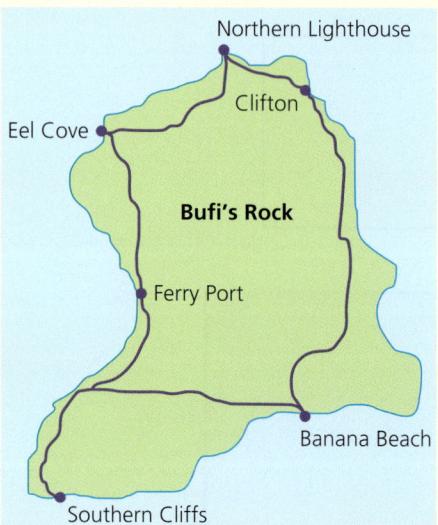

	From	To	Distance (km)	Time	Average speed (km per h)
Friday	Ferry Port	Southern Cliffs	18 km	1 h 15 mins	
	Southern Cliffs	Banana Beach		45 mins	25.3
Saturday	Banana Beach	Clifton	32.5 km		13
Sunday	Clifton	Northern Lighthouse		Half an hour	12.6
	Northern Lighthouse	Eel Cove	10.5 km		14
Monday	Eel Cove	Ferry Port	12 km	$\frac{3}{4}$ hour	

a   Copy and complete the table.

b   How far did Gwyneth and Alun cycle altogether?

c   What was their average cycling speed for the whole trip?

d   On which stretch of the journey did Gwyneth and Alun reach their highest speed?
    Why do you think their speed was higher on this stretch?

Consolidation 2: Chapters 4–5

Band 1 questions

1 Find these:

a 11% of 600 mm
b 5% of 480 cm^2
c $7\frac{1}{2}$% of 2000 kg
d 27% of 87 tonnes
e 41% of 94 km
f 7% of 56 cm
g 87% of 60 mm
h 17% of 99 g

2 Copy and complete this table.

Give all fractions in their simplest form.

The first row has been done for you.

Fraction	Decimal	Percentage
$\frac{17}{20}$	0.85	85%
$\frac{7}{10}$		
	0.55	
		14%
$\frac{21}{40}$		
	0.875	
		62.5%
$\frac{11}{5}$		

3 There were 15 000 women and 18 000 men living in the town of Drefach.

The population increased over the next few years.

The number of women increased by 10%.

The number of men increased by only 4%.

a How many men and how many women lived in Drefach after this increase?
b What was the total increase as a percentage of the original population?

Consolidation 2

4 Ollie leaves his house at 07:30.
He drives 52 miles to work at an average speed of 39 miles per hour.
At what time does Ollie arrive at work?

5 Harish runs 400 metres in 56 seconds.
Calculate Harish's average speed in metres per second. Round your answer to 2 decimal places.

6 The distance from New York to Los Angeles is 2457 miles.
A flight between the two cities takes 6 hours 30 minutes.
Find the average speed of the aeroplane.

7 A wind speed of 333 km per hour was recorded in Greenland in 1972.
How fast is this in metres per second?

8 a Find the volume of each of these cuboids in cubic centimetres.

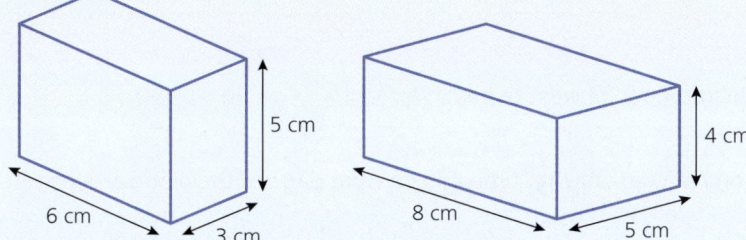

b Convert your answers to cubic millimetres.

9 Huw surveyed all the people in his class about how they get to school.
The results are shown in the pie chart.

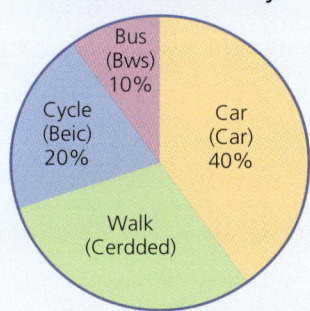

Travel to school survey

- Bus (Bws) 10%
- Cycle (Beic) 20%
- Car (Car) 40%
- Walk (Cerdded)

a What percentage of pupils said they walk to school?
b There are 30 pupils in Huw's class. How many of them said
  i car   **ii** walk   **iii** cycle   **iv** bus?

Band 2 questions

10 What multiplier would you use to
  a increase by 12%   **b** decrease by 20%   **c** increase by 7%
  d decrease by 80%   **e** increase by 400%?

11 Use a multiplier to find
  a 40 kg increased by 25%   **b** 120 cm increased by 30%   **c** 2200 metres increased by 22%
  d £310 decreased by 15%   **e** 670 decreased by 40%

Strategic competence

12. These items go down in value by the percentage shown. Find the new value of each item using a multiplier.
    a Exercise equipment £250, down 30%
    b Scooter £45, down 20%
    c Electric piano £640, down 25%

13. In a building supplies shop, all prices include VAT at 20%.
    i What is the price of these items before VAT is added?
    ii How much VAT is added?
    a Shovel: £42
    b Timber: £11.76 per metre
    c 100 nails: £4.80
    d Sand: £38.40 per tonne

Band 3 questions

Logical reasoning

14. A plane travels across 7 time zones in the direction of west to east.
    The flight time is 11 hours 15 minutes.
    If the plane departs at 13:00 local time on Monday, at what time and on what day will the plane arrive at its destination?

In the following four questions (15–18), either part **a** or part **b** is the correct equation of motion with constant acceleration.

These are important formulas used in physics:

s represents a length.

u and v both represent speed.

a represents acceleration whose units are metres per second^2.

t represents time.

Using dimensional analysis, decide whether **a** or **b** is the correct equation.

15. a $v = u + at$
    b $v = u + \dfrac{a}{t}$

16. a $s = \dfrac{u+v}{2t}$
    b $s = \left(\dfrac{u+v}{2}\right)t$

17. a $v = u + 2as$
    b $v^2 = u^2 + 2as$

18. a $s = ut + \dfrac{1}{2}at^2$
    b $s^2 = ut^2 + \dfrac{1}{2}at^2$

6 Using ratio and proportion and percentage

Coming up...
▶ Ratio and proportion review
▶ Profit and loss
▶ Simple and compound interest

Proportion crossword

Across

1. I need 2 eggs to make 14 cookies.
   How many cookies could I make with 4 eggs?
2. To cover 50 yards of footpath a workman needs 110 paving slabs. How many slabs would he need to cover 40 yards?
5. A piece of string is 99 cm long. It is cut into 9 equal sections. Four of them are put end to end.
   What is the total length of these four sections in centimetres?
7. Each jailer in a prison carries an identical bunch of keys. Two of the jailers are patrolling the prison. In total they have 42 keys. There are six jailers currently on their tea break.
   How many keys do they have between them?
9. Twenty tea bags need the leaves from 325 tea plants.
   How many tea plants must be plucked to make 8 tea bags?
11. There are 380 rooms in a hospital and usually 20 cleaners clean them all.
    Today only 8 cleaners are working. How many rooms can they clean?
13. The weight of five identical chocolate eggs is 315 g. How many grams do two eggs weigh?

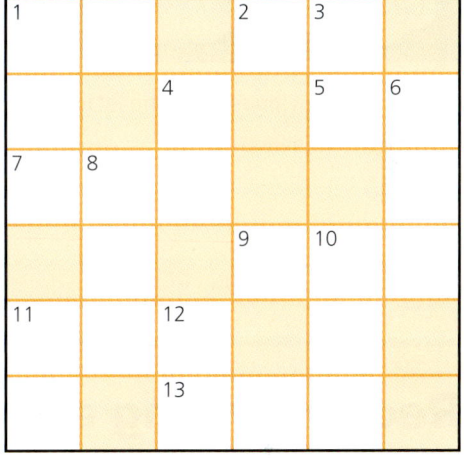

Down

1. I need 17 grams of sugar to make a single fondant fancy.
   How many grams would I need for 13 fondant fancies?
3. Seven children with exactly the same weight are chosen for a TV advert. Together they weigh 294 kg.
   How much, in kilograms, do two of them weigh?
4. Eleven identical pot plants cost £44. A restaurant buys four of these plants. How much do they spend?
6. A man has swum the English Channel 19 times, a total distance of 380 miles. By the end of next year he hopes to have swum the channel another 5 times. How many miles will that be **altogether**?
8. Ten identical parts for a machine cost £190. How much will fifteen of these parts cost?
10. It takes a robot 392 hours to build 28 electric cars. How long would it take the robot to build 24 of them?
11. Lena comes from Poland. She takes a holiday to Wales and changes 2880 Polish złoty into £600. By the end of her holiday she has just £2.50 left. How many Polish złoty would she get for this?
12. A runner runs a marathon at a steady pace. After 2 hours she has completed 14 miles.
    How many miles has she run after 3 hours?

6.1 Ratio and proportion review

Skill checker

Ratio clues in the clouds

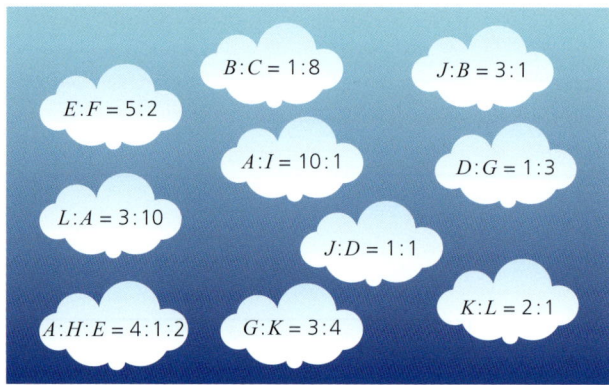

Can you find the values of all the letters using the clues in the clouds?
The value of I has been given.

$A=$ $B=$ $C=$ $D=$

$E=$ $F=$ $G=$ $H=$

$I=2$ $J=$ $K=$ $L=$

▶ Recap: finding an equivalent ratio

To find an equivalent ratio, multiply or divide each part of the ratio by the same number.

$4:1:15$ and $12:3:45$ are equivalent ratios. ← *Each part of the first ratio is multiplied by 3.*

Worked example

Complete these equivalent ratios by filling in the empty boxes.

a $3:7=\square:28$
b $30:40=3:\square=\square:80$
c $7:9=\square:45$
d $8:36=\square:9=\square:27$
e $10:\square=55:44$

Solution

a $3:7=12:28$
b $30:40=3:4=60:80$
c $7:9=35:45$
d $8:36=2:9=6:27$
e $10:8=55:44$

6 Using ratio and proportion and percentage

▶ Recap: splitting in a given ratio

Worked example

A charity divides £2600 between a rescue puppy centre and a donkey sanctuary in the ratio 8 : 5.
How much money does each of these animal centres receive?

Solution

Add the number of parts in the ratio:

$$8 + 5 = 13$$

Divide the total amount of money by this number of parts:

$$£2600 \div 13 = £200$$

The rescue puppy centre receives £200 × 8 = £1600
The donkey sanctuary receives £200 × 5 = £1000

As a check: £1600 + £1000 = £2600, the original amount of money being shared.

Activity

Proportion

Fflur the painter has been asked to paint the art room in a school.

She makes a paint colour called terracotta by mixing three different paint colours: orange, red and pink.

a The art room has a surface area of 96 m².
One litre of paint covers 12 m².
How much terracotta paint must Fflur make?

b The instructions for terracotta paint are:
Mix 5 litres of orange paint, 2 litres of red paint and 1 litre of pink paint.
What is the ratio of orange : red : pink paint Fflur must use?

c How many litres of each colour will Fflur need?   *Divide the total amount of paint needed in your ratio from part b.*

d One tin of paint contains 2 litres.
How many tins of each colour must Fflur buy?

6.1 Now try these

Band 1 questions

1 Simplify these ratios:

   a 21 : 3   **b** 6 : 24   **c** 3 : 27   **d** 14 : 35   **e** 63 : 56   **f** 100 : 65

   g 46 : 69   **h** 15 : 35 : 10   **i** 38 : 2 : 12   **j** 100 : 24 : 10

2 Find the number that goes in the box to make an equivalent ratio.

   a 4 : 5 = ☐ : 40   **b** 1 : 4 = 5 : ☐   **c** 1 : 9 = ☐ : 36

   d 5 : 4 = 45 : ☐   **e** 8 : 3 = ☐ : 27   **f** 9 : 7 = 81 : ☐

   g 25 : 6 = 75 : ☐   **h** 1 : 6 : 5 = ☐ : ☐ : 25   **i** 3 : 11 : 2 = ☐ : 44 : ☐

   j 41 : 2 : 6 = ☐ : 8 : ☐   **k** 12 : 22 = ☐ : 11 = 36 : ☐

Curriculum for Wales Mastering Mathematics: Book 3

3 For each shape:
  a Work out the fraction of the shape that is coloured.
  b Work out the ratio of the coloured area to the white area.

i 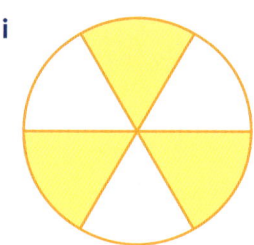 ii  iii  iv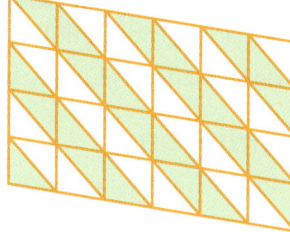

v vi  vii  viii

ix

4 Split £30 in these ratios:
  a 1:1   b 1:2   c 1:4   d 1:5   e 1:9   f 1:14
  g 2:1   h 2:3   i 2:13   j 3:2   k 3:7   l 4:1
  m 4:11   n 5:1   o 5:7

Band 2 questions

5 Write these ratios in their simplest form.
  a 4.5 kg : 250 g   b 750 ml : 3.5 litres   c 5 metres : 75 cm   d 48 cm : 1.68 metres

6 Write these ratios in their simplest form using whole numbers.
  a 1.1 : 1.2   b 1.5 : 2   c $7\frac{1}{2} : 10$
  d 9.5 : 10   e $6\frac{1}{4} : 8$ f $5\frac{3}{4} : 2\frac{1}{4}$

7 A carpenter buys 192 metres of timber to build 32 windows. How many metres of timber would the carpenter need to build
  a 1 window?
  b 19 windows?

8 Gireesha has 3 cats, 2 dogs and a tortoise.
Marika has 2 cats and a dog.
Find these ratios in their simplest form.
  a The number of cats Gireesha has to the number of cats Marika has.
  b The number of dogs Gireesha has to the number of dogs Marika has.
  c The total number of pets Gireesha has to the total number Marika has.

9 I have a photo measuring 4 cm by 6.4 cm.

The photo can be enlarged, for example by doubling the height and width to 8 cm by 12.8 cm.

In a shop, picture frames with the following dimensions are on sale:

a 10 cm by 16 cm
c 7.5 cm by 12 cm
b 25 cm by 40 cm
d 12 cm by 18 cm

Which of these frames would be suitable for my photo? There may be more than one answer.

10 Look at the three flags.

a

b

c

For each one work out
 i the fraction of the flag that is white
 ii the ratio of the areas green to white to black.

For the flag in part **c**, the heights of the green and black stripes are both 25 cm. The height of the white stripe is 10 cm.

11 Eira is making ice cream. She uses this recipe.

> 320 ml double cream
> 300 ml milk
> 140 g caster sugar
> 1 vanilla pod
> 4 egg yolks
> **Serves 4**

Find how much cream, milk and sugar Eira would need to make ice cream for

a 8 people
b 1 person
c 3 people
d 6 people.

12 Tajikistan is a country in central Asia. The currency is called somoni.

Two British pounds is roughly equivalent to 27 Tajik somoni.

Copy and complete this table to help you answer the following questions.

British pounds	1	2	100	
Tajik somoni		27		540

a How many Tajik somoni would you get for £1?
b How many Tajik somoni would you get for £100?
c How many British pounds could be bought for 540 Tajik somoni?
d Roughly how many pence is one somoni worth?

Band 3 questions

13 Sion's salary changes each month, but his expenses are fixed. Each month he has to spend
- £800 on rent
- £240 on food
- £100 on bills
- £200 on his train fare.

Anything that is left is Sion's **disposable income**.

a Find the ratio of his spending on rent to his spending on train fares, giving your answer in its simplest form.

b Find the ratio of his spending on rent to his spending on food, giving your answer in its simplest form.

c In January Sion earns £2340.
Find the ratio of his rent to his disposable income for January in its simplest form.

d In February Sion earns £1540.
Find the ratio of his rent to his disposable income for February in its simplest form.

e In March Sion moves to a different flat. His food bill remains at £240 per month. His total monthly expenses also remain the same. His monthly rent, bills and train fare are now in the ratio 14 : 3 : 5.
Find how much Sion now pays each month for his

    i rent   **ii** bills   **iii** train fare.

14 There are roughly 250 billion stars in the Milky Way galaxy.

On average, there is one planet orbiting each star.

Astronomers are interested in how many planets are similar to Earth, because intelligent life may live there.

They have estimated that there could be 40 billion planets similar to Earth.

a What is the ratio of Earth-like planets to non-Earth-like planets? Give your answer in its simplest form.

b Of these 40 billion Earth-like planets, 10 billion could be orbiting stars that are similar to the Sun.
What fraction of Earth-like planets are orbiting Sun-like stars?

c What is the ratio of Earth-like planets orbiting Sun-like stars to Earth-like planets orbiting non-Sun-like stars? Give your answer in its simplest form.

d Suppose that on three planets out of every one billion Earth-like planets orbiting Sun-like stars there is intelligent life.
On how many planets in the Milky Way galaxy can we expect intelligent life?

15 Look at this model of the letter H.

It is made from two cuboids standing upright and a smaller cube between them.

The base of each cuboid is a square with side length 2 cm.

The height of each cuboid is three times its width.

The cube in the centre has a side length of 2 cm.

Find

a the ratio of the volume of one cuboid to the cube

b the ratio of the total height of the model to its total width.

The model is placed into a tray of red paint so that the base of each cuboid is painted red. The rest of the surface area remains blue.

c What fraction of the surface area is now painted red?

d What is the ratio of red to blue surface area?

6.2 Profit and loss

Skill checker

Students from class 7LF have received their Welsh test scores, but only as a mark and not a percentage.
Calculate each of their test scores as a percentage.

- Bronwen's score is 20
- Aled's score is 17
- **Cymraeg test out of 32**
- Carwyn's score is 28
- Delyth's score is 9

Profit is when something is sold for more than it cost.
The actual profit is the selling price, subtract the buying price.
Actual profit = selling price − buying price
Loss is when something is sold for less than it cost.
The actual loss is the buying price, subtract the selling price.
Actual loss = buying price − selling price

Worked example

Find the actual profit or loss of the following.

a Buying price = £20, selling price = £45
b Buying price = £80, selling price = £30

Solution

a The selling price is greater than the buying price, so:
   Actual profit = £45 − £20 = £25
b The selling price is less than the buying price, so:
   Actual loss = £80 − £30 = £50

Profit and loss can be written as a percentage change from the original amount (the buying price). This is a useful way to make direct comparisons between different profits and losses.

$$\text{Percentage profit} = \frac{\text{actual profit}}{\text{original amount}} \times 100\%$$

$$\text{Percentage loss} = \frac{\text{actual loss}}{\text{original amount}} \times 100\%$$

Curriculum for Wales Mastering Mathematics: Book 3

> ### Worked example
>
> **a** A smartwatch is bought for £250 and sold for £300.
> What is the percentage profit?
>
> **b** A car is bought for £8000 and sold for £4500.
> What is the percentage loss?
>
> ### Solution
>
> **a** Actual profit = £300 − £250 = £50
>
> So, the percentage profit = $\dfrac{\text{actual profit}}{\text{original amount}} \times 100\%$
>
> $\dfrac{50}{250} \times \dfrac{100\%}{1} = 20\%$
>
> The smartwatch is sold at 20% profit.
>
> **b** Actual loss = £8000 − £4500 = £3500
>
> So, the percentage loss = $\dfrac{\text{actual loss}}{\text{original amount}} \times 100\%$
>
> $\dfrac{3500}{8000} \times \dfrac{100\%}{1} = 43.75\%$
>
> The car is sold at 43.75% loss.

6.2 Now try these

Band 1 questions

For questions 1 to 10, find the actual profit or loss. State whether it is a profit or loss.

	Buying price	Selling price
1	£17.00	£19.50
2	£21.00	£45.00
3	£7.00	£18.40
4	£19.00	£11.00
5	£35.00	£18.50
6	£45.75	£60.00
7	£25.50	£30.00
8	£9.50	£14.00
9	£21.80	£18.30
10	£100.00	£66.00

For questions 11 to 18, find the total profit or loss. State whether it is a profit or loss.

	Item	Buying price	Selling price
11	Pens	£4.00 for 10	50p each
12	Pencils	£15.00 for 100	20p each
13	Rulers	£10.00 for 20	65p each
14	Paper	£5.00 for 100 sheets	4p per sheet
15	Apples	£2.50 for 25	9p each
16	Oranges	£20.00 for 50	35p each
17	Pears	£8.00 for 20	43p each
18	Bananas	£30.00 for 60	40p each

Band 2 questions

Find the percentage profit or loss of the following:

	Buying price	Selling price
19	£4.00	£6.00
20	£2.50	£3.00
21	£9.00	£7.20
22	£6.50	£5.85
23	£62.00	£63.24
24	£0.90	£0.72
25	£37.00	£74.00
26	£18.00	£60.30
27	£19.00	£209.00
28	£23.00	£0.00

Remember

Percentages can be greater than 100%.

In this topic, for example, a 200% percentage profit means the selling price was three times the buying price.

29. Ffinian has a snack shop at school.

    He buys bars of chocolate for 20p and sells them for 28p.

    What is his percentage profit?

30. Hendy Electronics purchases televisions for £350 and sells them for £472.50

    What percentage profit does the shop make on each television?

31. Mari buys a painting for £40.

    She decides she does not like it and sells it for £33.

    What is her percentage loss for this?

Band 3 questions

32. Tomos owns a clothes shop.

    He buys designer coats for £250 and sells them at a profit of 35%.

    What is the selling price of the coats?

33. Constance buys a new mobile phone for £650.

    After a week, she decides to sell it and makes a loss of 18%.

    How much did she sell it for?

34. Scarlett buys a box of plates to sell at her market stall.

    She wants to make an overall profit of 20% from the box.

    She pays for 50 plates in the box, with each plate costing her £3.00.

    Scarlett accidentally breaks five plates before she can sell them.

    What should the selling price of each plate be for Scarlett to make an overall percentage profit of 20% on the money she paid for the box of plates?

35. Clementine buys Venus flytrap plants for £2.50 each.

    It is known that 20% of the plants will die before being sold.

    a   At what price must Clementine sell the plants to break even?

    b   At what price must she sell them to make a 20% profit?

Cross-curricular activity

The Venus flytrap is an example of a carnivorous plant. It traps its prey by snapping its claw-like leaves shut.

Find out about other types of carnivorous plants and the methods they use to trap their prey.

6.3 Simple and compound interest

Skill checker

A multiplier is a number you can multiply an amount by to increase or decrease it by a percentage. You looked at this in Chapter 4, section 2.

What would the multipliers be for the following percentage increases and decreases?

	5%	10%	25%	75%	100%	2.5%	0.5%	0.01%
Increased by								
Decreased by								

Interest is the amount of money paid at regular time periods when money is borrowed or lent.

Interest is paid when money is invested in a bank. Interest can be seen as a reward for being a loyal customer.

However, a customer is charged interest if money is borrowed from the bank.

Simple interest is when a percentage is worked out for an amount, and is then multiplied by the number of time periods the money is invested for.

Compound interest is when the interest is compounded. This means that when interest is added to an amount, in the next time period, the added interest is part of the new interest calculated.

Note

It is always 'borrow from' and 'lend to'.

Worked example

a Carwyn invests £500 on which 15% **simple interest** is paid per annum (each year).

How much money will he have after 3 years?

b Cerys invests £500 in another bank which offers 15% **compound interest** per annum.

How much money will she have in the bank after 3 years?

Discussion activity

Which language does the word *annum* originate from?

Do you know any other words, in English or Welsh, that are borrowed from this language?

Solution

a Interest after 1 year = 15% of £500 = $\frac{15}{100} \times £500 = £75$

Interest after 3 years = $3 \times 75 = £225$

Therefore, the amount Carwyn has after 3 years is:

£500 + £225 = £725

b After the first year, the amount she has is £500 + $\frac{15}{100}$ × £500 = £575

After the second year the amount she has is £575 + $\frac{15}{100}$ × £575
= £661.25

After the third year the amount she has (correct to the nearest penny)
is £661.25 + $\frac{15}{100}$ × £661.25 = £760.44.

The amount of interest received increases more quickly with compound interest over the same time period.

The interest Cerys receives is:
£760.44 − £500 = £260.44 (correct to the nearest penny)

> Using the multiplier method, this calculation would be:
> 500 × 1.15^3
> Refer to Chapter 4, sections 4.2 and 4.3 to see why this works.

6.3 Now try these

Band 1 questions

Calculate the total amount of money a person would have at the end of the time periods stated, based on simple interest.

	Investment	Simple interest % per annum	Time period in years
1	£80	5	3
2	£450	7	4
3	£900	3	8
4	£2000	11	3
5	£856	3	4

Band 2 questions

Calculate the total amount of money a person would have at the end of the time periods stated, based on compound interest.

Give your answers to the nearest penny.

Use the multiplier method to answer questions 8, 9 and 10.

	Investment	Compound interest % per annum	Time period in years
6	£60	5	2
7	£500	8	3
8	£800	3	4
9	£5000	11	7
10	£10 000	3	12

Band 3 questions

11 Jamil invests £2000 in a bank account.

The bank pays compound interest of 4% per annum.

Find the value of his investment, and the interest paid, after

a 3 years
b 5 years
c 11 years.

12 Avin invests £7000 in a building society for 11 years.

Find the value of his investment, and the interest earned, if the annual interest rate is

a 14%
b 6%
c 2.9%
d 0.7%.

13 A key worker's salary in 2021 is £40 000 a year.

If their pay is increased by 4% each year, what would their salary be in 5 years' time?

14 Dafydd has invested £3000 in a bank.

The bank offers (compound) interest of 14% per annum.

After how many years will Dafydd's money have doubled?

15 Angharad and Meirion invest their money in two different bank accounts.

Angharad invests £40 000 at 9% compound interest per annum.

Meirion invests £30 000 at 12.5% compound interest per annum.

Who has more money after 10 years, and by how much?

> **Key words**
>
> Here is a list of the key words you met in this chapter.
>
> | | | | |
> |---|---|---|---|
> | Compound interest | Loss | Percentage loss | Percentage profit |
> | Profit | Proportion | Ratio | Simple interest |
>
> Use the glossary at the back of this book to check any you are unsure about.

Review exercise: using ratio and proportion and percentage

Band 1 questions

1. If six teenagers can plant 108 trees in a day, how many trees can ten teenagers plant in a day?

2. Aunt Liyana decides to give her nephew and niece, Hari and Lili, £540 between them.
   Because Lili is older than Hari, Aunt Liyana decides that the money must be divided so that Lili gets £5 for every £4 that Hari gets.
   How much does each child get?

3. Eight workers can plaster a block of flats in six days.
   a   How many workers are needed to plaster the flats in four days?
   b   How many days would it take one worker to plaster the block of flats on their own?

4. The number of windows in a church, a school and a house are in the ratio 3 : 7 : 1.
   There are 154 windows altogether and of these, 42 are in the church.
   How many windows are there in the school and in the house?

5. Medwyn buys a second-hand car for £3150.
   He then sells it for £5000.
   What profit does he make?

6. It will take four workers five hours to repair a road.
   The boss wants it finished in two hours or less.
   How many workers are needed?

7. Niamh buys a house for £60 000 and sells it for £88 000.
   What is her percentage profit?

Band 2 questions

8. Irin invests £3000 in a bank which offers 21% simple interest per annum.
   What will her investment be worth in 3 years' time?

9. Mair invests £10 000 in a bank which offers 8% compound interest per annum.
   How much interest will she earn after 3 years?

10. A pump can empty a swimming pool in six hours.
    How long would it take to empty the pool using
    a   two pumps
    b   three pumps
    c   12 pumps?

Band 3 questions

11. Sami buys a mobile phone for £850.
    It loses value by 20% each year.
    What will it be worth in 3 years' time?

12. The population of an island is 70 000 and increases by 17% each year.
    After how many years will the population be trebled?

13. Nihar invests some money in a bank.
    The bank provides Nihar with the formula he should use to work out his investment.
    $A = 700 \times 1.04^n$
    a   How much did Nihar invest?
    b   What was the rate of interest?
    c   What does the letter n represent in the formula?
    d   How much will the investment be worth in 20 years' time?

14. A radioactive element has a mass of 150 g.
    Its mass reduces by 15% each year.
    How long will it take for the mass of the element to halve?

7 Equations, expressions, formulas and inequalities

Coming up...
- Solving equations (review)
- Using the laws of indices
- Expanding brackets
- Formulas
- Solving inequalities

Magic square

a Copy out the grid below, replacing each equation with its solution.

$c - 7 = 10$	$s - 9 = 15$	$6 + f = 7$	$p + 5 = 13$	
$k + 8 = 31$		$10x = 70$	$3n = 42$	
$5r = 20$			$g - 8 = 12$	$z + 15 = 37$
$g - 2 = 8$		$t - 10 = 9$	$10j = 210$	$14h = 42$
$12y = 132$	$\dfrac{p}{3} = 6$	$\dfrac{a}{3} = 5$		$4h = 36$

The solutions to these equations form a **magic square**. The rows, columns and diagonals all add up to the same total.

b What is this total?

c Can you make up some equations that could be used in the empty boxes to complete the magic square?

7.1 Solving equations review

Skill checker

Find the number that each symbol represents.

The numbers given are the row and column totals.

✓	🚲	🚲	23
✓	🛡	🛡	21
✓	🚲	🛡	22
33	17	16	

✦	●	🕷	🚌	23
●	🚌	🕷	⊘	16
✦	🕷	🕷	🕷	35
20	15	27	12	

▶ Solving equations with an unknown on one side

Remember that an **equation** says that one expression is equal to another. For example:

$12x + 6 = 30$

$32 - 5y = 22$

x and *y* are called 'unknowns'.

Solving an equation means finding the value of the **unknown** that makes the equation true – this is the **solution** to the equation.

You can use the **balance method** to solve an equation.

Apply the same operation to both sides of the equation to keep it balanced.

Worked example

Solve:

a $\quad 12x + 6 = 30$

b $\quad 32 - 5y = 22$

Solution

a

$-6 \begin{pmatrix} 12x + 6 = 30 \\ 12x = 24 \end{pmatrix} -6$

Imagine a pair of old-fashioned scales, ⚖ ; to keep them balanced you add or subtract the same amount to both sides.

$\div 12 \begin{pmatrix} 12x = 24 \\ x = 2 \end{pmatrix} \div 12$

You can check your answer by substituting it back into the equation.
$12 \times 2 + 6$
$= 24 + 6$
$= 30 \ ✓$

Communication using symbols

b Method 1

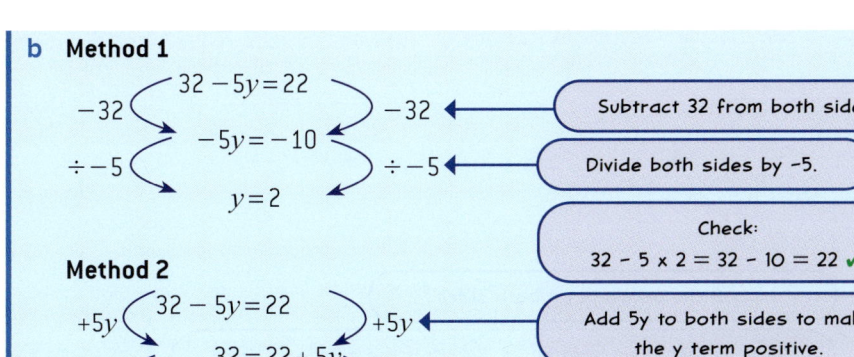

- Subtract 32 from both sides.
- Divide both sides by −5.
- Check: $32 - 5 \times 2 = 32 - 10 = 22$ ✓

Method 2

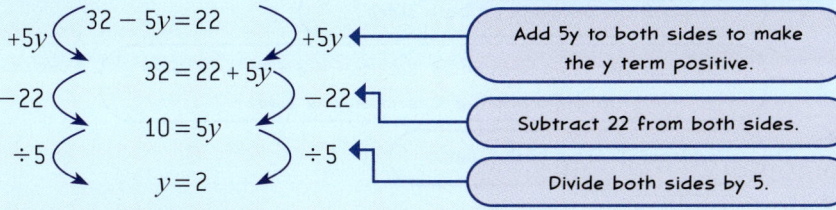

- Add 5y to both sides to make the y term positive.
- Subtract 22 from both sides.
- Divide both sides by 5.

Worked example

Solve:

a $\dfrac{15}{c} = 5$

b $\dfrac{x}{8} - 2 = 3$

Solution

a

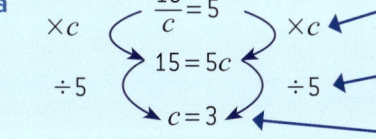

- Multiply both sides by c remove the unknown from the denominator.
- Divide both sides by 5.
- In the final answer, it is usual to write the unknown first.

Check: $\dfrac{15}{3} = 5$ ✓

b

$+2 \Big(\dfrac{x}{8} - 2 = 3 \Big) +2$

$\times 8 \Big(\dfrac{x}{8} = 5 \Big) \times 8$

$x = 40$

- Add 2 to both sides.
- Multiply both sides by 8.

Check: $\dfrac{40}{8} - 2$
$= 5 - 2$
$= 3$ ✓

Worked example

You can use bar model diagrams to help you solve equations.

Solve:

a $4x + 5 = 19$

b $4x + 7 = 2x + 23$

Solution

a

x	x	x	x	5
19				

x	x	x	x
14			

x
3.5

$x = 3.5$

b $4x + 7 = 2x + 23$

x	x	x	x	7
x	x	23		

x	x	x	x
x	x	16	

x	x
16	

x
8

$x = 8$

Worked example

Solve these equations.

a $5(7x+5) = 4(7x+8)$

b $\dfrac{3}{5x-1} = \dfrac{4}{2x+1}$

Solution

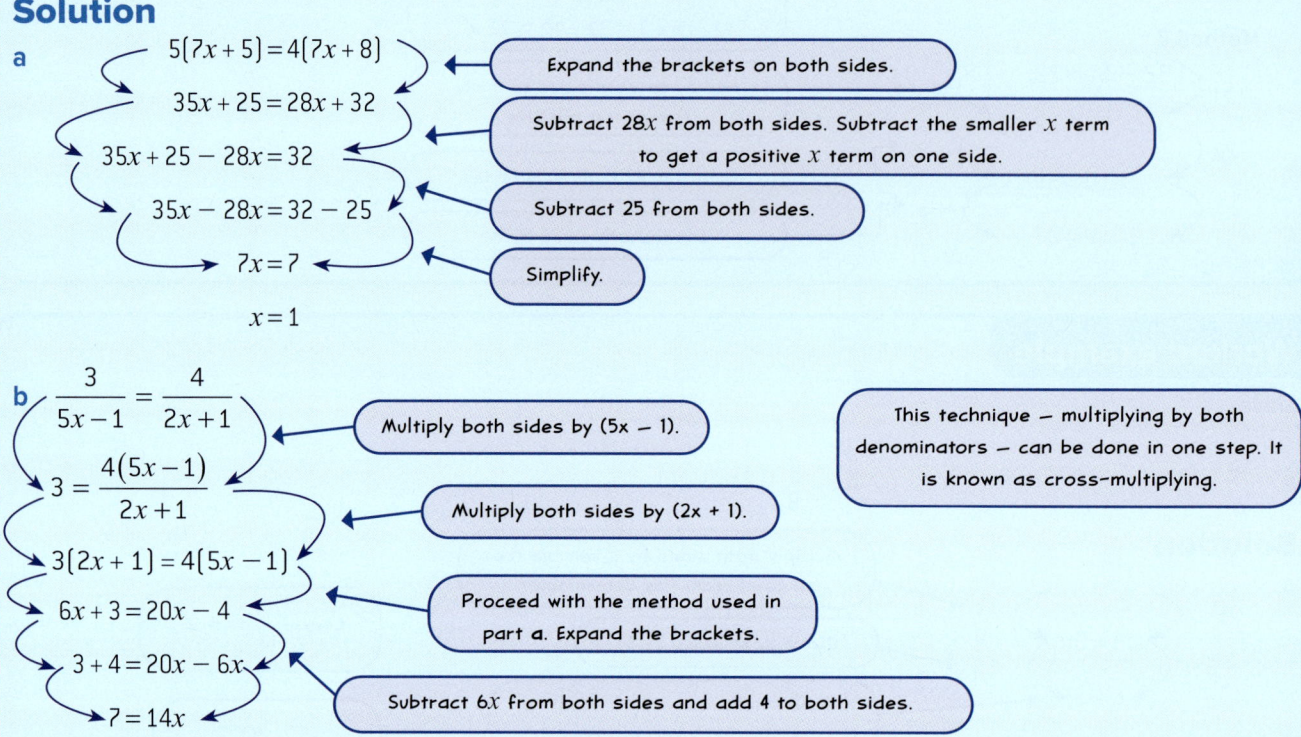

a
$5(7x+5) = 4(7x+8)$ — Expand the brackets on both sides.
$35x + 25 = 28x + 32$ — Subtract $28x$ from both sides. Subtract the smaller x term to get a positive x term on one side.
$35x + 25 - 28x = 32$
$35x - 28x = 32 - 25$ — Subtract 25 from both sides.
$7x = 7$ — Simplify.
$x = 1$

b
$\dfrac{3}{5x-1} = \dfrac{4}{2x+1}$ — Multiply both sides by $(5x - 1)$.
$3 = \dfrac{4(5x-1)}{2x+1}$ — Multiply both sides by $(2x + 1)$.
$3(2x+1) = 4(5x-1)$ — Proceed with the method used in part a. Expand the brackets.
$6x + 3 = 20x - 4$
$3 + 4 = 20x - 6x$ — Subtract $6x$ from both sides and add 4 to both sides.
$7 = 14x$
$x = \dfrac{7}{14} = \dfrac{1}{2}$

This technique — multiplying by both denominators — can be done in one step. It is known as cross-multiplying.

▶ Problem solving using equations

Worked example

One cinema ticket costs £c.

Bronwen buys five tickets and one large box of popcorn, which is £3.50.

She gets £4 change from £60.

a Write an equation for c and solve it to find the cost of one ticket.
b Check your answer.

Solution

a The price of one cinema ticket is c, so the cost of 5 tickets is $5c$.

The cost of the tickets and popcorn is $5c + 3.5$.

Bronwen gets £4 change from £60, so the total cost is £56.

$5c + 3.5 = 56$ — Subtract 3.5 from both sides.
$5c = 52.5$ — Divide both sides by 5.
$c = 10.5$

The final answer is an amount of money and must have 2 decimal places. In the working, however, this isn't necessary. The £ sign only needs to be given with the final answer.

The cost of each ticket is £10.50.

b Check that Bronwen bought 5 tickets for £10.50 each and popcorn for £3.50:

$5 \times 10.5 + 3.5$
$= 52.5 + 3.5$
$= 56$

The amount she spends is £56, which agrees with her getting £4 change from £60.

Worked example

At the start of the week, Jamie and Holly both have the same amount of credit on their school meal cards. This week:

Jamie has 5 school meals and has £1 credit left.

Holly has 2 school meals and has £8.80 credit left.

a Find the cost of one school meal.
b How much credit did Jamie and Holly start with?

Solution

a Let c represent the cost of one meal in pence. Work in pence because whole numbers are easier to work with.

Jamie's credit = Holly's credit
$5c + 100 = 2c + 880$
$3c + 100 = 880$
$3c = 780$
$c = 260$

So a school meal costs 260p or £2.60.

b Substitute $c = 260$ into the expression for Jamie's credit.
$5c + 100 = 5 \times 260 + 100$
$= 1400$
So Jamie's starting credit was 1400p or £14.
Check that Holly's credit is also 1400p:
$2c + 880 = 2 \times 260 + 880$
$= 1400$

Activity

Gulnaz has got her maths homework wrong.

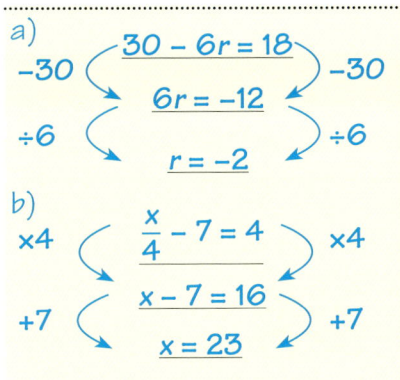

- How can you show Gulnaz that she has the wrong answers without solving the equations?
- Where has Gulnaz gone wrong?
- Work out the correct solutions to both equations.

Curriculum for Wales Mastering Mathematics: Book 3

7.1 Now try these

Band 1 questions

1 Solve these equations.

Show a check for **one** of your answers by substituting into the equation.

a $m + 4 = 11$ b $p - 6 = 10$

c $5 + d = 7$ d $y - 8 = 22$

e $4b = 12$ f $\dfrac{d}{2} = 3$

g $-4j = 12$ h $\dfrac{1}{2}x = 5$

i $\dfrac{x}{2} = \dfrac{5}{2}$ j $\dfrac{2}{3}x = 8$

2 Solve these equations.

Show a check for **one** of your answers by substituting into the equation.

a $2a + 3 = 5$ b $4d + 1 = 9$

c $2 + 2c = 8$ d $6c + 2 = 20$

e $6m + 4 = 16$ f $3d - 2 = 10$

g $2k - 10 = 8$ h $7 - 4l = 15$

i $6 - 2m = 0$ j $18 - 5p = 8$

3 Solve these equations.

Show a check for **one** of your answers by substituting into the equation.

a $\dfrac{x}{6} + 3 = 9$ b $3 = 4 - \dfrac{4}{p}$

c $4 + \dfrac{d}{3} = 19$ d $\dfrac{7}{q} - 24 = 4$

e $\dfrac{2}{f} - 7 = 3$ f $\dfrac{3}{a} + 4 = 10$

4 Solve these equations.

Show a check for **one** of your answers by substituting into the equation.

a $2r = 4.2$ b $s + 6.2 = 11$

c $5.4 = 8 - t$ d $4u - 5.6 = 10$

e $9.5 = 7.1 + 3y$ f $15 - 3n = 1.2$

Band 2 questions

5 a Match each of these equations with its solution.

     i $3x - 2 = 10$ ii $5x - 1 = 24$ iii $4x + 11 = 17$

     iv $2x + 5 = 10$ v $3x - 33 = 0$

| $x = 5$ | $x = 1\frac{1}{2}$ | $x = 3$ | $x = 4$ | $x = 11$ | $x = 2\frac{1}{2}$ |

b Which value of x is not used?

Write an equation with this value of x as its solution.

7 Equations, expressions, formulas and inequalities

6 Jo has a pencils and Humza has b pencils.
Write in words the meaning of each of these equations.
The first one has been done for you.
   a $a + b = 12$ In total Jo and Humza have 12 pencils
   b $a - b = 4$
   c $a = 8$
   d $a = 2b$

7 Tomos is buying some fish and chips in Calais.

He pays €13 for three portions of fish and four portions of chips.
   a Write down an equation to show this information.
   Use f to represent the cost of one portion of fish.
   b Solve your equation to work out the price of one portion of fish.
   c Check your answer.

8 Steffan buys five pens in a shop for £1.20.
   a Write an equation for this information using p pence as the cost of one pen.
   b Solve your equation to find the cost of one pen.

Band 3 questions

9 Three prizes are given from a prize fund, with these conditions.
   - The first prize is twice the third prize.
   - The second prize is £5 more than the third prize.
   a Find the values of the prizes when the total prize money is
      i £65   ii £37.
   b It is decided that the first prize must be at least £5 more than the second prize.
      Investigate the total prize money needed to meet all three conditions.

10 Both of these rectangles have a perimeter of 30 cm.

   a 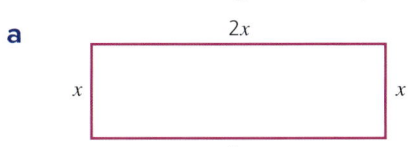   b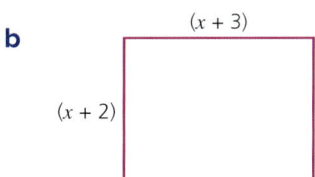

   Find the length and width of each rectangle.

11 Each obtuse angle of this parallelogram is 32° more than each acute angle.
   Let one obtuse angle be x.
   Form an equation for x and solve it.
   Find all the angles of the parallelogram.

12 Sunita has 11 metres of wire netting.
   She says, 'I'm making a run for my guinea pig. I'll make it against the garden wall.'
   She makes the run 3 metres wide.
   The length of the run is x metres.
   Form an equation for x and solve it.

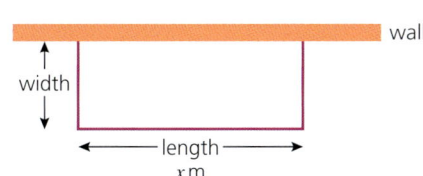

13 Solve these equations by filling in the blanks on the bar diagrams.

a $3x + 7 = 31$

| x | x | x | 7 |

| x | x | x |

| x |

b $5x + 8 = 23$

| x | x | x | x | x | 8 |

| x | x | x | x | x |

| x |

14 Write down using algebra all the steps in the solution of the equation below.

$5x + 7 = 3x + 29$

| x | x | x | x | x | 7 |
| x | x | x | 29 | | |

| x | x | x | x | x |
| x | x | x | 22 | |

| x | x |
| 22 | |

| x |
| 11 |

15 Solve these equations.

Show a check for **one** of your answers by substituting into the equation.

a $2n + 1 = n + 3$ b $3b - 2 = 4 + b$ c $5k - 1 = 2k + 2$
d $4m - 3 = 5 + 2m$ e $10 + 2f = 15 - 3f$

16 Solve these equations by expanding the brackets first.

Choose **one** of your solutions to check by substituting back into the original equation.

a $4(x + 1) = 28$ b $4(a + 2) = 18$ c $5(b - 7) = 16$
d $10(7 - c) = 26$ e $13 = 5(2d - 4)$ f $6(3e + 5) = 14$

17 Anwen and Seimon are both thinking of numbers.

> Anwen says:
> I think of a number.
> I multiply by 6.
> I add 7.
> My answer is 19.

> Seimon says:
> I think of a number.
> I double it.
> I subtract 6.
> My answer is 18.

a i Write down an equation for Anwen's number.
   ii Solve your equation.
b i Write down an equation for Seimon's number.
   ii What number was Seimon thinking of?

18 Jamie and Heulwen are allowed the same amount of screen time one afternoon.

Jamie spends 1 hour playing computer games and watches some TV.

Heulwen only watches TV. She watches TV for four times longer than Jamie.

a How long does Jamie spend watching TV?
b How long does Heulwen spend watching TV?

19 Jac says:

> One small tile weighs $\frac{1}{2}$ kilogram.
> Eight small tiles and 6 large tiles weigh 16 kg.
> What is the weight of one large tile?

Write an equation to solve Jac's puzzle.

20 Solve these equations by expanding the brackets first.

Show a check for **one** of your answers by substituting into the equation.

a $7x + 4 = 4(3x + 1)$
b $2(x + 4) = 5(x + 1)$
c $5(7x + 7) = 7(4x + 6)$

21 Solve these equations by cross-multiplying first.

a $\dfrac{1}{2x + 4} = \dfrac{1}{x + 1}$

b $\dfrac{3}{6x + 9} = \dfrac{4}{7x + 7}$

c $\dfrac{6}{4x + 1} = \dfrac{5}{5x - 5}$

> These are harder to check. For one of the equations, substitute your answer into both sides of the equation.

7.2 Using the laws of indices

Skill checker

Write the following in index form.
The first one has been done for you.

a $6 \times 6 \times 6 \times 6 \times 6 = 6^5$
b $5 \times 5 \times 5 \times 5 \times 5 \times 5$
c 4×4
d $3 \times 3 \times 3$
e $8 \times 8 \times 8 \times 8$
f $10 \times 10 \times 10 \times 10 \times 10 \times 10$

▶ The laws of indices

$a^n = a \times a \times a \times \ldots \times a$ (n times)

So, for example:

$a^5 = a \times a \times a \times a \times a$

Here are some **laws of indices**:

Law 1 $a^x \times a^y = a^{x+y}$

> For example $a^2 \times a^3 = a^{2+3} = a^5$
> $a^2 \times a^3 = (a \times a) \times (a \times a \times a) = a \times a \times a \times a \times a = a^5$

Law 2 $a^x \div a^y = a^{x-y}$

> For example $a^5 \div a^3 = a^{5-3} = a^2$
> $a^5 \div a^3 = \dfrac{a \times a \times a \times a \times a}{a \times a \times a} = \dfrac{a \times a}{1} = a^2$

Law 3 $(a^x)^y = a^{xy}$

> For example $(a^2)^3 = a^{2 \times 3} = a^6$
> $(a^2)^3 = (a \times a) \times (a \times a) \times (a \times a) = a^6$

Law 4 $a^1 = a$

Law 5 $a^0 = 1$

> For example $5^1 = 5$ and $3^0 = 1$
> $3^3 = 3 \times 3 \times 3 = 27$
> $3^2 = 3 \times 3 = 9$ Subtracting 1 from the index means dividing by 3.
> $3^1 = 3$
> Following this pattern:
> $3^0 = 1$

Worked example

Simplify these expressions.

a $2a^3b^7 \times 4ab^5$

b $\dfrac{12a^4b^3}{4ab^2}$

c $(2c^2)^3$

Solution

a   First, multiply the numbers: $2 \times 4 = 8$
Next use the rules of indices: $a^x \times a^y = a^{x+y}$
$a^3 \times a = a^3 \times a^1 = a^4$
$b^7 \times b^5 = b^{12}$
So
$2a^3b^7 \times 4ab^5 = 8a^4b^{12}$

b   First, divide the numbers: $12 \div 4 = 3$
Next use the rules of indices: $a^x \div a^y = a^{x-y}$
$\dfrac{a^4}{a} = \dfrac{a^4}{a^1} = a^3$
$\dfrac{b^3}{b^2} = b^1 = b$
So
$\dfrac{12a^4b^3}{4ab^2} = 3a^3b$

c   First, cube the number: $2^3 = 8$
Next use the rules of indices: $(a^x)^y = a^{xy}$
$(c^2)^3 = c^6$
So $(2c^2)^3 = 8c^6$

7.2 Now try these

Band 1 questions

1. Write these in index form.
   The first one has been done for you.
   - a $a \times a = a^2$
   - b $b \times b \times b \times b$
   - c $c \times c \times c \times c \times c$
   - d $d \times d \times d$
   - e $f \times f \times f \times f \times f \times f$
   - f $g \times g \times g \times g \times g \times g$

2. Write these expressions as single powers.
   - a $a^6 \times a^3$
   - b $b^7 \times b^2$
   - c $c^5 \times c^4$
   - d $d^{50} \times d$
   - e $e^6 \div e^4$
   - f $f^6 \div f^2$
   - g $g^{10} \div g^{10}$
   - h $h^{50} \div h^{49}$
   - i $i^3 \times i^4$
   - j $j^2 \times j^2$
   - k $k^{15} \times k^{20}$
   - l $l^5 \times l$
   - m $m^5 \div m^2$
   - n $n^5 \div n$
   - o $p^{50} \times p^{49}$
   - p $q^{100} \div q^{100}$

3. Copy and complete these algebra pyramids.
   The value in each brick is found by multiplying the values in the two bricks underneath it.

   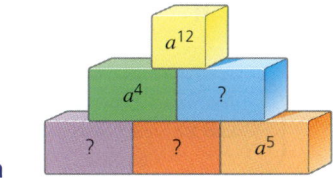
   a
   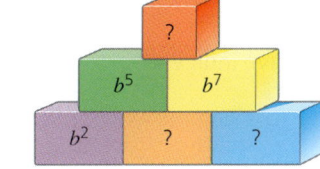
   b

4. Copy and complete these algebra pyramids.
   The value in each brick is found by multiplying the values in the two bricks underneath it.

   a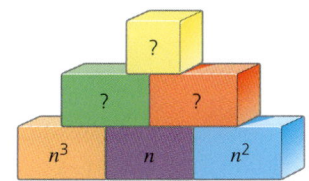
   b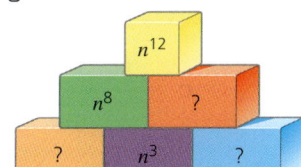
   c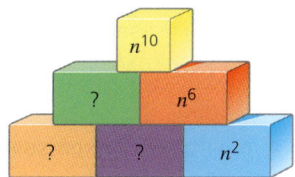

Band 2 questions

5. Jake simplifies $4s^3 \times 5s^2$ like this:

   $4s^3 \times 5s^2 = 4 \times 5 \times s^3 \times s^2$
   $= 20s^5$

   Use Jake's method to simplify these expressions.
   - a $6a^3 \times 3a^5$
   - b $9b^7 \times 6b^5$
   - c $3c^5 \times 2c^9$
   - d $30d^{40} \times 20d^{60}$

6. Simplify these expressions.
   - a $2a^5 \times 5a^7$
   - b $3b^4 \times 2b$
   - c $6c^3 \times 2c^{10}$
   - d $5d^4 \times 2d^0$

7. Simplify these expressions.
   - a $5a^2b \times 3ab$
   - b $7bc^2 \times 6b^5c^3$
   - c $4c^3d^2 \times 3cd$
   - d $3de \times 2d^2e^6$
   - f $10f^{100}g^{50} \times 7f^{75}g^{25}$
   - e $9ef^6 \times 4e^2f^3$

Band 3 questions

8 Laila simplifies $\dfrac{12b^5}{4b^3}$ like this:

$$\dfrac{12}{4} = 3$$

$$\dfrac{b^5}{b^3} = b^2$$

So $\dfrac{12b^5}{4b^3} = 3b^2$

Use Laila's method to simplify these expressions.

a $\dfrac{15a^6}{3a^4}$ b $\dfrac{24b^8}{8b^3}$ c $\dfrac{150c^{27}}{15c^9}$ d $\dfrac{125d^{10}}{5d}$

e $\dfrac{3ef^3}{ef^2}$ f $\dfrac{12fg^2}{3fg}$ g $\dfrac{8g^{10}}{4g^{10}}$ h $\dfrac{4gh^6}{2gh^0}$

9 Simplify these expressions involving negative indices.

a $a^{-3} \times a^5$ b $b^7 \times b^{-2}$ c $c^{-3} \times c^{-2}$

d $d^{-4} \div d^5$ e $e^6 \div e^{-2}$ f $f^{-5} \div f^{-3}$

> The usual rules of indices can be used with negative indices.

10 Simplify these expressions.

a $3a^{-2}b \times 3ab$ b $7bc^2 \times 3b^5c^{-3}$ c $-4c^{-3}d^2 \times 5c^{-2}d^{-3}$

d $10d^{-2}e^{-1} \times 2d^{-4}e^{-7}$ e $2ef^{-6} \times 2e^{-2}f^{-3}$

11 Simplify these expressions.

a $\dfrac{18a^{-6}}{3a^2}$ b $\dfrac{40b^6}{8b^{-2}}$ c $\dfrac{50c^{-7}}{5c^{-3}}$ d $\dfrac{12d^{-8}}{2d}$ e $\dfrac{6ef^{-1}}{3ef^{-2}}$

12 Simplify these expressions.

a $(x^3)^4$ b $(a^{-2})^3$ c $(2c^2)^3$ d $(y^3)^{-5}$

e $(z^{-2})^{-6}$ f $(3d^{-2})^2$ g $(2c^2)^3 \times 7c^2$ h $(3e^4)^{-2} \times 2e^2f^2$

7.3 Expanding brackets

Skill checker

A	D	E	G	I
$16x - 8$	$25 - 15x$	$5x + 10$	$12x + 32$	$8x + 12$

L	N	P	R	S
$36x + 24$	$24x + 12$	$3 - 3x$	$3x - 18$	$7x + 7$

Expand the brackets and write down the corresponding letter from the table.
What countries do you spell?

① $5(2 + x)$, $3(4 + 8x)$, $4(3x + 8)$, $2(18x + 12)$, $2(8x - 4)$, $4(3 + 6x)$, $5(5 - 3x)$

② $4(2x + 3)$, $3(x - 6)$, $5(2 + x)$, $3(8 + 12x)$, $8(2x - 1)$, $6(2 + 4x)$, $-5(3x - 5)$

③ $7(1 + x)$, $3(1 - x)$, $4(4x - 2)$, $2(4x + 6)$, $12(2x + 1)$

④ $2(6x + 16)$, $-3(6 - x)$, $-5(-x - 2)$, $-5(-2 - x)$, $2(12x + 6)$, $3(12x + 8)$, $-8(1 - 2x)$, $-6(-2 - 4x)$, $5(-3x + 5)$

7 Equations, expressions, formulas and inequalities

▶ Expanding a single set of brackets

Worked example

Expand $x(x + 5)$.

The phrases 'expand brackets' and 'multiply brackets out' mean the same thing.

Solution

$x(x + 5)$
$= (x \times x) + (x \times 5)$
$= x^2 + 5x$

Multiply x outside the brackets by everything inside.

▶ Expanding double brackets

When you expand a pair of brackets you multiply every term in the second bracket by every term in the first bracket and then simplify your answer.

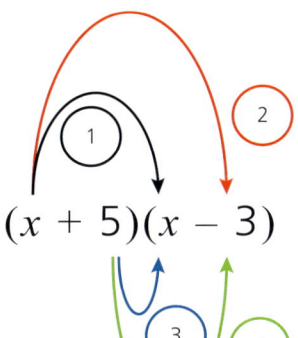

So $(x + 5)(x - 3) = (x \times x) + (x \times -3) + (5 \times x) + (5 \times -3)$
$\qquad = x^2 - 3x + 5x - 15$
$\qquad = x^2 + 2x - 15$

You can remember FOIL:

F stands for First: Multiply the first term in each set of brackets: $x \times x = x^2$

O stands for Outer: Multiply the outer terms: $x \times -3 = -3x$

I stands for Inner: Multiply the inner terms: $5 \times x = 5x$

L stands for Last: Multiply the last terms: $5 \times -3 = -15$

Add these four terms:

$x^2 - 3x + 5x - 15$
$= x^2 + 2x - 15$

Alternatively, you can use a multiplication grid:

	x	5
x	x^2	$5x$
-3	$-3x$	-15

$x^2 - 3x + 5x - 15$
$= x^2 + 2x - 15$

Worked example

Expand the following:

a $(x-4)(x-2)$

b $(x+4)^2$

c $(x-4)(x+2)(x+3)$

Solution

a **Method 1:** Using FOIL to expand $(x-4)(x-2)$.

First: $x \times x = x^2$

Outer: $x \times -2 = -2x$

Inner: $-4 \times x = -4x$

Last: $-4 \times -2 = 8$

Add the four terms:

$(x-4)(x-2) = x^2 - 2x - 4x + 8$
$= x^2 - 6x + 8$

Method 2: Using a grid.

	x	-4
x	x^2	$-4x$
-2	$-2x$	8

Put the contents of one bracket along the top and the other down the side then multiply each pair.

When you expand a pair of brackets, take care with the signs.

$(x-4)(x-2) = x^2 - 4x - 2x + 8$
$= x^2 - 6x + 8$

Add the results.

b $(x+4)^2$ means $(x+4)$ multiplied by itself.

$(x+4)^2 = (x+4)(x+4)$
$= x^2 + 4x + 4x + 16$
$= x^2 + 8x + 16$

You can use FOIL or a grid to help.

c **Step 1:** Expand $(x-4)(x+2)$.

$(x-4)(x+2) = x^2 + 2x - 4x - 8$
$= x^2 - 2x - 8$

Step 2: Multiply the result by $(x+3)$.

$(x^2 - 2x - 8)(x+3)$

You can use a grid to help you.

	x^2	$-2x$	-8
x	x^3	$-2x^2$	$-8x$
3	$3x^2$	$-6x$	-24

$(x^2 - 2x - 8)(x+3) = x^3 + 3x^2 - 2x^2 - 6x - 8x - 24$
$= x^3 + x^2 - 14x - 24$

Remember

FOIL only works when there are two terms inside each set of brackets, so it can't be used here.

7.3 Now try these

Band 1 questions

1 Expand the brackets.

a $x(x-5)$ b $2x(x+2)$ c $x(3x-5)$ d $4x(2x+5)$
e $x(5-x)$ f $2x(2+x)$ g $x(1-3x)$ h $4x(11+2x)$

2 You can use a grid to help you expand (multiply out) a pair of brackets.

a i Copy and complete this grid to expand $(x+4)(x+5)$.

	x	4
x	x^2	
5		20

  ii Add together the four terms inside the grid and simplify your expression.

b Use these grids to help you expand the expressions.

  i $(x+2)(x+3)$

	x	2
x		
3		

  ii $(x+2)(x+1)$

	x	2
x		
1		

  iii $(x+1)(x-2)$

	x	1
x		
-2		

  iv $(x-3)(x+2)$

	x	-3
x		
2		

3 The expression in the top brick of each of these algebra pyramids is found by multiplying together the expressions in the two bricks beneath it.

Copy and complete each algebra pyramid.
Expand and simplify the expression in each of the top bricks.

a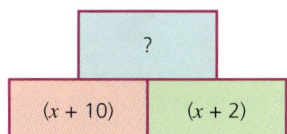

b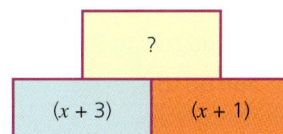

c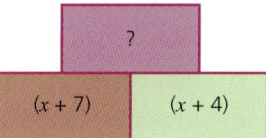

Curriculum for Wales Mastering Mathematics: Book 3

Fluency

4 Expand these expressions.
   a $(x+2)(x+6)$
   b $(x+3)(x+7)$
   c $(x+2)(x-6)$
   d $(x-3)(x-7)$

5 Expand the brackets in these expressions.
   Simplify your answers.
   a $(a+2)(a+3)$
   b $(b+1)(b+4)$
   c $(c-3)(c+5)$
   d $(d+4)(d-6)$
   e $(e+7)(e-1)$
   f $(f-3)(f-4)$
   g $(g-2)(g+2)$

Band 2 questions

Logical reasoning

6 Vaya and Sara have both expanded the expression $(b+3)^2$.
   Vaya says, 'The answer is $b^2 + 9$'.
   Sara says, 'No, it is $b^2 + 6b + 9$'.
   a Who is right?
   b How would you show the other person that she is wrong?
   c Expand and simplify these expressions.
      i $(n+3)^2$
      ii $(x-3)^2$
      iii $(y-30)^2$

7 a Expand $(x+5)^2$.
   b Use your answer to part **a** to work out 25^2.
      What value of x should you use?
   c Use this method to calculate
      i 15^2
      ii 105^2
      iii 1005^2
   d Use a similar method to work out 52^2.

8 Meena's Dad asks her a question.
   He says, 'Meena, what is $\left(3\frac{1}{2}\right)^2$?'

   Meena replies, 'That is easy, I have a special trick for squaring halves.
   One more than 3 is 4. You do 3 times 4 and then add $\frac{1}{4}$. The answer is $12\frac{1}{4}$.'
   a Is Meena correct?
   b Check Meena's method for these fractions.
      i $\left(1\frac{1}{2}\right)^2$
      ii $\left(2\frac{1}{2}\right)^2$
      iii $\left(5\frac{1}{2}\right)^2$
      iv $\left(10\frac{1}{2}\right)^2$
   c By writing $\left(1\frac{1}{2}\right)^2$ as $\left(1+\frac{1}{2}\right)^2$ and multiplying out the brackets, show how the rule works.
   d By using x to represent any positive integer, show this rule always works.

Band 3 questions

Logical reasoning

9 a Follow these instructions.
      Step 1: Think of two numbers that add up to 1.
      Step 2: Square the larger number and add this to the smaller number.
      Step 3: Square the smaller number and add this to the larger number.
   b Repeat for other pairs of values that add up to 1.
   c Use algebra to explain your results.

> **Note**
> If one number is x, the other is $1 - x$.

7 Equations, expressions, formulas and inequalities

10 a Expand these expressions.
   i $(a+3)(a-3)$
   ii $(a+b)(a-b)$
   iii $(2c+7)(2c-7)$
   iv $\left(d-\dfrac{1}{2}\right)\left(d+\dfrac{1}{2}\right)$

   What do you notice?

  b Expressions in the form $a^2 - b^2$ are often called the difference of two squares.
     Why do you think this is?

  c Write these as a product of two brackets. This is known as **factorising**.
   i $a^2 - 16$
   ii $a^2 - 36$
   iii $a^2 - 64$
   iv $a^2 - 1$
   v $a^2 - \dfrac{1}{9}$

  d Show how you would use the difference of two squares to work out these calculations without using a calculator.
   i $26^2 - 16^2$
   ii $378^2 - 376^2$
   iii $1025^2 - 1024^2$

11 Find as many different expressions as you can for the blue area in this square.
    Show algebraically that your expressions are equivalent.

12 Look at the diagram.

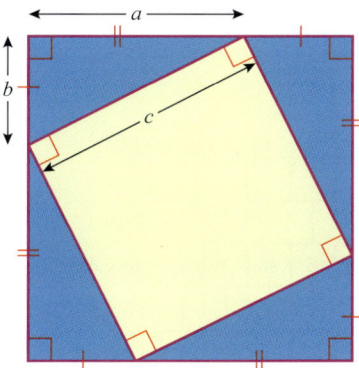

  a What is the area of the large square, in terms of a and b?
     Expand the brackets in your expression for the area.
  b i   What is the area of the yellow square in terms of c?
     ii What is the area of one of the blue triangles, in terms of a and b?
     iii Using your answers from **bi** and **bii**, find a second expression for the area of the large square in terms of a, b and c.
  c Now you should have two expressions for the area of the large square.
     Can you prove Pythagoras' theorem?

13 Copy and complete these algebra pyramids.
    The expression in the top brick is found by multiplying the expressions in the two bricks underneath it.
    What rules should you use to find the expressions in the bottom bricks?

  a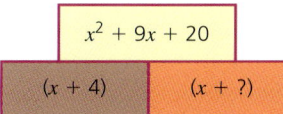
  b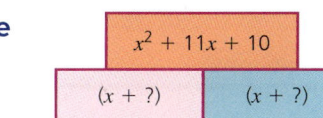
  c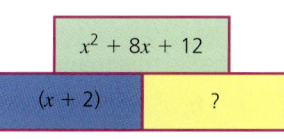

  d $x^2 + 13x + 40$ $(x + ?)$ $(x + ?)$
  e $x^2 + 11x + 10$ $(x + ?)$ $(x + ?)$
  f $x^2 + 2x + 1$ $?$ $?$

137

14. Look at this shape.

    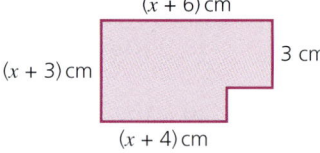

    a i   Divide the shape into two rectangles with a horizontal line.
       ii Find the area of each rectangle in terms of x.
       iii Add these two expressions to find the total area of the shape in terms of x, simplifying your answer.
    b i   Divide the shape into two rectangles with a vertical line.
       ii Find the area of each rectangle in terms of x.
       iii Add these two expressions to find the total area of the shape in terms of x, simplifying your answer.
    c Check that the simplified answers you got in parts **a** and **b** are the same.

15. Expand the following sets of brackets and simplify your answers.

    a $x(x+1)(x+2)$
    b $x(x-2)(x+4)$
    c $(x+6)(x-1)(x+2)$
    d $(2x-1)(x+1)(x-1)$
    e $(x-1)(x+1)(2x-1)$

16. This pattern is called Pascal's triangle.

    Each number in Pascal's triangle is the sum of the two numbers above it.

    Pascal's triangle can be used to help you expand brackets.

    For example, to help expand $(x+1)^2$ take the numbers from Row 2:

    $(x+1)^2 = 1x^2 + 2x + 1$
    $ = x^2 + 2x + 1$

    To help expand $(x+1)^3$ take the numbers from Row 3:

    $(x+1)^3 = 1x^3 + 3x^2 + 3x + 1$
    $ = x^3 + 3x^2 + 3x + 1$

    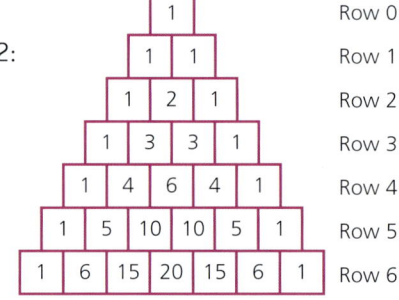

    Use Pascal's triangle to expand $(x+1)^4$.

7.4 Formulas

Skill checker

The letters a to j have the values shown in the table.

a	b	c	d	e	f	g	h	i	j
1	2	3	4	5	6	7	8	9	10

1. What is this part of the body?
   $bc; g-f; a+b; b^2+a$

2. What is this thing to wear?
   $j-h; a^2; \sqrt{f+j}; c+d; \dfrac{j}{b}$

3. What is this type of food?
   $\sqrt{d}; a+d; g-b; \sqrt{3j+f}$

7 Equations, expressions, formulas and inequalities

▶ Substituting into a formula

You can substitute numbers into formulas. These may include negative numbers and decimals.

Worked example

v^2 is the subject of the formula $v^2 = u^2 + 2as$.

Find v when $u = -3$, $a = 9.8$ and $s = 20$.

Solution

$v^2 = u^2 + 2as$
$v^2 = (-3)^2 + 2 \times 9.8 \times 20$ ← Substitute the values given into the formula.
$v^2 = 9 + 392$
$v^2 = 401$
$v = 20.02$ or -20.02 to 2 d.p. ← Square root both sides of the formula to find v. Remember that there are two answers when you square root a number, one positive and one negative.

▶ Rearranging a formula

You can rearrange a formula to make another letter the subject.

You must do the same thing to both sides of the formula to get the variable you want by itself on one side of the formula. ← This is similar to solving an equation.

$v^2 = u^2 + 2as$

Make s the subject.

Subtract u^2 → $u^2 + 2as = v^2$ ← Subtract u^2
$2as = v^2 - u^2$
Divide by $2a$ → $s = \dfrac{v^2 - u^2}{2a}$ ← Divide by $2a$

Make u the subject.

Subtract $2as$ → $u^2 + 2as = v^2$ ← Subtract $2as$
$u^2 = v^2 - 2as$
Square root → $u = \pm\sqrt{v^2 - 2as}$ ← Square root

Remember the $\pm$ sign when taking the square root. On its own, the $\sqrt{}$ sign means only the positive root.

▶ Substituting negative numbers into a formula

Worked example

Work out the value of a when $b = -2$ using the formula

$a = \dfrac{3b^2 - 6}{b + 4}$

Solution

$a = \dfrac{3(-2)^2 - 6}{(-2) + 4}$ ← Substitute -2 in place of b.

$a = \dfrac{3(4) - 6}{2}$ ← $(-2)^2$ means -2×-2 which is equal to 4.

$a = \dfrac{6}{2} = 3$

Remember

Take care when substituting a negative number into a formula.

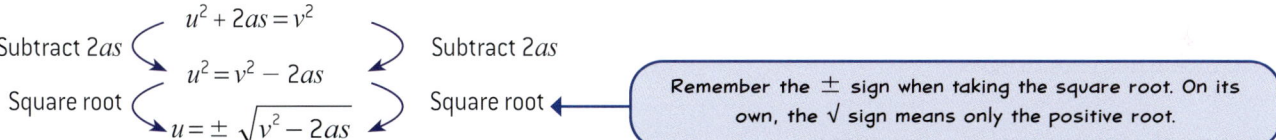

Be careful squaring a negative number on the calculator.
Always keep the negative number in brackets, with the power of 2 outside the brackets.
Try these on the calculator:

The wrong way: -2^2 ✗ (Gives the answer -4)
The correct way: $(-2)^2$ ✓ (Gives the answer 4)

Curriculum for Wales Mastering Mathematics: Book 3

▶ Working with formulas

Worked example

The area of this trapezium is 48 cm².
Work out the height of the trapezium.

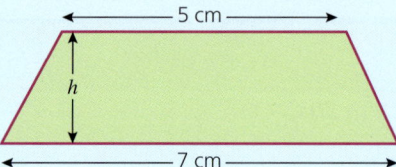

Solution

The formula for the area of a trapezium is $A = \dfrac{1}{2}(a+b)h$ where

▶ a and b are the lengths of the parallel sides; and
▶ h is the distance between the parallel sides.

$$48 = \frac{1}{2}(5+7)h$$ ← Substitute the values for A, a and b into the formula.

$$\frac{1}{2} \times 12 \times h = 48$$ ← It is usual to work with the variable on the LHS of the equation.

$$6h = 48$$
$$h = 8$$

So the height of the trapezium is 8 cm.

Worked example

The formula for the perimeter of a rectangle is

$$P = 2(l + w)$$

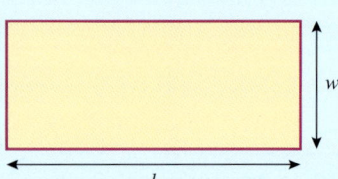

Rearrange the formula to make l the subject.

Solution

Look at how Beth and David rearrange the formula. These methods are both correct.

Beth

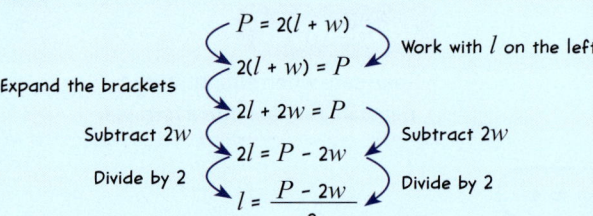

David

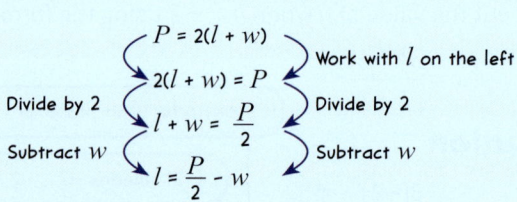

140

7 Equations, expressions, formulas and inequalities

Activity

Substitute the numbers below into the formulas given and complete the cross-number puzzle.

a	b	c	d	e	f	g	h	i	j
1	2	3	4	5	6	7	8	9	10

Across
1. $(de)^2 - 9$
3. $(j+e)^2$
5. eg
6. $a + 2a + 3a + 4a + 5a + 6a$
7. gh
9. $5d^2$
10. $gj^2 + 8$
11. e^2
12. $4(13+j)$
14. $10d + 1$
15. $e(d+f+1)$
17. $j(e+j) + c^2$
18. $2j^2 + 2j + 2$

Down
2. $100j - ei$
4. $b^2 e(j+1)$
8. $100f + 10g + e$
9. $111h + 1$
11. $6f^2 - 1$
13. $2(e^3 + 1)$
14. $14c - 1$
16. $7h - d$

7.4 Now try these

Band 1 questions

1 Find the value of $4a + 9b - c$ when
  a $a = 3, b = 3$ and $c = 4$
  b $a = 4, b = -4$ and $c = 5$
  c $a = 0.2, b = 0.3$ and $c = 0.4$
  d $a = 0.1, b = -0.2$ and $c = 0$.

2 In rugby union, the formula to calculate the total score is
$S = 5t + 2c + 3g$
where t is the number of tries scored, c is the number of conversions and g is the number of goal kicks or drop goals.
  a   Find the total score when
    i $t = 2, c = 1$ and $g = 8$
    ii $t = 10, c = 7$ and $g = 12$.
  b   In one match, Wales score two tries, one conversion and four goal kicks.
Tonga score three tries, three conversions and one goal kick.
Who wins?

3. This swimming pool is a rectangle l metres long and w metres wide.

a Copy and complete this formula for the perimeter of the pool:
$P = $ _____ + _____

b Copy and complete this formula for the area of the surface of the pool in terms of l and w.
$A = $ _____

c Find the perimeter and surface area of the pool when
   i $l = 11$ and $w = 2$
   ii $l = 9$ and $w = 4$
   iii $l = 7$ and $w = 6$.

4. Make x the subject in each of these formulas.

a $y = 2x$
b $y = \dfrac{x}{2}$
c $y = x + 3$
d $y = x - 3$

Band 2 questions

5. a Change the subject of this formula to B:
$A = 180 - B - C$

b Change the subject of this formula to w:
$P = 2(l + w)$

c Change the subject of this formula to R:
$V = IR$

6. Jerin lets out holiday flats during the summer.
The formula she uses to work out the charge is
Cost = rate × number of weeks + deposit + £25 per person.
The rate is £275 per week and the deposit is £150.

a Write out the formula using C to stand for the cost, W for the number of weeks and N for the number of people.

b Find the cost when
   i   four people stay for two weeks
   ii two people stay for three weeks.

c Three people have £1050 to spend.
How long can they stay?

7. Joe, Mari, Bryn and Myra are trying to find the value of $5b^2 - 3c$ when $b = 4$ and $c = 2$.
They all get different answers.
Joe says, 'The answer is 34.'
Mari says, 'No, it's 48!'
Bryn says, 'I think it's 74.'
Myra says, 'You are all wrong! It's 2884.'
Who has the right answer?
Explain where the others have gone wrong.

8 An electrician charges £15 per hour plus a call-out charge of £25.
   a Write a formula for the total charge, £c, for a job lasting h hours.
   b Calculate c when
      i $h = 2$ ii $h = 3\frac{1}{2}$
   c Rearrange your formula to make h the subject.
   d Find how many hours were worked when the total charge is
      i £115   ii £145   iii £77.50.

9 A Greek mathematician named Hero showed that the area of a triangle of sides a, b and c could be calculated using this formula:
$$A = \sqrt{s(s-a)(s-b)(s-c)}$$
where $s = \frac{1}{2}(a+b+c)$.

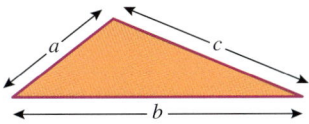

   a Look at these two triangles.
      The side lengths are in centimetres.

      a 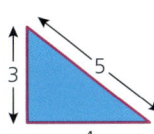   b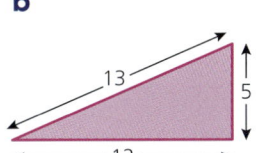

      i   Use Pythagoras' theorem to check that they are right-angled.
      ii Use the formula $A = \frac{1}{2}bh$ to find their areas.
      iii Check that Hero's formula gives the same answers.
   b Now use Hero's formula to find the areas of these non-right-angled triangles:
      i $a = 4$ cm, $b = 6$ cm and $c = 8$ cm   ii $a = 6.7$ cm, $b = 9.3$ cm and $c = 12.9$ cm

10 The picture shows a model of a garden swing seat.

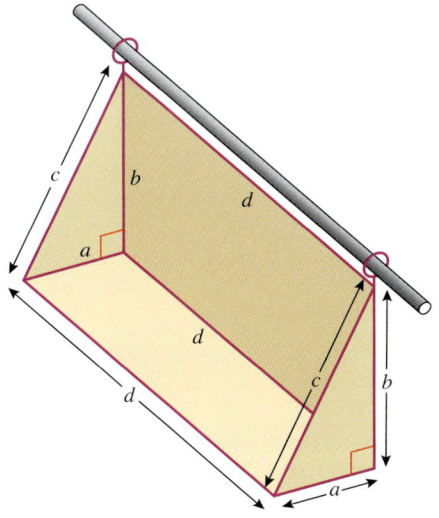

The frame is made out of thin wire.
The lengths are in centimetres.
Each end is in the shape of a right-angled triangle.
   a Write down a formula for the perimeter P of each triangle.
   b Write down a formula for the total length T of wire. Simplify your expression.
   c The back, sides and seat are made out of wood.
      Write a formula for the area A of wood.

Band 3 questions

11 The path of an object orbiting the Sun is an ellipse, not a circle.

The formula for the area of an ellipse is $A = \pi ab$ where a and b are the lengths shown in the diagram.

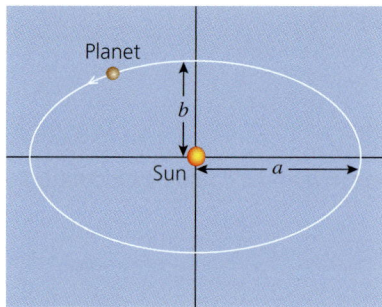

For Pluto, $a = 5.9 \times 10^9$ km and $b = 5.7 \times 10^9$ km.

Pluto is no longer classed as a planet, but is a 'dwarf planet'.

a Find the area inside Pluto's orbit.

b Compare this with the area of a circle of radius 5.8×10^9 km.

12 This tent frame is made from aluminium tubing.

The measurements are in metres.

The length x is called the 'size' of the tent.

a Write down the formula for the total length L of tubing needed.

b **i** The size of a tent is 1.8 m.

How much tubing is needed?

ii A tent uses 25 m of tubing.

What is its size?

iii A tent uses 30 m of tubing.

What is its size?

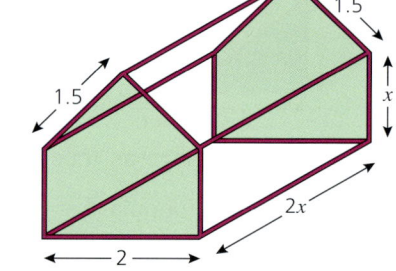

13 a Change the subject of this formula to h:

$V = \pi r^2 h$

b Change the subject of this formula to b:

$A = \dfrac{1}{2}(a + b)h$

c Change the subject of this formula to l:

$A = l^2$

14 Look at these three dot patterns.

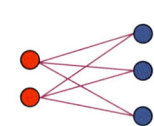 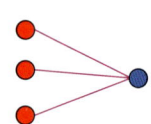 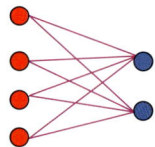

Find a formula for the number of lines, l, joining m red dots to n blue dots.

How many lines come from each dot?

Hint

Look at the red dots.

7.5 Solving inequalities

Skill checker

Remember, the inequality signs are as follows:

The symbol ⩾ means greater than or equal to.

The symbol > means greater than.

The symbol ⩽ means less than or equal to.

The symbol < means less than.

In the inequality $x > 3$, x can be replaced with any number greater than 3, *but not 3*.

This can also be represented on a number line.

On a number line you can represent < or > by an open circle: ○

⩽ and ⩾ are represented by a solid circle: ●

$x > 3$ is:

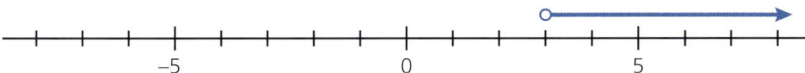

How would you represent the following inequalities on a number line?

a $x \leqslant -4$
b $x \geqslant -1$
c $x < 0$
d $-3 \leqslant x < 2$

Many of the skills you have developed to solve equations in Section 7.1 can be applied to solving inequalities.
For example, work out the following problem using algebra:

Worked example

I am thinking of a number.
I multiply it by 4 and then subtract 7.
The answer is less than 13.
What numbers could I be thinking of?

Solution

Write the question using algebra: $4x - 7 < 13$
Now solve it like an equation.

$$\begin{aligned} 4x - 7 &< 13 \\ 4x &< 20 \\ x &< 5 \end{aligned}$$

(+7 to both sides, ÷4 both sides)

This means the number can be any number which is less than 5.

Activity

Think of a positive number.

Multiply your number by 5. Is your answer greater than or less than the number you thought of?

Now start again with the number you were thinking of.

Multiply your number by -5. Is your answer greater than or less than the number you thought of?

What happens if the number you thought of was a negative number?

In the previous example, the inequality was divided by a positive number.

Inequalities behave differently to equations if both sides are **multiplied or divided** by a **negative number**.

Consider this example:

Worked example

I am thinking of a number.
I multiply it by -3.
The answer is greater than or equal to 12.
What numbers could I be thinking of?

Solution

$$\begin{array}{c} -3x \geqslant 12 \\ 0 \geqslant 12 + 3x \\ -12 \geqslant +3x \\ -4 \geqslant x \\ \text{or } x \leqslant -4 \end{array}$$

(with $+3x$, -12, $\div 3$ applied to both sides)

As you can see, the inequality sign has changed from $\geqslant$ to $\leqslant$.

Note

Use knowledge acquired from the previous activity — multiplication of both sides of an inequality by a negative number leads to a false statement. Therefore the inequality sign needs to be swapped around to make the statement true.

Therefore, the **rule** for **multiplying or dividing** an inequality by a **negative number** is, you must **reverse the inequality** sign. For example, $<$ becomes $>$, $\geqslant$ becomes $\leqslant$ and so on.

Worked example

Solve the inequality $7 - 2x < 3$.

Solution

$$7 - 2x < 3$$
$$7 - 2x - 7 < 3 - 7$$
$$-2x < -4$$
$$\frac{-2x}{-2} > \frac{-4}{-2}$$
$$x > 2$$

7.5 Now try these

Band 1 questions

Solve the following inequalities.

1. $x - 4 > 10$
2. $x + 5 \leq 11$
3. $x + 7 < 2$
4. $x - 6 \geq -4$
5. $3x + 2 \geq 14$
6. $5x - 6 < 29$
7. $-2x > 10$
8. $30 - 4x \leq 2$

9. I am thinking of a number.
   I multiply it by 3 and then add 4.
   The answer is greater than 22.
   What numbers could I be thinking of?

Band 2 questions

Solve the following inequalities.

10. $3x < x + 10$
11. $5 + x > -11$
12. $7x \leq 3x - 16$
13. $2x \geq 6x + 20$
14. $3(x + 1) > 12$
15. $5(x - 2) < 20$
16. $3(5 - x) \leq 9$
17. $\dfrac{x}{5} + 7 \geq 10$

18. I am thinking of a number.
    I subtract 4 from it and then multiply it by 7.
    The answer is less than or equal to 42.
    What numbers could I be thinking of?

Band 3 questions

Solve the following inequalities.

19 $7x + 2 > 3x + 14$

20 $10 - 2x < 3x + 25$

21 $5(x + 2) \geq 2x - 11$

22 I am thinking of a number.

I multiply it by 3 and then subtract this number from 15.

I then double this new number and the answer is greater than 18.

What numbers could I be thinking of?

23 Catherine has a collection of books.

Ann had 5 times as many books as Catherine, but she has now lost 17 of them.

Ann now has fewer books than Catherine.

By using n as the number of books Catherine has, write down an inequality in terms of n to show this information.

Now solve this inequality to find the greatest number of books that Catherine may have.

Key words

Here is a list of the key words you met in this chapter.

Equation	Expand	Expression	Formula
Index	Indices	Inequality	Substitute

Use the glossary at the back of this book to check any you are unsure about.

7 Equations, expressions, formulas and inequalities

Review exercise: equations, expressions, formulas and inequalities

Band 1 questions

1 Solve these equations.
   Choose **one** of your answers to check by substituting into the equation.
   - **a** $a + 11 = 15$
   - **b** $6 + b = 8$
   - **c** $x + 5 = 15$
   - **d** $d - 7 = 13$
   - **e** $a - 4 = 4$
   - **f** $q - 5 = 15$

2 Solve these equations.
   Choose **one** of your answers to check by substituting into the equation.
   - **a** $999z + 999 = 999$
   - **b** $5x + 6 = 16$
   - **c** $3a - 7 = 14$
   - **d** $10 + 3f = 19$
   - **e** $12 + 7g = 33$

3 Solve these equations.
   Choose **one** of your answers to check by substituting into the equation.
   - **a** $\dfrac{x}{3} = 20$
   - **b** $\dfrac{3}{4}x = 100$
   - **c** $\dfrac{g}{5} = 5$
   - **d** $\dfrac{t}{9} = 4$
   - **e** $\dfrac{x}{2} = 5$
   - **f** $10 + \dfrac{f}{2} = 14$
   - **g** $\dfrac{m}{3} - 11 = 21$
   - **h** $2 - \dfrac{a}{3} = 6$
   - **i** $3 - \dfrac{m}{2} = 15$
   - **j** $4 + \dfrac{a}{5} = 16$

4 a Change the subject of this formula to h:
   $V = lwh$

   b Change the subject of this formula to a:
   $$m = \dfrac{a + b + c + d}{4}$$

5 Simplify these expressions.
   - **a** $a^3 \times a^4$
   - **b** $3b^5 \times 4b$
   - **c** $2c^3 \times 5c^4$

6 Simplify these expressions.
   - **a** $a^4b \times ab$
   - **b** $2bc^3 \times 5b^2c^5$

7 Solve the following inequalities:
   - **a** $x + 7 > 15$
   - **b** $2x + 11 < 23$
   - **c** $13 - x \geq 4$
   - **d** $14 - 3x \leq 2$

Band 2 questions

8 Expand the brackets in these expressions.
   - **a** $(a + 3)(a + 5)$
   - **b** $(b + 5)(b + 5)$
   - **c** $(c + 3)(c - 2)$
   - **d** $(d - 2)(d - 1)$
   - **e** $(e - 1)(e + 1)$
   - **f** $(f - 3)(f - 3)$

9 Look at the isosceles triangle shown.

   a Two of the angles are $75°$.

   Form an equation and solve it to find the size of the third angle marked x.

   b The perimeter of the triangle is 11 cm.

   The two equal sides are 4 cm in length.

   Form an equation and solve it to find the length of the base.

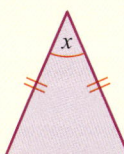

Curriculum for Wales Mastering Mathematics: Book 3

10 Kate buys two apples, at 24p each, and four bananas.

She spends £1.28 altogether.

  a Write an equation for this information, using b pence as the cost of one banana.

  b Solve your equation to find the cost of one banana.

11 The length of a rectangle is 4 times its width.

The perimeter is 50 cm.

Find the area of the rectangle.

12 Peter is four times as old as his daughter Tara.

In five years' time their ages will total 50 years.

  a Copy and complete this table

	Peter's age	Tara's age
Now	$4x$	x
In 5 years		

  b Form an equation and solve it.

  c State Peter's and Tara's present ages.

13 Solve the following inequalities.

  a $5x > x - 8$ **b** $3x + 1 < 2x$ **c** $3(x + 2) \geq 15$ **d** $10 - 2x \leq 2$

Band 3 questions

14 Paul thinks of a number, x.

> I think of a number.
> I multiply it by 4 and add 2.
> The answer is four more than twice my number.

  a Write down expressions for

  i 'multiply my number by 4 and add 2'

  ii 'four more than twice my number'.

  b Form an equation.

  c Solve your equation to find Paul's number.

15 Solve these equations by filling in the blanks on the bar model diagrams.

  a $2x + 7 = 19$

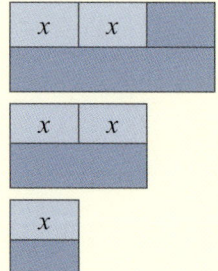

  b $4x + 6 = 2x + 39$

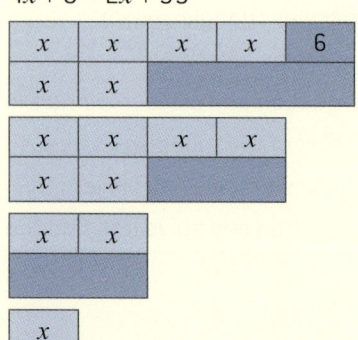

150

16 Solve these equations.
- a $3p - 2 = 2p + 1$
- b $12 - 3c = 18 - 5c$
- c $7x - 8 = 4x + 4$
- d $2x + 9 = 3x - 5$
- e $15 - x = 23 - 2x$

17 Simplify these expressions.
- a $\dfrac{12a^2}{2a}$
- b $\dfrac{b^8 c^4}{b^5 c^2}$
- c $\dfrac{24 b^6 c^3}{4 b^4 c^3}$

18 Solve these equations.
- a $6(5x + 4) = 3(4x + 2)$
- b $2(5x + 7) = 4(2x + 7)$
- c $3(6x + 2) = 10(5x + 7)$
- d $2(6x + 10) = 5(4x + 8)$
- e $\dfrac{6}{5x + 9} = \dfrac{10}{9x - 3}$
- f $\dfrac{1}{8x + 1} = \dfrac{1}{10x + 1}$
- g $\dfrac{10}{4x + 10} = \dfrac{8}{3x + 9}$

19 Find the value of $2x^2 - y$ when
- a $x = 1$ and $y = 2$
- b $x = 2$ and $y = 6$
- c $x = 5$ and $y = 10$
- d $x = 10$ and $y = 1$.

20
- a   Change the subject of this formula to r:
$$A = \pi r^2$$
- b   Change the subject of this formula to V:
$$x = \sqrt[3]{V}$$
- c   Change the subject of this formula to h:
$$t = \sqrt{\dfrac{h}{5}}$$

Consolidation 3: Chapters 6–7

Band 1 questions

1 Solve these equations.
Choose **one** of your answers to check by substituting your answer into the equation.
- **a** $11j - 11 = 22$
- **b** $5d + 3 = 18$
- **c** $2s - 6 = 4$
- **d** $2n - 4 = 8$
- **e** $6b + 5 = 17$

2 Solve these equations.
Choose **one** of your answers to check by substituting your answer into the equation.
- **a** $4c - 5 = 11$
- **b** $2x + 5 = 45$
- **c** $3y - 8 = 28$
- **d** $9z + 35 = 125$
- **e** $2b - 7 = 7$
- **f** $100c + 500 = 1500$
- **g** $3t - 12 = 3$
- **h** $4k + 8 = 24$
- **i** $6h - 5 = 1$

3 Solve these equations.
Choose **one** of your answers to check by substituting your answer into the equation.
- **a** $\frac{r}{4} = 10$
- **b** $\frac{y}{7} = 1$
- **c** $\frac{n}{5} + 7 = 15$
- **d** $8 - \frac{a}{3} = 12$
- **e** $\frac{3}{s} - 1 = 17$

4 Solve the following inequalities.
- **a** $x - 6 > 14$
- **b** $x + 18 \leq 10$
- **c** $x + 9 < 3$
- **d** $x - 4 \geq -8$

Band 2 questions

5 Rhiannon is thinking about putting some money into a savings account.
Here is an advert for the account.

> **Drefynydd Building Society**
> *Live Savers Savings Account*
> Balances below £500:
> - 1.5% simple interest per annum
>
> Balances £500 or above:
> - 1.75% simple interest per annum

Rhiannon is trying to decide whether to put £450 or £500 into the account.
- **a** If Rhiannon puts £450 into the savings account, how much simple interest would she receive each year?
- **b** If Rhiannon puts £500 into the savings account, how much simple interest would she receive each year?

Rhiannon plans to keep her money in the account for 3 years.
- **c** Find the amount of interest she would receive over 3 years on both £450 and £500.
- **d** What is the difference in the total amount of interest she would receive over 3 years?

6 Find the value of $5a(a - b)$ when
- **a** $a = 2$ and $b = 5$
- **b** $a = 0$ and $b = -20$
- **c** $a = 4$ and $b = 3$
- **d** $a = 1$ and $b = -4$.

7 £7000 is invested at 5% compound interest.
Calculate the value of the investment and the interest earned after
- **a** 3 years
- **b** 20 years.

Consolidation 3

8 Solve this equation by filling in the blanks on the bar diagram.

$3x + 6 = 39$

| x | x | x | 6 |

| x | x | x |

| x |

9 Solve the following inequalities.

a $5(x - 2) < 25$

b $2(3 - x) \leqslant 10$

c $\dfrac{x}{4} + 6 \geqslant 10$

10 A carton of orange juice says it contains 20% extra free.

The carton contains 396 ml of juice.

What volume of juice is in a carton that does not have the extra 20%?

Band 3 questions

11 Look at this model of a letter L.

The front face is painted blue, while the rest is yellow.

What is the ratio, in its simplest form, of blue to yellow paint used?

12 Solve these equations.

a $5x - 5 = 2x + 4$ b $5x + 1 = 6x - 1$ c $1 + 4r = 2r + 5$

d $6x - 10 = 4x$ e $3 + 3w = 10 - 4w$

13 Solve these equations.

a $10(8x + 7) = 10(6x + 6)$ b $6(x + 7) = 10(x + 3)$ c $9(9x + 9) = 8(4x + 4)$

d $\dfrac{10}{9x + 3} = \dfrac{4}{2x - 6}$ e $\dfrac{8}{3x + 10} = \dfrac{10}{4x - 3}$

153

14. Look at this shape.

    a i   Divide the shape into two rectangles with a horizontal line.
       ii Find the area of each rectangle in terms of x.
       iii Add these two expressions to find the total area of the shape in terms of x, simplifying your answer.
    b i   Divide the shape into two rectangles with a vertical line.
       ii Find the area of each rectangle in terms of x.
       iii Add these two expressions to find the total area of the shape in terms of x, simplifying your answer.
    c   Check that the simplified answers you got in parts **a** and **b** are the same.
    d   If the shape is divided into two rectangles using a vertical line, there is a larger rectangle on the left and a smaller one on the right.

        The area of the smaller rectangle is 10 cm².
        Find the value of x.

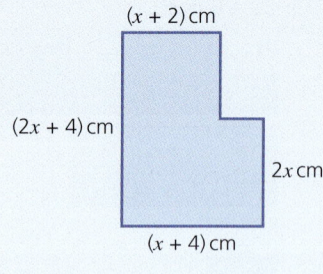

15. a Change the subject of this formula to v:
       $s = vt$
    b Change the subject of this formula to r:
       $C = 2\pi r$

16. a Change the subject of this formula to l:
       $t = 2\pi\sqrt{\dfrac{l}{g}}$
    b Change the subject of this formula to a:
       $h^2 = a^2 + b^2$

17. Solve the following inequalities:
    a $5x - 7 > 2x + 11$
    b $3(2x + 1) < 11 + 8x$

18. A radioactive element's mass reduces by 13% each year.
    How long will it take for the mass to halve?

8 Graphs

Coming up...
- The gradient of a line
- The equation $y = mx + c$

Who wants to be a straight-lines pro?

Answer all eight questions correctly to show you are an expert on straight lines.

① What is the gradient of this line?

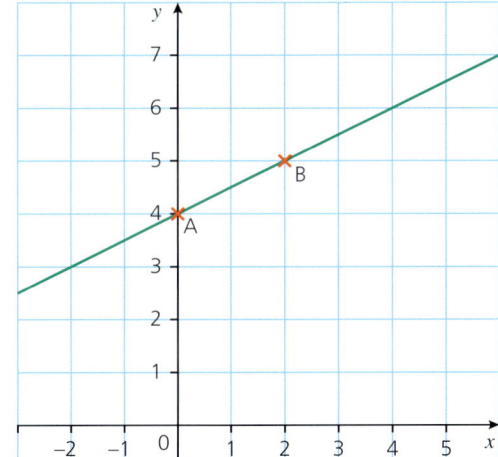

| 2 | 0 | $\frac{1}{2}$ | 4 |

② What can you say about these two lines?

$y = 2x + 3$

$y = 2x - 1$

- They are both curves.
- They both have the same gradient.
- They both have the same y-intercept.
- They have nothing in common.

③ Point A lies on the line $y = 2x - 3$.

A is the point (1, ?).

| −2 | 2 | −3 | −1 |

④ Which of these points lies on the line $y = 2x + 1$?

(0, 3)   (1, 4)   (2, 5)   (3, 6)

5. What is the equation of this line?

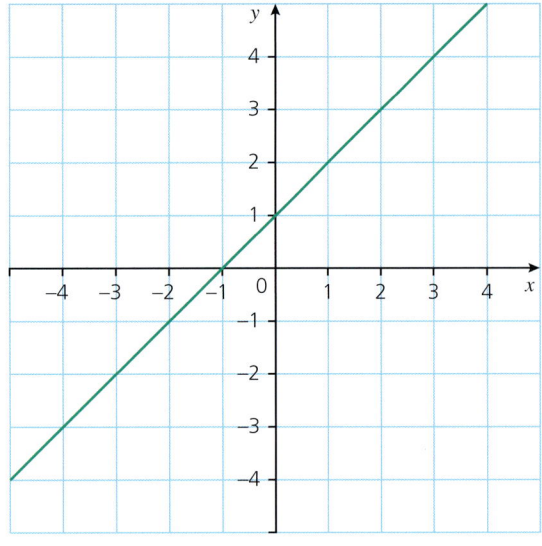

$y = x - 1$ $y = x + 1$ $y = -x - 1$ $y = -x + 1$

6. What is the y-intercept of this line?

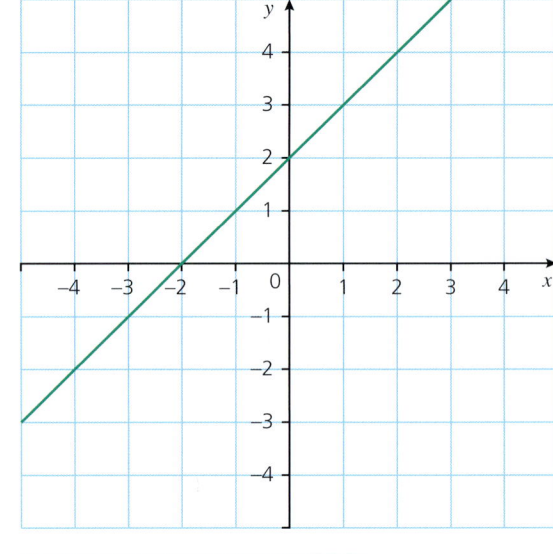

−2   2   1   −1

7. Which of the following is NOT a property of this line?

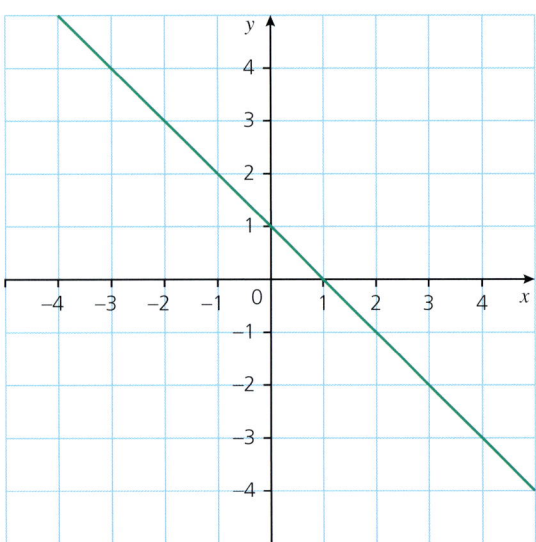

It has a negative gradient.

It has a y-intercept of 1.

It has a gradient of −1.

It has the equation $y = x + 1$.

8. Which of these could be the equation of this line?

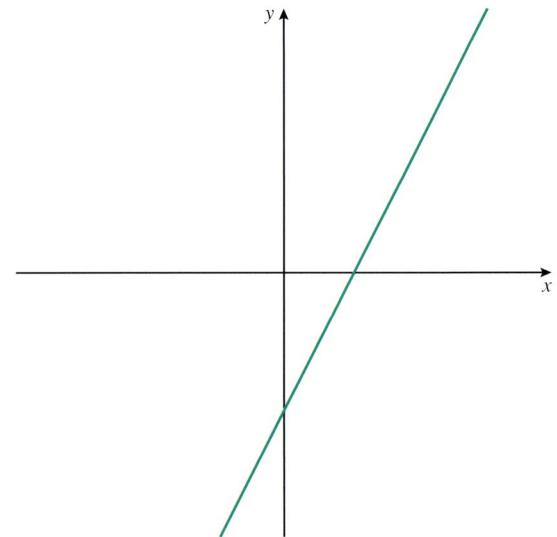

$y = 2x + 2$ $y = 2x - 2$ $y = -2x + 2$ $y = -2x - 2$

8 Graphs

8.1 The gradient of a line

Skill checker

a   Copy and complete this table of values for the two straight lines with these equations:

Line 1: $y = x + 1$

Line 2: $y = -x + 1$

Line	x	−2	−1	0	1	2
1	$y = x + 1$				$1 + 1 = 2$	
2	$y = -x + 1$					

b   Copy this grid and plot the two straight lines.

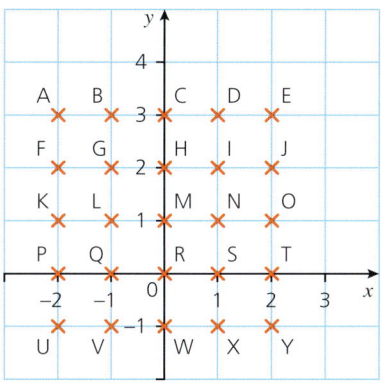

c   Also plot these lines on your graph:

Line 3: $x = -1$

Line 4: $x = 1$

Line 5: $y = 1$

Line 6: $y = 3$

d   Now crack this code to find the secret word.

i   Where do lines 1 and 6 intersect?
ii   Where do lines 4 and 5 intersect?
iii   Where do lines 1 and 4 intersect?
iv   Where do lines 2 and 3 intersect?
v   Where do lines 1 and 2 intersect?
vi   Where do lines 2 and 6 intersect?

▶ Calculating the gradient

The gradient of a line is the steepness of the line.

To find the gradient of a straight line, choose any two points that lie on the line.

❶ Find the change in the y coordinates.

❷ Find the change in the x coordinates.

❸ The gradient is the change in y divided by the change in x.

The letter m is used for the gradient.

Worked example

Find the gradient of the line shown.

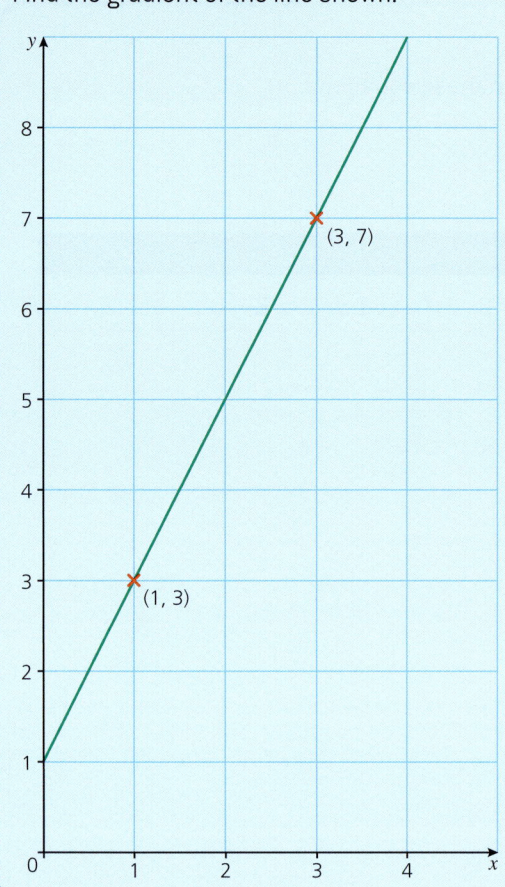

Solution

The points (1, 3) and (3, 7) on the line have been chosen.

Subtracting the y coordinates gives $7 - 3 = 4$.

Subtracting the x coordinates gives $3 - 1 = 2$.

$$\text{Gradient} = \frac{\text{change in } y}{\text{change in } x}$$

$$= \frac{4}{2}$$

$$= 2$$

Be consistent here: use
- the second y coordinate take away the first; and
- the second x coordinate take away the first.

If the line slopes down from left to right the gradient is negative.

Worked example

Find the gradient of the line shown.

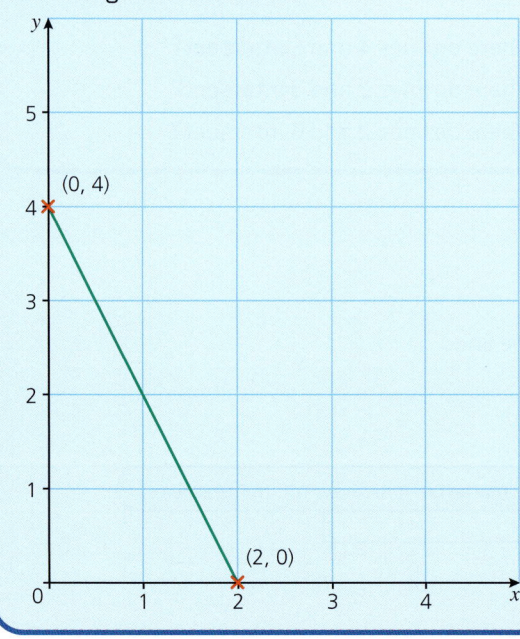

Solution

The two points marked are (0, 4) and (2, 0).

Subtracting the y coordinates gives $0 - 4 = -4$.

Subtracting the x coordinates gives $2 - 0 = 2$.

$$\text{Gradient} = \frac{\text{change in } y}{\text{change in } x}$$

$$= \frac{-4}{2}$$

$$= -2$$

▶ Parallel lines

Parallel lines have the same gradient.

Worked example

a The line L_1 shown in the diagram passes through the points A(18, 32) and B(1, 1).

The line L_2 passes through the points C(9, 11) and D(3, 0).

Are the two lines parallel?

b The line L_3 shown in the diagram passes through the points E(0, 20) and F(16, 16).

The line L_4 passes through the points G(3, 22) and H(7, 21).

Are the two lines parallel?

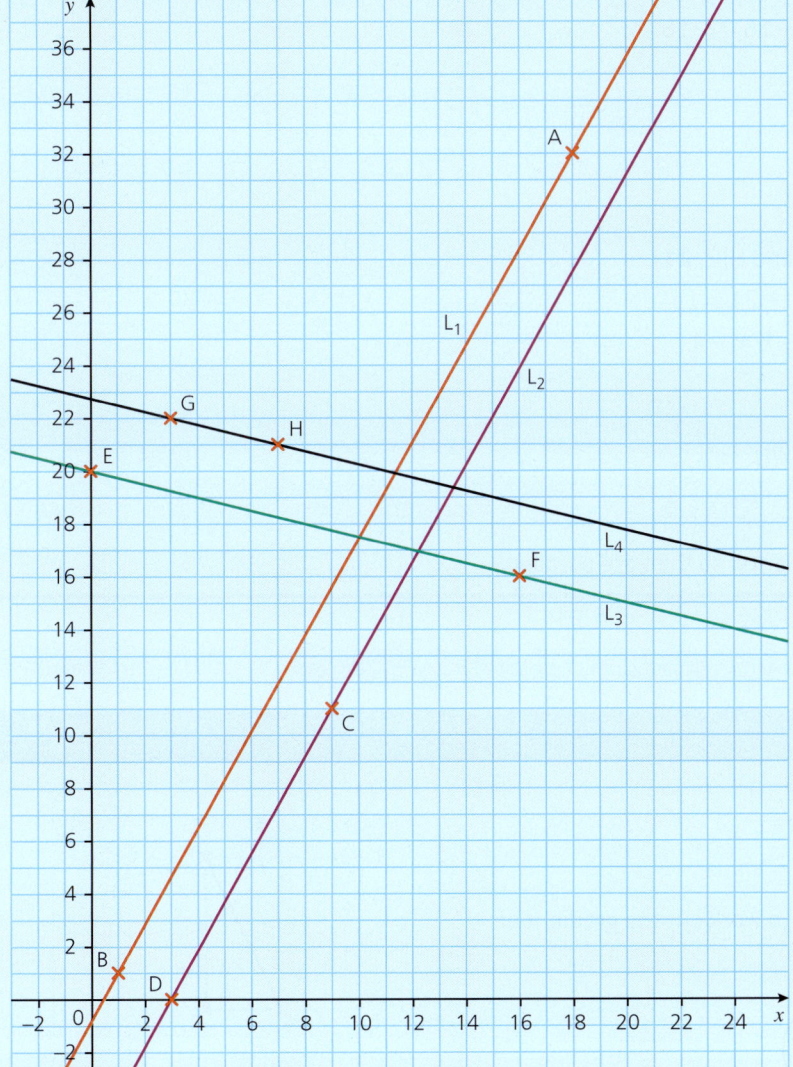

Solution

a For the line L_1, use the points A(18, 32) and B(1, 1).

Subtracting the y coordinates gives $32 - 1 = 31$.

Subtracting the x coordinates gives $18 - 1 = 17$.

Gradient $= \dfrac{\text{change in } y}{\text{change in } x} = \dfrac{31}{17} = 1.82$ (2 d.p.)

For the line L_2, use the points C(9, 11) and D(3, 0).
Subtracting the y coordinates gives $11 - 0 = 11$.
Subtracting the x coordinates gives $9 - 3 = 6$.

$$\text{Gradient} = \frac{\text{change in } y}{\text{change in } x}$$

$$= \frac{11}{6} = 1.83 \text{ (2 d.p.)}$$

The lines L_1 and L_2 do not have the same gradients, so although they appear to be parallel lines, they are not.

b For the line L_3, use the points E(0, 20) and F(16, 16).
Subtracting the y coordinates gives $20 - 16 = 4$.
Subtracting the x coordinates gives $0 - 16 = -16$.

$$\text{Gradient} = \frac{\text{change in } y}{\text{change in } x}$$

$$= \frac{-4}{16} = -0.25$$

For the line L_4, use the points G(3, 22) and H(7, 21).
Subtracting the y coordinates gives $22 - 21 = 1$.
Subtracting the x coordinates gives $3 - 7 = -4$.

$$\text{Gradient} = \frac{\text{change in } y}{\text{change in } x}$$

$$= \frac{1}{-4} = -0.25$$

The lines L_3 and L_4 have the same gradients, so they are parallel.

8.1 Now try these

Band 1 questions

1 P is the point (1, 1) and Q is the point (2, 3).

    a Write down the change in y, RQ.

    b Write down the change in x, PR.

    c Use your answers to parts **a** and **b** to work out the gradient of the line PQ in the diagram.

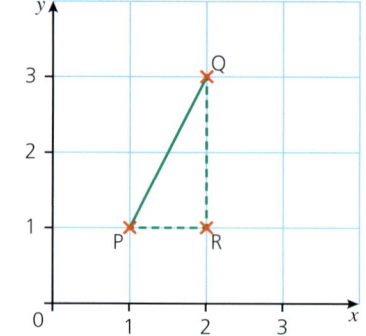

Strictly, this is a line segment, not a line!

2 Work out the gradient of the line joining each of these pairs of points.

    a (5, 7) and (7, 11)

    b (−4, 2) and (2, 4)

    c (7, 9) and (12, −1)

    d (−8, −2) and (−3, −7)

Again, these are line segments, not lines. A line continues forever in both directions.

8 Graphs

3 Work out the gradient of the line PQ shown in each of these diagrams.

a 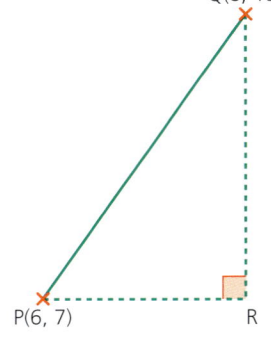
b 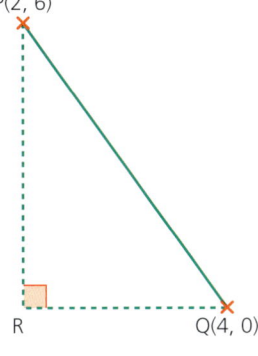
c

4 For each of the lines **a** to **g** in these diagrams

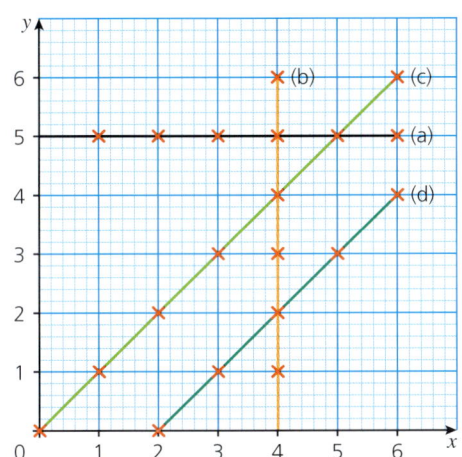 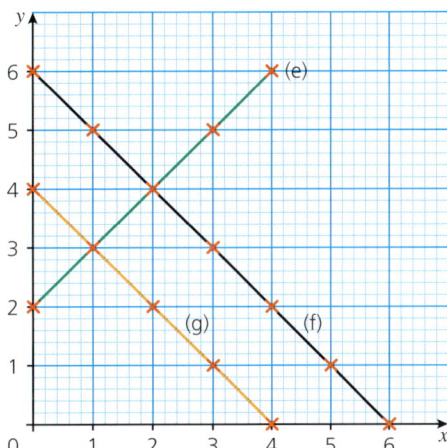

  i write down the coordinates of two points marked on the line
  ii calculate the gradient of the line.

Band 2 questions

5 a Find the coordinates of two points which lie on each of these lines:

> You can use a table of values to help.

  i $y = x$
  ii $y = 2x$
  iii $y = 3x$

  b Use your coordinates to draw all the lines on the same axes. Label each line.
  c Calculate the gradient of each line.
  d What do you notice about the gradients of the lines?
  e What is the gradient of the line $y = 20x$?

6 a Copy and complete these tables:

  i for the equation $y = 3x - 2$

x	−1	0	1	2	3
$y = 3x - 2$		$3 \times 0 - 2$ = ___			$3 \times 3 - 2$ = 7

161

ii for the equation $y = 2x - 2$

x	−1	0	1	2	3
$y = 2x - 2$	$(2 \times -1) - 2$ $= -4$	$2 \times 0 - 2$ $= \_\_\_$			

b Draw the two lines on the same axes.

c Find the gradient of each line.

7 For each line segment in this diagram

a write down the coordinates of two points on the line segment

b calculate the gradient.

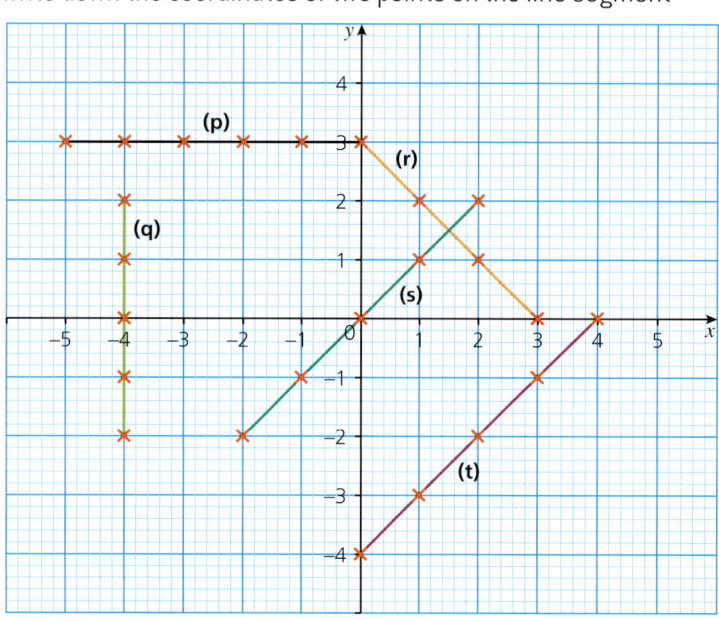

8 a Copy and complete these tables for the lines $y = 2x + 2$ and $y = 2x - 1$.

x	−2	−1	0	1	2	3
$y = 2x + 2$	$(2 \times -2) + 2$ $= -2$					

x	−2	−1	0	1	2	3
$y = 2x - 1$	$(2 \times -2) - 1$ $= -5$					

b Draw and label these two lines on the same axes.

c What can you say about the gradients of these two lines?
How can you tell this from your graph?

d Draw and complete similar tables for $y = 2x$ and $y = 2x + 4$.

Band 3 questions

9 Look at these road signs.

a Which hill is steepest?

b Which hill is least steep?

10 Draw x- and y-axes.

Use the same scale on each axis and take values from -2 to 6.

a Plot the points A$(-2, -2)$, B$(-1, 5)$, C$(3, 6)$ and D$(2, -1)$.

b Work out the gradient of the lines AB, BC, DC and AD.

c What type of quadrilateral is ABCD?

11 The shape shown is a rectangle.

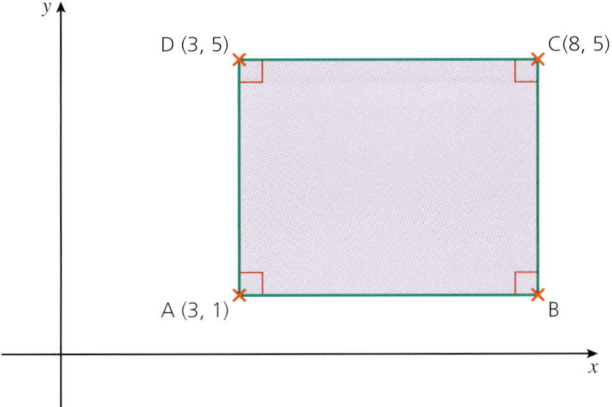

a What are the coordinates of point B?

b What is the gradient of the line AB?

c What is the gradient of the line AC?

12 Fill in the blanks in each sentence with the words on one of these cards:

a A square has ____ of sides that have the same gradient.

b A rectangle has ____ of sides that have the same gradient.

c A trapezium has ____ of sides that have the same gradient.

d A rhombus has ____ of sides that have the same gradient.

e A kite has ____ of sides that have the same gradient.

f A parallelogram has ____ of sides that have the same gradient.

| one pair | two pairs | no pairs |

8.2 The equation of a straight line

Skill checker

a Draw x- and y axes using values from -6 to 6.

Plot these lines.

$x = 1$	$x = -1$	$x = 1.5$	$x = -1.5$
$y = -x - 1$	$y = x - 1$	$y = -x + 0.5$	$y = x + 0.5$
$y = 1$	$y = -1$	$y = -x - 0.5$	$y = x - 0.5$
$y = 1.5$	$y = -1.5$	$y = -x + 1$	$y = x + 1$

b Which famous flag have you drawn?

c Can you shade it in with the correct colours?

▶ The equation of a straight line

Straight lines have an equation of the form $y = mx + c$.

- The value of m represents the gradient or 'steepness' of the line.
- The value of c tells you where the line crosses the y-axis. This is called the **y-intercept**.

Special cases are:

- Horizontal lines which have an equation of the form $y = a$ where a is a number.
- Vertical lines which have an equation of the form $x = b$ where b is a number.

Worked example

Here are the equations of ten lines.
Write down which ones are parallel.

a $\quad x = 7$
b $\quad y = 3 + 7x$
c $\quad y = 2x + 3$
d $\quad y = -5$
e $\quad y = 2x - 4$
f $\quad x = -2$
g $\quad y = 5 - x$
h $\quad y = -x$
i $\quad y = 7x + 5$
j $\quad y = 4$

Solution

a and **f** are parallel   These are both vertical lines. ◀ *Vertical lines have an infinite gradient!*
b and **i** are parallel   These both have a gradient of 7.
c and **e** are parallel   These both have a gradient of 2.
d and **j** are parallel   These are both horizontal lines.
g and **h** are parallel   These both have a gradient of −1.

Worked example

Match each line in this diagram with the correct equation.

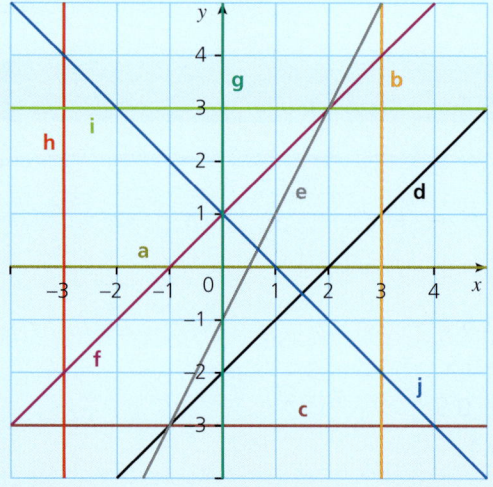

| $x = 3$ | $x = -3$ | $x = 0$ | $y = 3$ | $y = -3$ |
| $y = 0$ | $y = x + 1$ | $y = x - 2$ | $y = 2x - 1$ | $y = 1 - x$ |

Solution

Line **a** is the x-axis. Its equation is $y = 0$

Line **b** is a vertical line. Every point on the line has an x-coordinate of 3. The equation of the line is $x = 3$

Line **c** is a horizontal line. Every point on the line has a y-coordinate of -3. The equation of the line is $y = -3$

Line **d** has a gradient of 1 and a y-intercept of -2. The equation of the line is $y = x - 2$

Line **e** has a gradient of 2 and a y-intercept of -1. The equation of the line is $y = 2x - 1$

Line **f** has a gradient of 1 and a y-intercept of 1. The equation of the line is $y = x + 1$

Line **g** is the y-axis. Its equation is $x = 0$

Line **h** is a vertical line. Every point on the line has an x-coordinate of -3. The equation of the line is $x = -3$

Line **i** is a horizontal line. Every point on the line has a y-coordinate of 3. The equation of the line is $y = 3$

Line **j** has a gradient of -1 and a y-intercept of 1. The equation of the line is $y = -x + 1$ or $y = 1 - x$

Activity

Links to sequences

① A sequence has the position-to-term formula $2n - 1$.

   a Copy and complete this table.

n	1	2	3	4	5
$2n - 1$	$(2 \times 1) - 1$ $= 1$				

   b Copy this graph and plot the remaining points.

   c The sequence with the formula $2n - 1$ is an arithmetic sequence. Arithmetic sequences are sometimes called **linear** sequences. Why?

   d Draw a line through the points on your graph.
   What is the equation of the line you have drawn?

   > Think about the pattern made by the points on the graph.

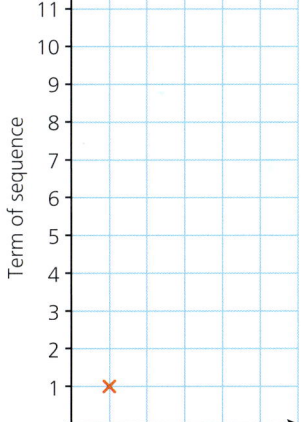

② Repeat for the sequence with the position-to-term rule $\frac{1}{2}n + 2$.

   a Construct a table to find the first five terms of the sequence.

   b Plot a graph of term against n for the first five terms.

   c Draw a line through the points.
   Find the equation of this line.

Curriculum for Wales Mastering Mathematics: Book 3

8.2 Now try these

Band 1 questions

Logical reasoning

1. For each of the following sets of points
   a. plot them on graph paper and draw a straight line through them
   b. find the equation of the line.
      i. $(-5, -5), (-4, -4), (-3, -3), (-2, -2)$
      ii. $(9, 6), (8, 5), (7, 4), (6, 3)$
      iii. $(4, 3), (3, 4), (2, 5), (1, 6)$

2. a. Draw the lines with these equations on the same axes.
      i. $y = 2x + 2$
      ii. $y = x + 4$
      iii. $y = 2x - 3$
      iv. $y = 4 + 2x$
      v. $y = x + 2$
      vi. $y = x - 3$
   b. Which lines have the same gradient?
      What do you notice about their equations?
   c. Which pairs of lines have the same y-intercept?
      What do you notice?
   d. A line has a gradient of 3 and y-intercept of -5.
      What is its equation?

Fluency

3. a. Draw the graphs of these lines on the same axes.
      Take values of x and y from 0 to 5.
      i. $x = 4$
      ii. $x = 2$
      iii. $y = 5$
      iv. $y = 3$
   b. Which of the lines are parallel to the x-axis?
      Which of the lines are parallel to the y-axis?
   c. What are the equations of the x-axis and y-axis?

Logical reasoning

4. a. Find the missing coordinates on these straight lines.
      i. $(0, 0), (1, 1), (2, 2), (3, \square), (\square, 4), (5, \square)$
      ii. $(6, 0), (5, 1), (4, 2), (3, \square), (\square, 4), (1, \square)$
      iii. $(-1, 1), (0, 2), (1, 3), (2, \square), (\square, 5), (4, \square)$
      iv. $(-2, 5), (-1, 5), (0, 5), (1, 5), (\square, 5), (3, \square)$

   > You can do this by writing the x-coordinates as one sequence of numbers and the y-coordinates as another.

   b. Find an equation for each line.

   > Begin by taking any two points on the line and finding the gradient between them.

5. a. Draw x- and y-axes from 0 to 8.
      On your axes draw the pattern formed by these lines.
      i. $x = 2$
      ii. $x = 5$
      iii. $y = 2$
      iv. $y = 5$
      v. $y = x$
      vi. $x + y = 7$
   b. Draw more lines on your diagram to form eight triangles of the same size.
   c. Write down the equations of the lines you drew in part **b**.

166

6 Write down the gradient and the coordinates of the y-intercept for each of these lines.

a $y = 2x + 3$
b $y = 3x - 4$
c $y = x + 4$
d $y = x - 1$
e $y = 4x$
f $y = \frac{1}{2}x$
g $y = 7 - 2x$

Band 2 questions

7 Find the gradient and y-intercept for each of these lines.
Then write down the equation of the line.

a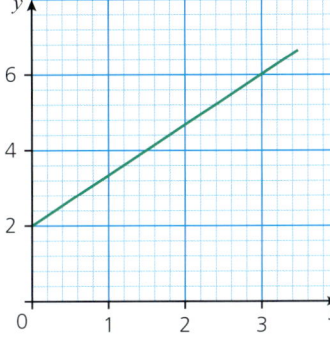
b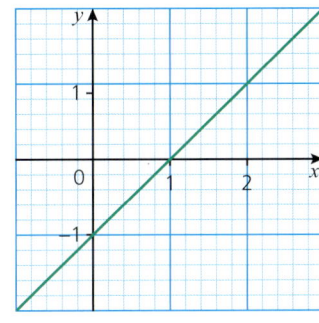
c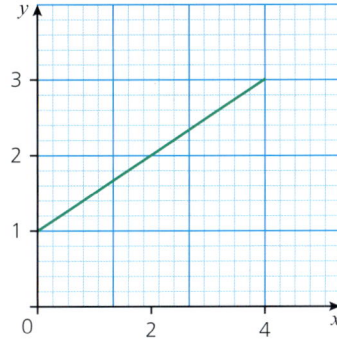
d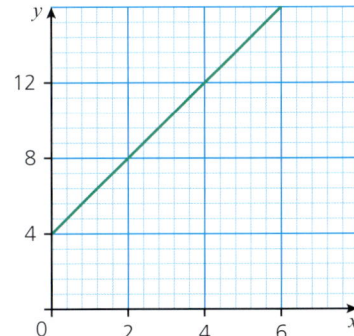
e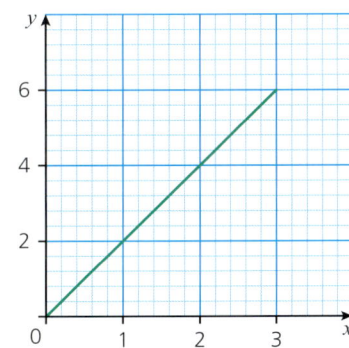

8 For each of these equations of straight lines
   i   rearrange the equation into the form $y = mx + c$
   ii write down the gradient and the coordinates of the y-intercept.

a $y = 2(x + 1)$
b $y = \frac{1}{2}(2x - 5)$
c $2y = 6x - 5$
d $x + y = 2$
e $2x - y = 7$

9 Here are the equations of three lines:
$y = x$ $\qquad$ $y = 3 - x$ $\qquad$ $y = x + 3$

Answer these questions **without** plotting the lines on a graph.

a Which of these lines are parallel?
b What is the gradient of the two parallel lines?
c Which of the lines have the same y-intercept?

Band 3 questions

10 Ella and Iwan are doing a physics experiment.

They are investigating how forces stretch springs.

Iwan hangs a weight of 1 newton (N) on the end of a spring and measures its length in millimetres.

They then take it in turns to add more weights, each time measuring the length, L mm, of the spring and the force, F N, until there is a total weight of 8 N hanging on the spring.

Ella and Iwan display their results in a table.

Force, F (N)	0	1	2	3	4	5	6	7	8
Length of spring, L (mm)		36	48	59	71	83	100	106	117

a Plot the graph of force against length, with force along the horizontal axis.

Draw a line of best fit.

b One of the results is wrong.

Which one do you think it is?

Suggest the correct value for this result.

c Iwan forgot to measure the length of the spring before he started hanging weights on it.

Use your graph to find the length of the spring.

d Calculate the gradient of the graph.

e Write the equation of the line in the form $L =$ _____ $F +$ _____

11 Lowri and Jac are doing an electricity experiment in their physics lesson.

They set up a circuit to measure the current, I, flowing through a resistor and the voltage, V, across it.

Their circuit is shown in the diagram.

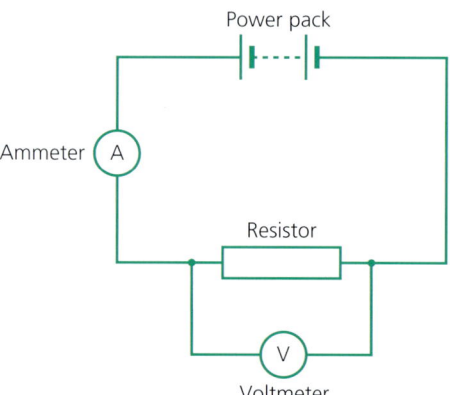

They set their power pack at 0 volts and record the readings shown on the voltmeter and the ammeter.

Lowri turns the control on the power pack to increase the voltage.

Jac records the new readings on both the voltmeter and the ammeter.

They repeat this several times.

Here are their results.

Voltage, V (volts)	Current, I (amps)	$V \div I$
0.0	0.0	—
1.0	0.2	
2.0	0.4	
3.0	0.5	
4.0	0.8	
5.0	1.0	
6.0	1.2	

a Draw a graph of V against I, with I along the horizontal axis.

b Jac wrote an incorrect value for one of the results.
 i Which result is it?
 ii What should you do about it?

c What is the current when the voltage is 4.5 V?

d What is the voltage when the current is 1.1 A?

e The resistance of the resistor is measured in ohms.

 The resistance equals the gradient of the line on the graph.

 Calculate the resistance.

12 On the graph below a point has been plotted and five lines drawn through it.

Draw a graph and plot a point of your choice.

Investigate the equations of lines which pass through your point.

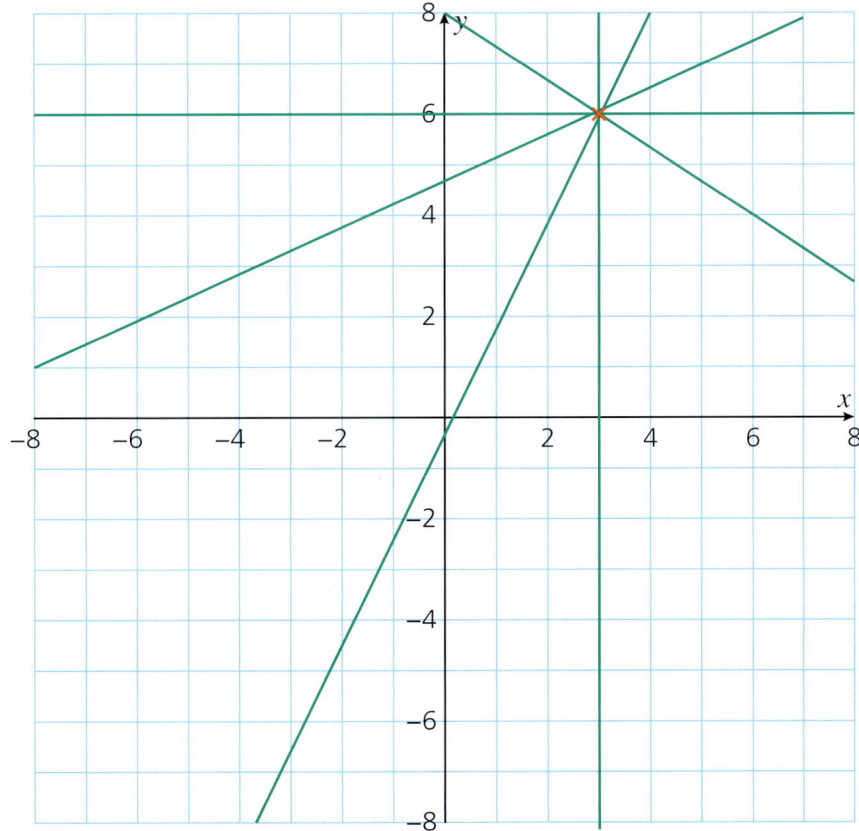

169

13 Aber Autos is a car-hire company.

The basic charge is £20 plus a further 30p per mile.

  a Show that $C = 0.3x + 20$, where C is the cost of driving x miles.
  b Plot a graph of this equation for values of x from 0 to 200.
  c Use your graph to find the cost of hiring a car and driving 80 miles.
  d Use your graph to find how many miles are driven when the hire cost is £56.

14 Gas bills have a standing charge of £15 plus 10p per gas unit used.

The bill is £B.

  a Show this is related to x, the number of gas units used, by the equation
     $B = 15 + 0.1x$
  b Plot a graph of this equation taking x values from 0 to 800.
  c Use your graph to find Mr Smith's bill when he uses 275 units of gas.
  d The Butterworths' bill is £89.
     Use your graph to find how many units of gas they used.

15 A sequence has the position-to-term rule $5 - 2n$.

  a Find the first 5 terms of the sequence.
  b Plot a graph of the term on the vertical axis against n on the horizontal axis for the first five terms.
  c Draw a line through the points.
     Find the equation of this line.

> **Key words**
>
> **Here is a list of the key words you met in this chapter.**
>
> Curve   Gradient   Linear y-intercept
>
> **Use the glossary at the back of this book to check any you are unsure about.**

8 Graphs

Review exercise: graphs

Band 1 questions

1. Write down the coordinates of the y-intercept and the gradient of each of these lines.
   - a $y = 2x + 5$
   - b $y = 3x + 2$
   - c $y = 4x - 5$
   - d $y = 7 + 3x$
   - e $y = 10x$
   - f $y = -2x$
   - g $y = \frac{2}{3} + \frac{1}{2}x$

2. Match each line in this diagram with the correct equation.

   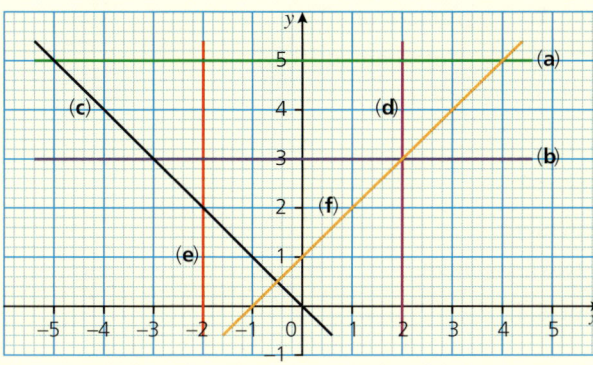

   - i $x = 2$
   - ii $y = 3$
   - iii $y = -x$
   - iv $y = x + 1$
   - v $y = 5$
   - vi $x = -2$

3. Here are the graphs of the lines $y = x - 2$ and $y = 2x$.

   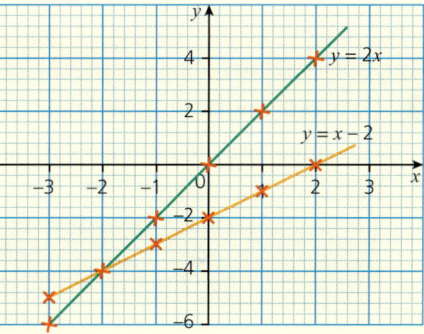

   - a Find the gradient of each line.
   - b Which is the steeper line?
   - c Find the y-intercept of each line.
   - d Write down the coordinates of the point where the two lines cross.

4. a Draw x- and y-axes with values from -2 to 8 on each axis.

   Plot the line with gradient 2 and with a y-intercept of 1.

   Write down the equation of this line in the form $y = mx + c$.

   b On the same axes, plot the line with gradient 3 and with a y-intercept of -2.

   Write down the equation of this line in the form $y = mx + c$.

   c Extend your lines so that they cross.

   Write down the coordinates of the point where the lines cross.

171

Band 2 questions

5 Here are the equations of eight lines.

$y = 2x - 3$	$x + y = 5$	$3y = 6x + 10$	$y + 3x = 0$
$y = -3$	$y = 3 - x$	$y = 3x + 5$	$y = 4x - 7$

Choose the equations of the lines which have

a a y-intercept at $(0, 0)$
b a gradient of 0
c a negative gradient
d a gradient of 3
e a gradient of 2
f a gradient of -3
g the steepest slope
h a y-intercept at $(0, 5)$.

6
a Draw x- and y-axes.
   Take values from -2 to 8 on each axis.
b Plot the line whose gradient is 1 and whose y-intercept is 2.
c Plot the line whose gradient is 2 and whose y-intercept is -2.
d Write down the coordinates of the point where these two lines meet.
e Plot the line whose gradient is $\frac{1}{2}$ and whose y-intercept is 4.
   What do you notice about this line?

Band 3 questions

7 Liz and Danny do an experiment with an elastic string.

They measure the extension of the string each time they add a weight to it.

Here are their results.

Force, F (N)	0	2	4	6	8	10	12	14
Extension of string, E (mm)	0	19	39	56	75	95	115	133

a Plot the graph of force, F N, against extension, E mm, with force along the horizontal axis.
b Calculate the gradient of the line and use it to write down the equation of the line.
c What does your graph tell you about the relationship between E and F?

8 A company makes rectangular metal strips to be used as machine components.

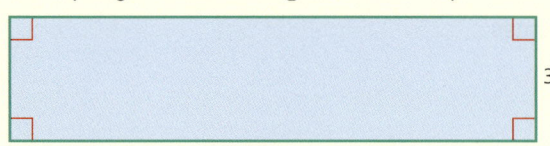

a Show that the area, A cm², of a strip is $A = 3x + 60$.
b Copy and complete this table:

x	2	4	6	8	10
$A = 3x + 60$					

c Plot a graph of A against x.

9 Here are the equations of eight lines.

| $y = 6x + 3$ | $y = 4x - 3$ | $y = 7$ | $y = 3 - 4x$ | $y = 2 - x$ |
| $y = 3x + 2$ | $y = 5x + 2$ | $y = 3x$ | $y = 3x - 4$ | $y = 8x - 5$ |

Choose the equation or equations that give(s)
- **a** a line that passes through the origin
- **b** a line that has a gradient of 5
- **c** a line that has a gradient of −1
- **d** a line that passes through the point (0, 3)
- **e** a line that is parallel to $y = 4x$
- **f** a line that passes through the point (0, −4)
- **g** a line that is parallel to $y = 5 - 4x$
- **h** a line that passes through the point (1, 5)
- **i** a line with a zero gradient
- **j** the steepest line.

10 A straight line joins the points (2, 2) and (5, 8).
- **a** What is the gradient of this line?
- **b** What is the equation of this line?

11 What is the equation of the line which passes through the points (0, 0) and (3, 18)?

12 A line passes through the point (2, 7).
The gradient of the line is −3.
What is the equation of the line?

13 Aled is making a table in his workshop.
The **sum** of the length and the width of his table is 10 m.

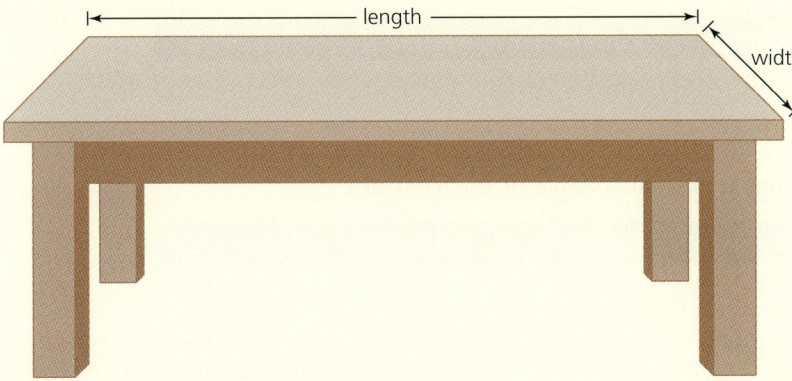

Draw a graph to show the possible lengths and widths of his table.

9 Real-life graphs

Coming up...
- ▶ Distance–time graphs (travel graphs)
- ▶ Reading from real-life graphs

Fractal triangles

The diagrams show a sequence of patterns made up of equilateral triangles.

To move from one pattern to the next, each triangle is divided into four smaller equilateral triangles.

The sides of the equilateral triangle in Pattern 1 are 1 metre in length.

In Pattern 2 there are four triangles each with sides $\frac{1}{2}$ metre in length.

In Pattern 3, there are 16 triangles each with sides of length $\frac{1}{4}$ metre.

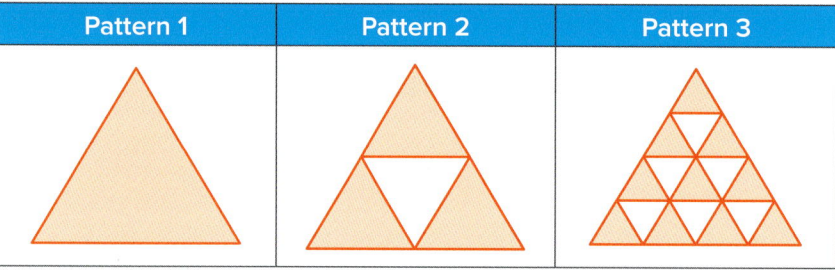

a Copy and complete this table for the **side length** of the small triangles in each pattern.

	Pattern 1	Pattern 2	Pattern 3	Pattern 4	Pattern 5
Length of side (m)	1	$\frac{1}{2}$	$\frac{1}{4}$		

b Copy and complete this table for the **number** of small triangles in each pattern.

	Pattern 1	Pattern 2	Pattern 3	Pattern 4	Pattern 5
Number	1	4	16		

c From one pattern to the next the **side length** of a small triangle halves.

   This is known as **exponential decay**.

   When a quantity halves and then halves again in the same space of time, this is known as 'exponential decay'.

   In which pattern would the small triangles have a side length of $\frac{1}{256}$ metres?

d From one pattern to the next the **number** of small triangles quadruples.

   Quadruples means it is multiplied by 4.

   This is known as **exponential growth**.

   When a quantity doubles and then doubles again in the same space of time, this is known as 'exponential growth'.

   In which pattern would there be 256 small triangles?

9.1 Distance–time graphs

Skill checker

a For each of these travel graphs, work out the speed in kilometres per hour.

Remember

In Chapter 6 on Measure you learnt that speed = distance ÷ time.

i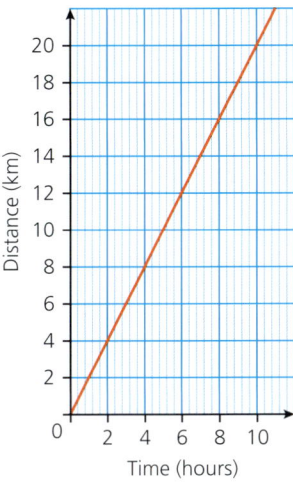

ii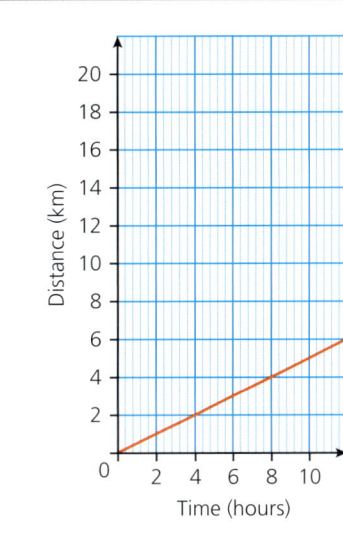

iii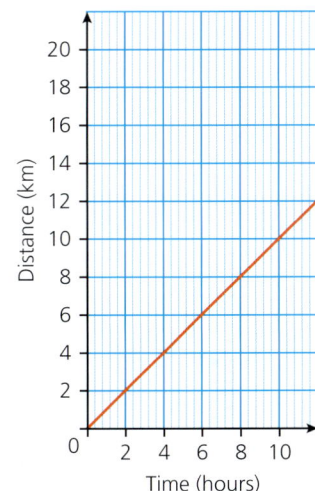

iv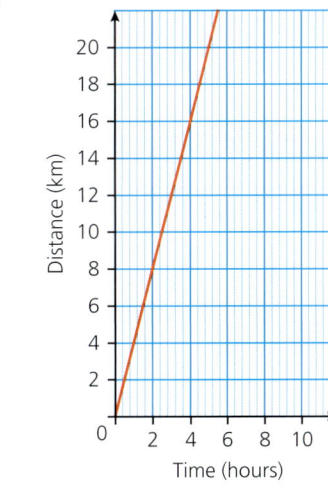

b In which of the travel graphs is the speed greatest?

c Copy and complete these sentences:

The steepest graph shows the object travelling at the _____ speed.

The graph with the _____ gradient shows the object travelling at the lowest speed.

Curriculum for Wales Mastering Mathematics: Book 3

Activity

Each of the distance–time graphs below represents a journey made by Alwyn.

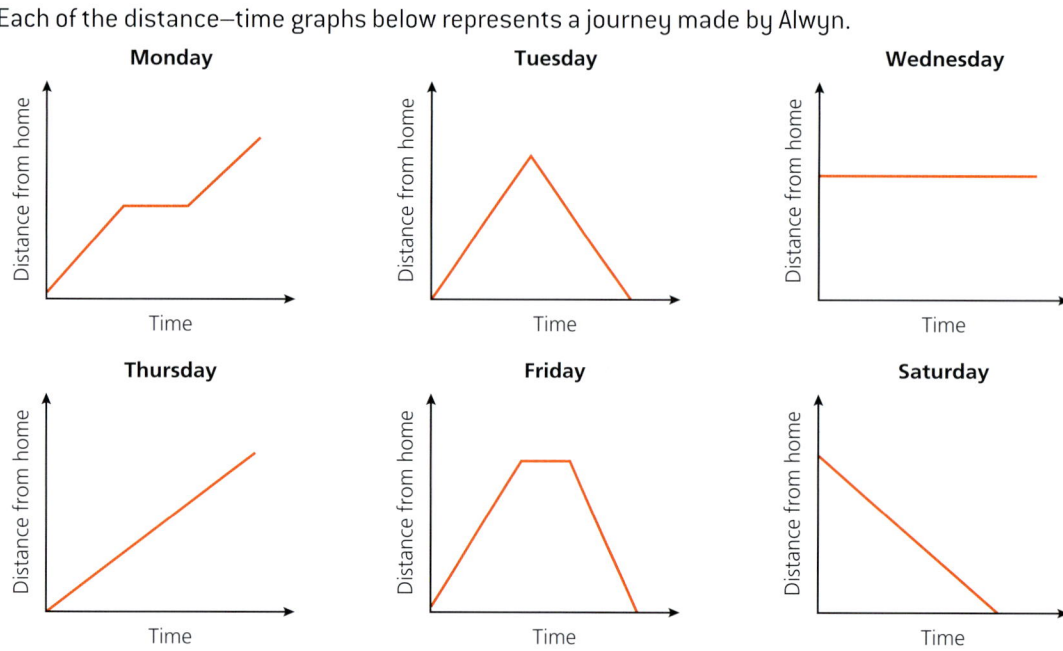

In pairs, discuss journeys which Alwyn could have made to create each day's distance–time graph.

Cross-curricular activity

Discuss with your PE teacher what kind of data you could log over a period of time, in order to record improvements in your fitness levels. Draw graphs to illustrate your improvement.

▶ Reading distance–time graphs

Worked example

Look at this graph showing the height of a hot air balloon.
It is called a travel graph or a distance–time graph.

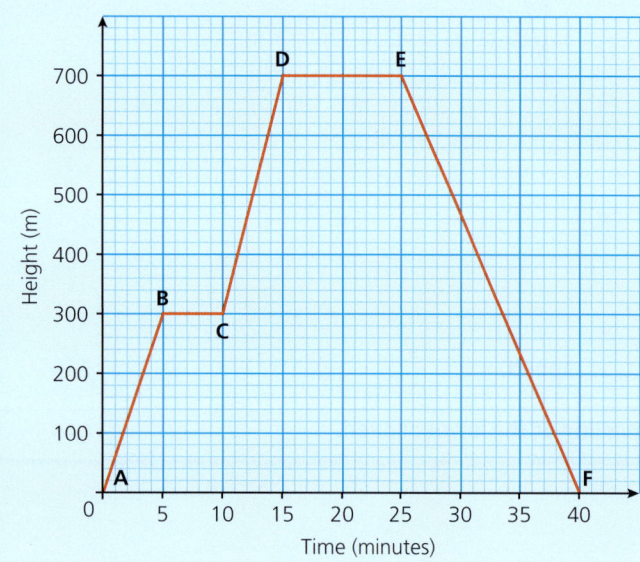

a How high does the balloon go?
b What is shown by the lines BC and DE?
c For how long is the balloon at a height of 300 m?
d How long does the balloon take to land?
e When is the balloon at a height of 500 m?

Solution

a The balloon reaches a maximum height of 700 m.

> The maximum height is the largest value of y that the graph reaches.

b The lines BC and DE represent times when the balloon remains at the same height.

> The lines BC and DE are horizontal lines.

c The balloon is at 300 m for 5 minutes.

> This is given by the length of line BC.

d It takes 15 minutes to come down.

> The line EF begins 25 minutes into the flight and finishes at 40 minutes.

e It is at a height of 500 m at $12\frac{1}{2}$ minutes on the way up and at 29 minutes on the way down.

> Draw a line horizontally from 500 m on the height axis and read off the time of the point where it crosses the graph.

▶ Drawing a distance–time graph

Worked example

Lloyd, Tegan and Raj leave Swansea and travel in the same direction.
▶ Lloyd sets off at 9:45 a.m. and travels 40 km by electric car.
   He drives at a constant speed and it takes him 1 hour.
▶ Tegan leaves at 9:15 a.m. She walks 3 km at a steady pace in one hour. Then she catches a bus which travels another 15 km at a constant speed. The bus journey takes 1 hour.
▶ Raj cycles, setting off at 9 a.m.
   He travels 20 km in 2 hours at a constant speed.
a Draw axes from 9 a.m. to 12 noon on the horizontal axis and from 0 km to 45 km on the vertical axis.
  Plot the travel graphs for all three people.
b At what time does Lloyd pass Tegan?
c At what time does Lloyd pass Raj?
d Does Raj pass Tegan? Explain your answer.
e At what speed is Raj cycling?
f At what speed is Lloyd driving?

Solution

a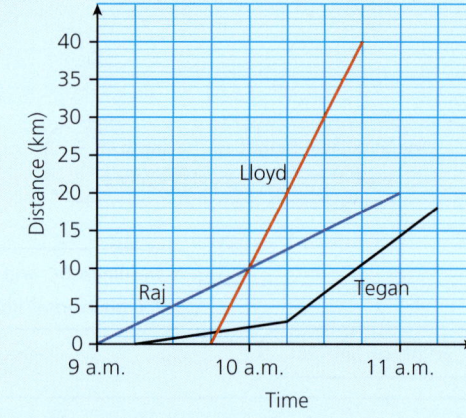

b Lloyd's red line and Tegan's black line cross at about 9:50 a.m.
This is the time at which Lloyd passes Tegan.

c Lloyd passes Raj at about 10 a.m.

d Raj's blue line and Tegan's black line do not meet, so Raj does not pass Tegan. Raj leaves Swansea earlier than Tegan and stays in front of her the whole time.

e Raj cycles 20 km in 2 hours.
$$\text{speed} = \frac{\text{distance}}{\text{time}} = \frac{20}{2} = 10 \text{ km/h}$$

f Lloyd drives 40 km in 1 hour.
$$\text{speed} = \frac{\text{distance}}{\text{time}} = \frac{40}{1} = 40 \text{ km/h}$$

Note

Sometimes when you read from a graph it is not possible to get exact values. Here it is only possible to read an approximate time. There may be a range of answers that are acceptable.

Activity

Make a poster all about your journey to school.

① Include a map of your journey on your poster.

② Show a travel graph for your journey.

Remember to include every part of the journey (e.g. walking, cycling, bus and car).

You will need to research how far you travel for each part of the journey.

Try to be accurate about what time you begin and end each part.

③ Make up two questions for a friend to answer about your travel graph.
For example:
  ▶ How far have I travelled at 8:30 a.m?
  ▶ At what time have I cycled 1 kilometre?
  ▶ How fast do I travel in the car?

Remember

For the faster parts of your journey the graph will be steeper.

9 Real-life graphs

9.1 Now try these

Band 1 questions

1 Seren walks to her local shop to buy some eggs and then returns home.

Here is a distance–time graph for her journey.

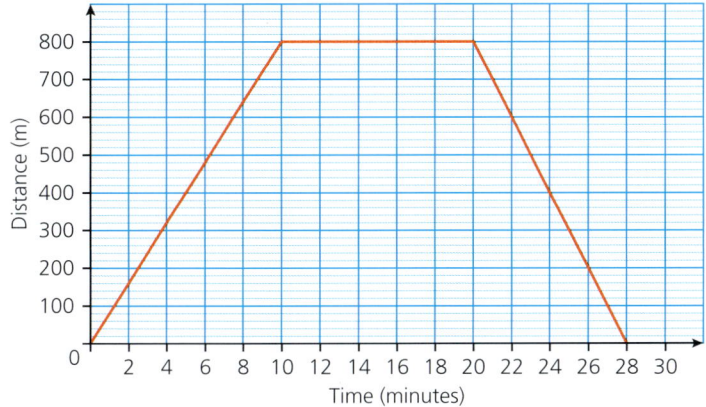

- **a** What does the shape of the graph tell you about Seren's journey?
- **b** How far is the shop from Seren's home?
- **c** How long does Seren spend in the shop?
- **d** How long does it take Seren to walk home?

2 A chef flips a pancake. The graph shows the pancake's distance above the ground at different times, measured in seconds.

- **a** How high is the pancake after one second?
- **b** The chef flips the pancake too high and it hits the ceiling.

    How many seconds does the pancake take to hit the ceiling after being flipped?
- **c** How high is the room?
- **d** When the pancake hits the ceiling, it sticks there. How long is it stuck for?

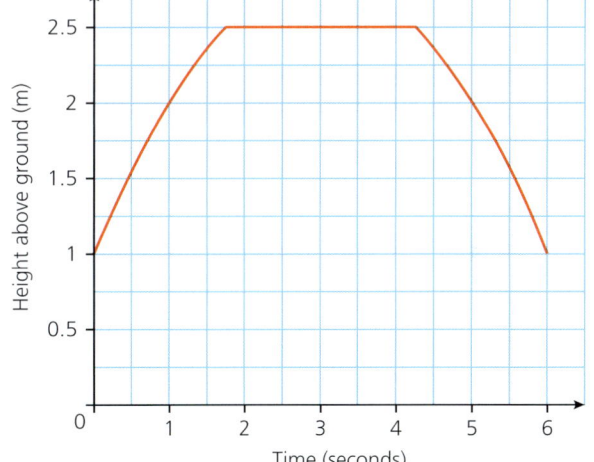

3 A bear is climbing a very tall tree to find some honey in a bees' nest at the top. The graph shows the height of the bear above the ground throughout its climb.

- **a** How high is the tree?
- **b** How long does it take the bear to climb the tree?
- **c** How high is the bear after 6 minutes?
- **d** How long does the bear spend eating the honey at the top of the tree?
- **e** How long does it take the bear to climb down the tree?

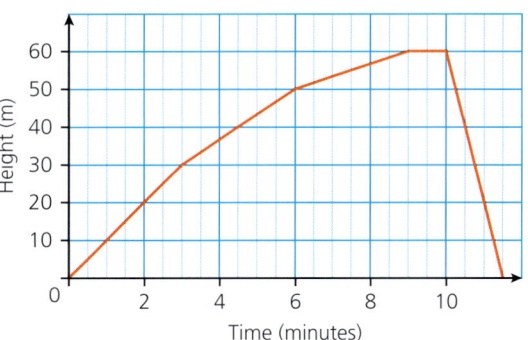

4 Mr Jones has a busy day delivering fruit and vegetables. The graph below shows his distance from home at certain times through the day.

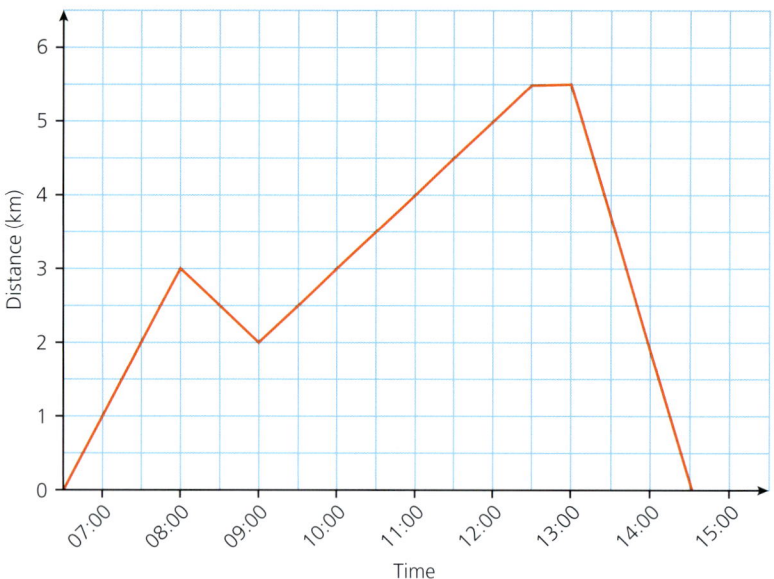

a Mr Jones' first delivery was 3 km away from his home. What time did he make this delivery?
b How far was Mr Jones from his home at 11:00 a.m.?
c Mr Jones made his third delivery of the day at 12:30 p.m.
   This delivery was to a shop 5.5 km from his home.
   How long did he stay at the shop?

Band 2 questions

5 Sketch a graph to represent the following situation:

From rest you start running until you reach flat-out speed.

Then you slow down gradually until you collapse from exhaustion. Your distance is plotted against time.

6 A cat jumps from ground level onto a fence.

The graph shows the height of the cat plotted against the time after it jumped.

a Roughly what is the cat's greatest height?
b After roughly how many seconds does it reach this height?
c Give a reason for the shape of the graph
   i   at 0.8 seconds
   ii after 1 second.

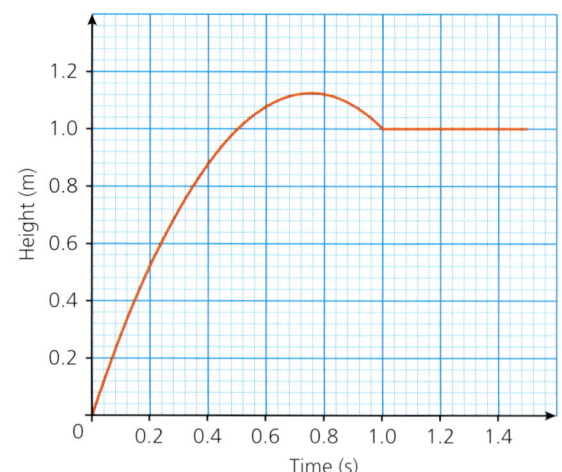

7 Majella runs the Newport Marathon. The graph shows her distance from the start line at certain times throughout the race.

a   At what time does the marathon start?
b   How far has Majella run by 11:00 a.m.?
c   The marathon is 40 km long. At what time does Majella finish the race?
d   Between which times does Majella run fastest?
    Explain why she might run faster between these times.
e   Majella's friend Paula is running in the same race. She runs at a steady speed of 10 km/h.
    Copy the graph. Draw a line representing Paula's journey on the same diagram.
f   At what times are the two women running together?
g   When do they next meet?
h   How long does it take Paula to complete the race?

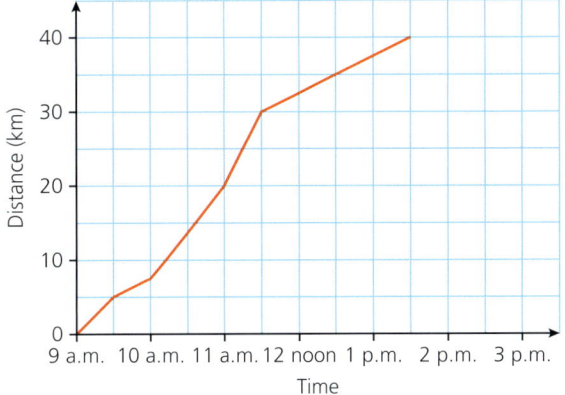

8 Suresh walks to the bakery to buy some scones.

The bakery is 1800 m from his home.

It takes Suresh 10 minutes to get there.

Suresh spends 12 minutes in the bakery.

On his way home, he covers the first 1000 m in 10 minutes.

He then sits down for 5 minutes to eat one of the scones.

The final 800 m home takes him 5 minutes.

The graph shows the first stage of Suresh's journey.

Copy the diagram and draw lines to show the rest of his journey.

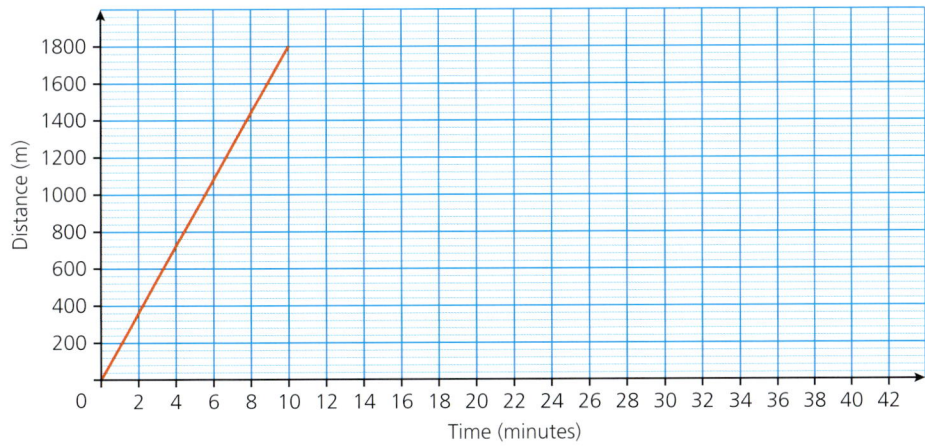

9 Parthi travels to work.

He walks 0.4 km from his home to the bus stop in 5 minutes.

He then waits 3 minutes for the bus.

The bus travels 2.1 km in 10 minutes to a bus stop near his workplace.

Parthi walks the last 0.3 km to work in 4 minutes.

   a Draw a horizontal axis from 0 minutes to 25 minutes.

      Draw a vertical axis from 0 km to 3 km.

      Plot a travel graph to represent Parthi's journey to work.

   b What is the average speed of the bus in kilometres per hour?

Band 3 questions

10 Steffan cycles to school every day.

On some days he gets to school more quickly than on others.

The distance–time graph below shows his journeys each day this week.

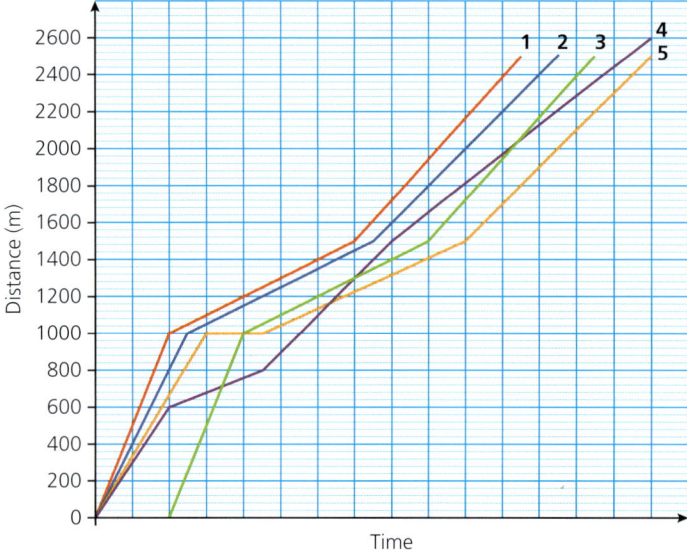

   a On Friday Steffan took a different route, which was slightly longer.

      Which number line shows Friday's journey?

   b How far does Steffan **usually** travel when he cycles to school?

   c The red (number 1) and blue (number 2) graphs have roughly the same shape.

      There is a steeper part of the graph, followed by a less steep part, followed by another steeper part.

      Can you explain this?

   d Steffan arrived at school at the earliest time on Tuesday.

      Which number line shows Tuesday's journey?

   e On Thursday Steffan stopped at the shop on the way to school.

      Which number line shows Thursday's journey?

   f On Wednesday Steffan left his house a bit later than usual.

      Which number line shows Wednesday's journey?

9 Real-life graphs

11 A bus is travelling at 10 metres per second towards a bus stop 50 metres away.

Here is a possible distance–time graph.

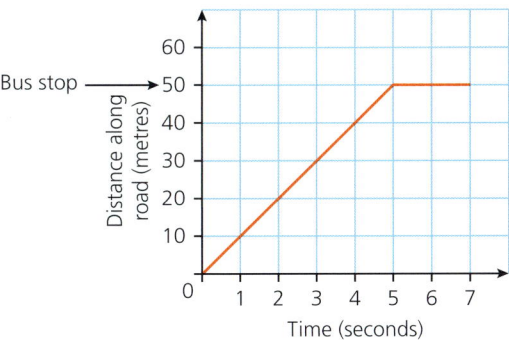

During the first 5 seconds, the bus is travelling at a constant speed of 10 m/s.

Then, on arrival at the bus stop, it suddenly stops dead in its tracks!

This is shown by the way in which the graph makes a sudden turn.

a What would happen if a real bus behaved in this manner?

b In the real world, a bus begins to slow down before it reaches the bus stop, and gradually comes to a halt.

This is represented using smooth curves.

Redraw the graph to show what it might look like in real life.

c Use your graph to find

  i when the driver begins to brake

  ii when the bus actually stops

  iii for how long the bus decelerates (slows down).

12 For each of the situations below, sketch a distance–time graph which will represent the events shown.

a
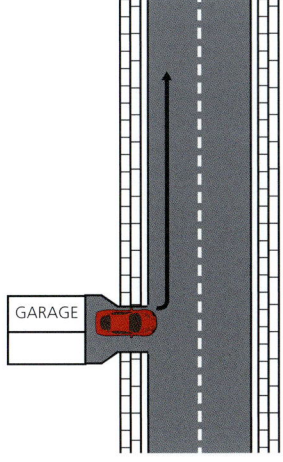

b
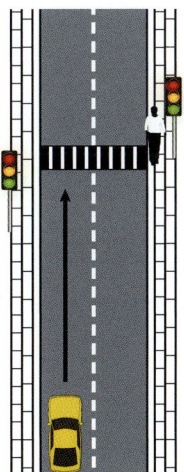

c 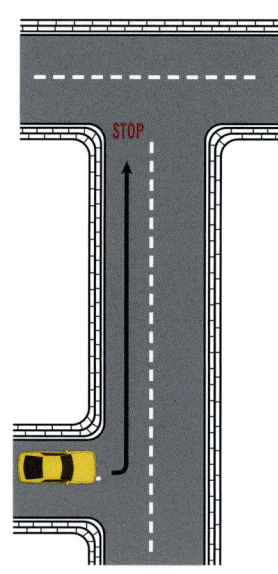 d 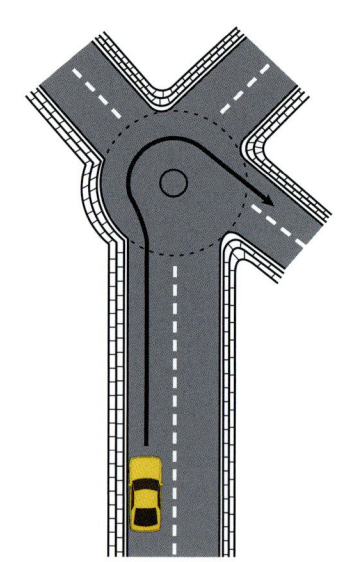 e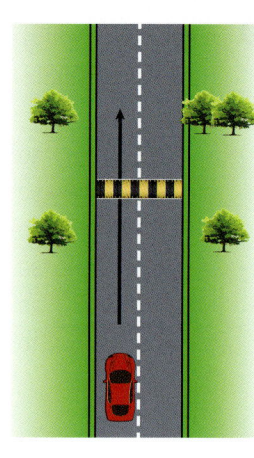

9.2 Reading from real-life graphs

Skill checker

In Chapter 8 you learnt how to plot a linear graph, for example for the straight line $y = 2x + 5$.

If y is the subject of the equation, you can use a table of values to plot any straight line.

Ioan draws up a table of values for the linear equation above.

Here is his table of values.

x	0	1	2	3	4	5	6
$y = 2x + 5$	5	7	9	10	13	15	17

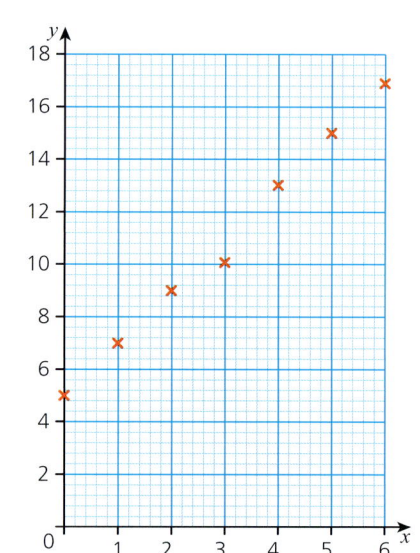

Ioan plots his points on a graph, shown here.

a One of the points looks wrong.

Copy the table of values, correcting the mistake.

b Copy the graph, plotting all the points correctly.

c Join the points with a straight line.

Use your graph to answer these questions:

d Find the value of y when $x = 1.5$.

e Find the value of x when $y = 14$.

Cross-curricular activity

Does insulation of water work by slowing down heat loss?

Globally, we need to slow down heat loss.

In your Science lesson, collect data by recording the temperature of 100 ml of hot water as it cools under different circumstances. Share the work and data with your friends, by thinking of different circumstances in which to test the water cooling, such as by standing a beaker of water in a bucket of ice, or by first pouring the hot water into a vacuum flask. Brainstorm different circumstances and set consistent timings for recording the temperatures. With all the results, draw graphs to clearly compare the results. In your English or Welsh lessons write a report with a conclusion using your graphs to support your findings.

▶ Reading information from a graph

Worked example

Rhian has solar panels on the roof of her house to generate electricity.

The solar panels work best during the middle part of the day when the sun is higher in the sky. They also work less well when it is cloudy.

The graph shows the amount of electricity the solar panels generate throughout one day in April. Solar energy is measured in kilowatts (kW).

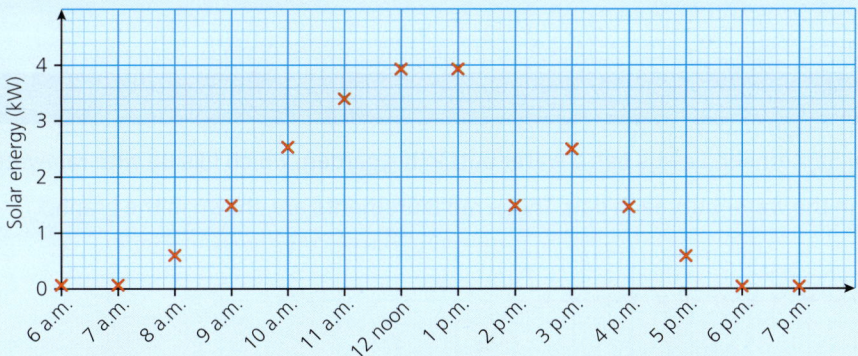

a The day was mainly sunny, but at one time it became partly cloudy. What time was this?

b How much electricity (in kW) were Rhian's solar panels making at 11 a.m.?

c Is it possible to be completely accurate when making this reading from the graph?
   Fill in the blanks in this sentence:
   The true reading is between _____ and _____.

d Estimate at what time the solar panels were making the most electricity.
   Roughly how much electricity (in kW) were the solar panels making at this time?

e At roughly what times were the solar panels making 1 kW of electricity? There is more than one answer.

Solution

Joining the points with a smooth curve allows you to answer this question more accurately.

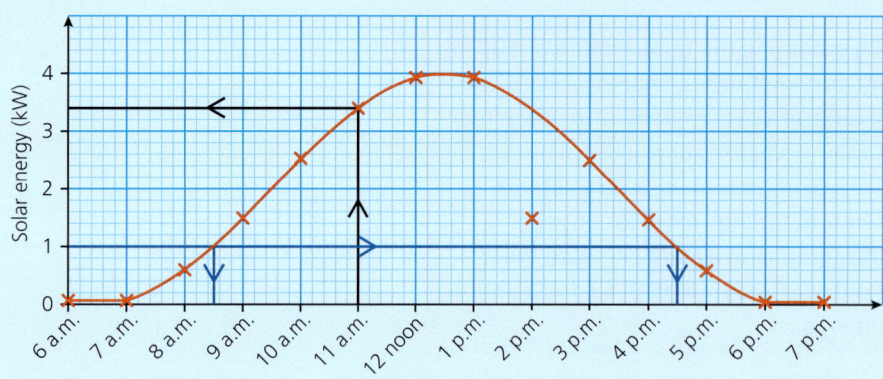

a At 2 p.m. the reading is much lower than the curve. This must be the time at which it was cloudy.
b The black lines show that at 11 a.m. the solar panels were generating roughly 3.4 kW of electricity.
c It is not possible to be completely accurate.
  The true reading is between 3.3 kW and 3.5 kW.
d The solar panels were making the most electricity at around 12:30 p.m.
  They were making roughly 4 kW at this time.
e The blue lines show that the solar panels were generating 1 kW of electricity at roughly 8:30 a.m. and 4:30 p.m.

9.2 Now try these

Band 1 questions

1. It rains every day for the first part of September. Lewys has a water butt in his garden, which fills up with rainwater.
   He measures the level of the water in the water butt each day. These measurements are shown in the graph.

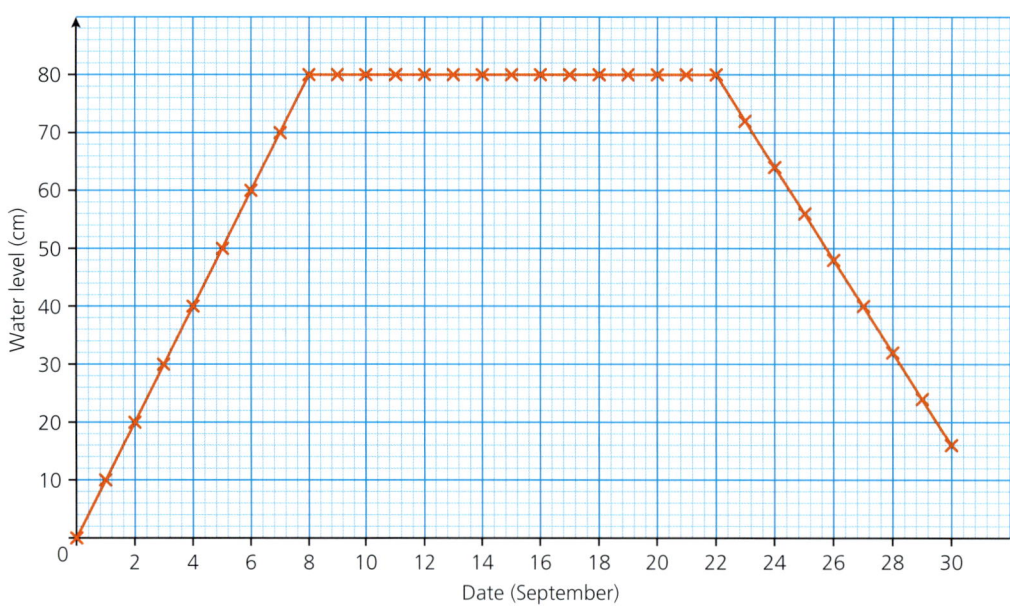

a   On which day did it stop raining?

b   After a while, Lewys begins to use the water for his garden.
    On which date did Lewys begin using the water?

c   Does Lewys use up all the water by the end of September?
    If not, how much is left?

2   The diagram shows the shape of an Olympic ski jump.

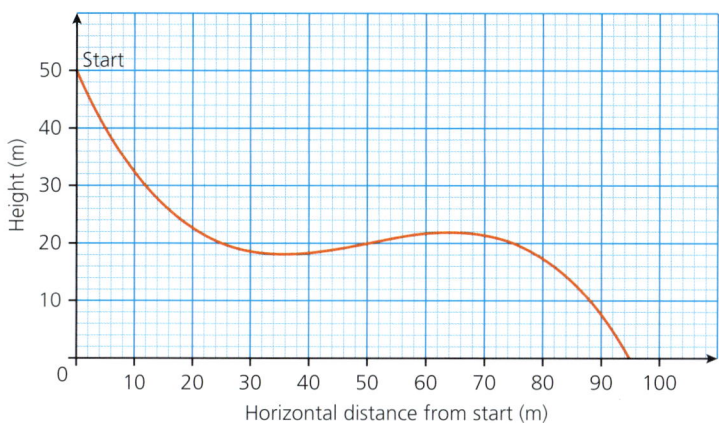

a   How high above ground are the skiers at the start of the slope?

b   How far from the start is the slope 20 metres high?

> **Hint**
> There is more than one answer to part b.

3   Satwinder cycles from his house to the nearest post box.

Starting from rest he increases his speed steadily until he reaches a top speed of 10 m/s.
This takes him 20 seconds.

He then stays at this speed for 70 seconds, until he is close to the post box.

He brakes and slows down steadily, coming to rest in 10 seconds.

a   Using the information above, copy and complete this table of values.

Time (seconds)	0	20	30	50	70	90	100
Speed (m/s)	0		10				

b   Draw axes with
    • time on the x-axis running from 0 to 100 seconds
    • speed on the y-axis running from 0 to 10 m/s.

    Use different scales on the x- and y-axes to fit the graph on your page.

c   Plot the points from your table on the graph.
    Join the points using straight lines.
    Use your graph to answer the following questions.

d   What was Satwinder's speed 60 seconds after leaving home?

e   At what times after leaving home was Satwinder travelling at 5 m/s ?

> **Note**
> This graph should be made up of three straight lines. It is called a 'piece-wise linear graph'.

Band 2 questions

4 Sara throws a ball from an upstairs window of her house.

Later she plots a graph of the ball's height against its distance from the house, as shown.

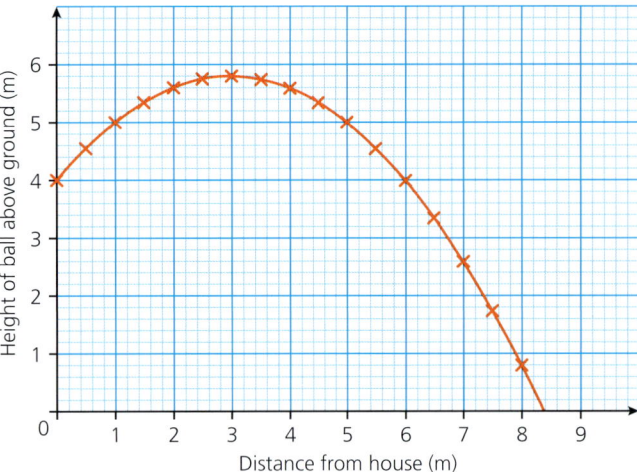

The height of the ball can be calculated using this equation:

$$h = -\frac{1}{5}x^2 + 1.2x + 4$$

where h is the ball's height and x is its horizontal distance from the house.

a How high above the ground is the window from which the ball was thrown?

b Using the graph, estimate the maximum height the ball reaches.

c Roughly how far from the house does the ball land?

5 Cerys is a mechanic in a garage in Brisbane, Australia.

For every car (of the same type) that comes in, she keeps a record of its age and an estimate of its value.

She plots a graph of the data she has collected this week.

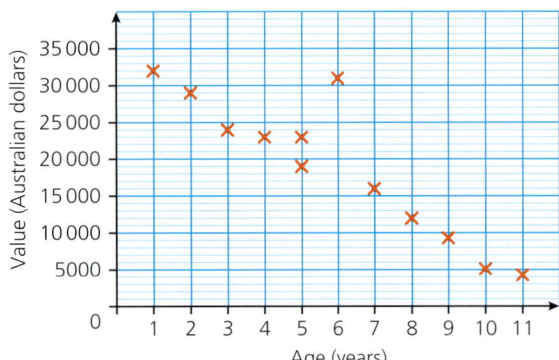

a Cerys thinks one of the points on her graph looks wrong.

She knows she recorded the car's age correctly, so she thinks she has estimated its value wrongly.

What is the age of the car?

b Estimate the value of this car more accurately, based on the information in the graph.

c Cerys wants to draw a straight line that passes as close as possible to all the points on her graph.

What should she do about the point that is wrong?

6 An offshore wind turbine is 120 metres tall. It casts a shadow across the sea. The length of the shadow varies throughout the day, as shown in the graph.

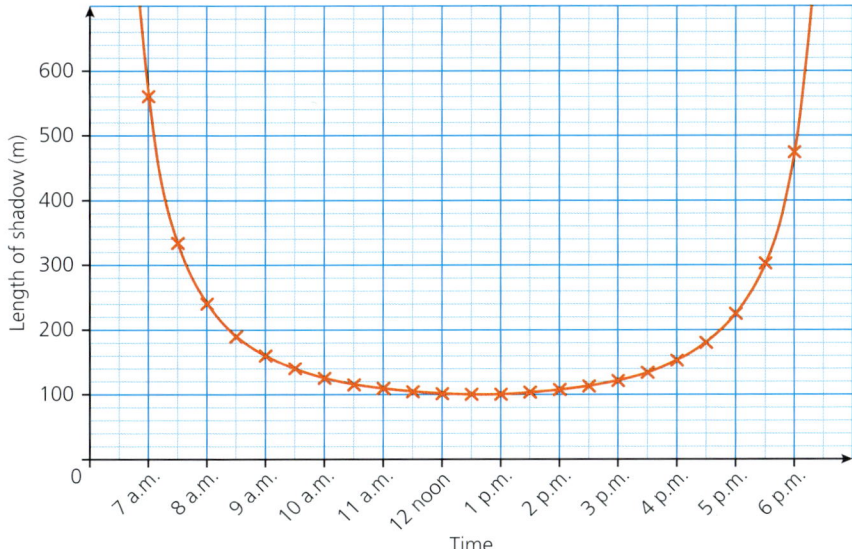

a Use the graph to find the length of the shadow at 3 p.m.
b Use the graph to estimate the length of the shadow at 9:15 a.m.
c At roughly what times of the day is the shadow 200 metres long?
   There are two answers.

7 If you climb a tall tower on a clear day you can see for miles.

The horizon is the imaginary line between the earth and the sky.
When you are higher up, the horizon appears to be further away.

The two diagrams show the distance to the horizon (the red curve) for two different towers.
From the taller blue tower, it is possible to see further.

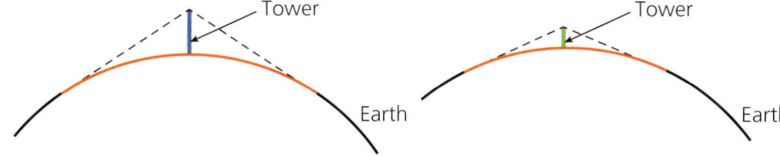

The graph shows the distance to the horizon for different size towers.
The distance d can be calculated from the height h using the formula

$$d = 1.22\sqrt{h}$$

a How far can you see from a tower 50 feet high?
b How high a tower would be needed to see at least 10 miles?

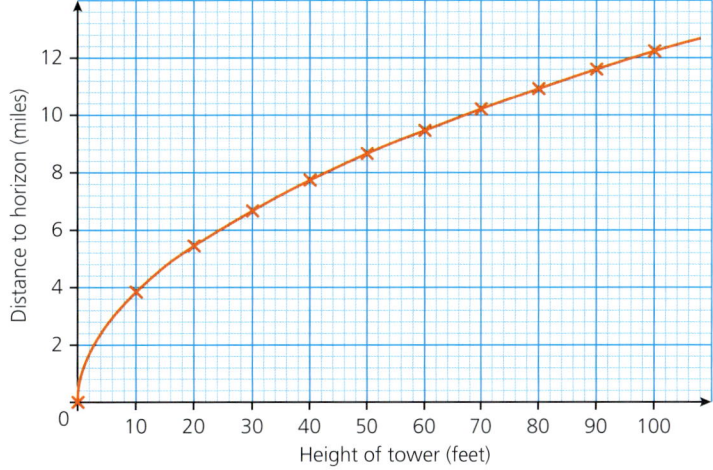

Curriculum for Wales Mastering Mathematics: Book 3

Band 3 questions

Logical reasoning

8 A farmer needs a rectangular field with an area of 360 m² for his sheep.

There are many ways to do this. Some of them are shown in the table below.

Width (m)	5	10	15		20		36	72
Length (m)	72		24	20		15		

a Copy and complete the table.

b Can you think of any other ways to make an area of 360 m²?

c Plot a graph of the length against the width, with the width on the x-axis.

Use values from 0 to 80 metres on both axes.

Use a scale of 1 cm for 10 metres.

Use your graph to answer these questions:

d How long is the field if the width is 30 metres?

e Estimate the width of the field if its length is 25 metres.

> **Note**
> The graph you have drawn is called a 'reciprocal curve'.

Strategic competence

9 Seema and Rhodri work in a university laboratory.

They are each studying a different colony of bacteria and they begin their experiments at the same time.

The graph shows the growth in the two populations of bacteria during the first 20 minutes of the experiments.

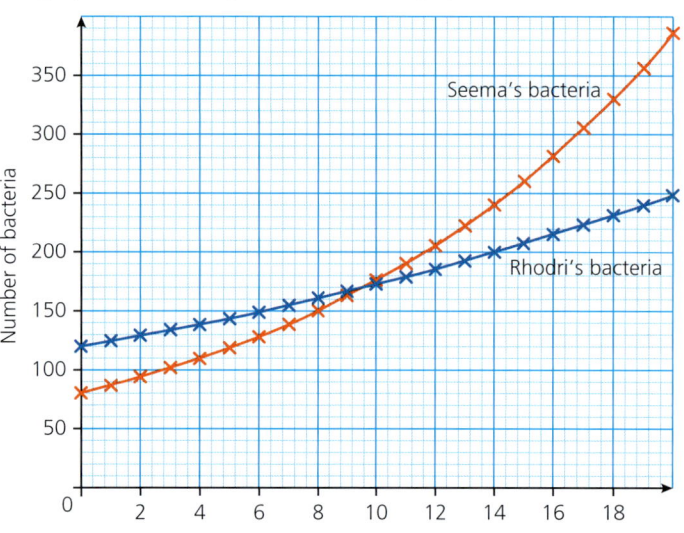

a What is the initial number of bacteria in Seema's colony? ← 'Initial' means at the start.

b Find approximately the number of bacteria in Seema's colony after four minutes.

Why do you think this is only an approximate figure?

c How long does it take for the number of bacteria in Seema's colony to reach 240?

d Which colony grows more quickly: Seema's or Rhodri's?

How can you tell from the graph?

e The number of bacteria in Seema's colony is initially lower than the number in Rhodri's colony. After how many minutes is Seema's colony larger?

f Roughly how long does it take for Seema's population of bacteria to double in size from its initial size?

Does the population double again in the same amount of time?

What is the name for this type of growth?

10 Mrs Hughes is a busy teacher. She makes a cup of tea in the staff room. She then places it on a table and forgets about it.

The temperature of the tea every minute is shown in the graph.

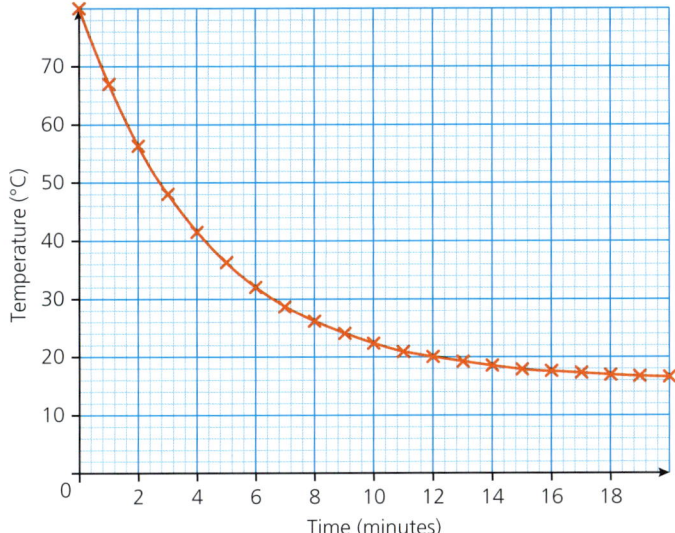

a What was the temperature of the tea when it was first placed on the table?
b How long does it take for the tea to cool to 20 °C?
c After a long time, the temperature of the tea becomes the same as the temperature of the room.
   Using the graph, what is the temperature of the room?
d What is the difference between the tea's temperature and the room temperature at the start?
e How long does it take for this temperature difference to half?
f How long does it take for this temperature difference to half again?
g What is the name for this type of curve?

Key words

Here is a list of the key words you met in this chapter.

Distance–time graph   Exponential graph   Piece-wise linear   Real-life graph
Reciprocal graph   Travel graph

Use the glossary at the back of this book to check any you are unsure about.

Curriculum for Wales Mastering Mathematics: Book 3

Review exercise: real-life graphs

Band 1 questions

Fluency

1. The Thomas family are going camping. They leave home in the morning and the graph below shows their distance from home during the journey.

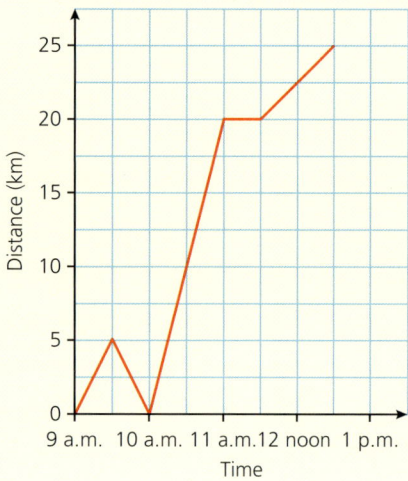

   a. After they had been travelling some time, Mrs Thomas realises they have forgotten to bring the tent. They return home to pick it up.

   At what time did Mrs Thomas realise this?

   b. Later, the Thomas family stopped to buy some drinks.

   At what time did they do this?

   c. The Thomas family arrived at the campsite at 12:30 p.m.

   How far from their home is the campsite?

Strategic competence

2. A walker climbs to the top of a mountain, then back down to ground level.
The graph below shows their height above ground level throughout the day.

   a. At what time did the walker start climbing the mountain?
   b. When the walker reached the top, for how long did they stop to look at the view?
   c. How high above ground level was the walker at 4 p.m.?

Band 2 questions

3 Tomos lives in Cardiff. He walks to a house called Glasfryn in the countryside outside Cardiff.

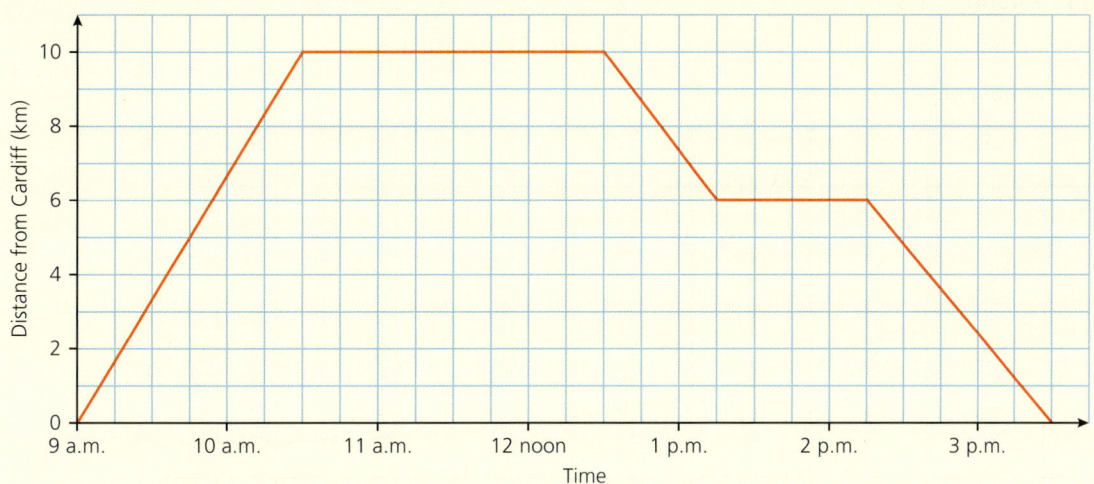

- **a** When does Tomos leave Cardiff?
- **b** How long does he spend at Glasfryn?
- **c** On the way back Tomos stops for a rest.

  How far is Tomos from Cardiff when he stops?
- **d** How long does he stop for?
- **e** Charles drives from Cardiff to Glasfryn, following the same route as Tomos.

  He leaves at 1:15 p.m. and arrives at 1:45 p.m.

  How far does he drive?
- **f** Copy Tomos's graph onto graph paper.

  Add a line to your graph representing Charles's journey.
- **g** At what time does Charles pass Tomos?

  What is Tomos doing at that time?
- **h** How far is Charles from Glasfryn when he passes Tomos?

4. A scientist investigates the number of bees in a hive.

   She finds that during the summer season their population N can be modelled by this exponential equation:

   $$N = 640 + 4360 \times 1.3^{-t}$$

   where t is the number of days into the summer season beginning on 1st June.

   A graph of the number of bees is shown.

   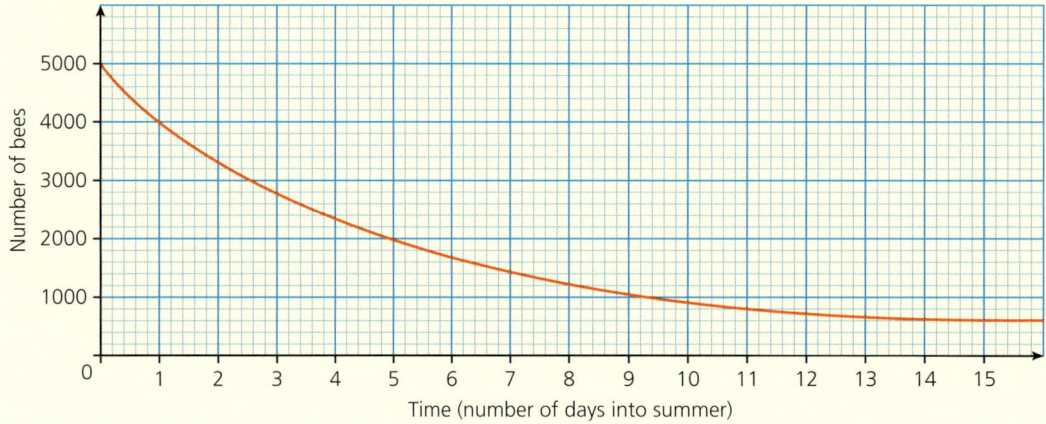

   a Find the population of the bees at midnight on 1st June. ← This is time $t = 0$.

   b Find the population of the bees at the beginning of 10th June.

   c On which date does the population first fall below 1000?

   d According to the equation, is it possible for the number of bees in the hive to fall to zero?

   If not, what is the smallest possible number of bees?

5. A, B and C are three stops on a bus route.

   The travel graph shows the journey of a bus from A to C via B.

   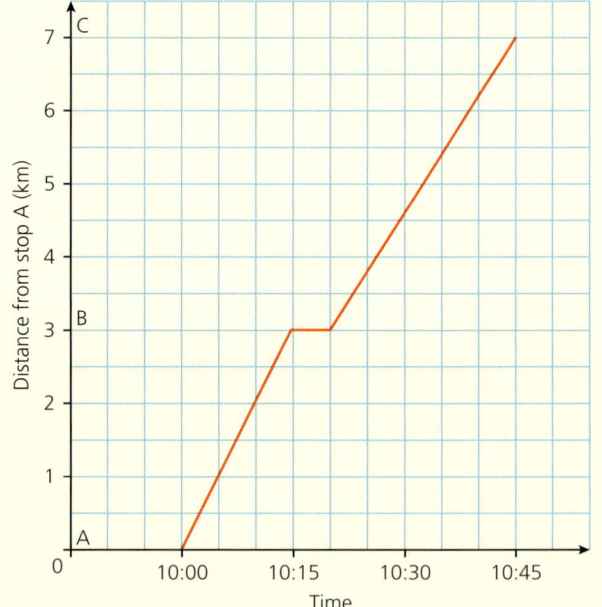

   a
     i When does the bus leave stop A?
     ii When does the bus arrive at stop B?
     iii For how long does the bus stop at B?
     iv What is the average speed of the bus between A and B?
     v What is the distance between stops B and C?

   b A second bus leaves C at 10:10.

   It travels at a constant speed to B, arriving at 10:45.

     i Copy the travel graph and show the second bus's journey on it.
     ii When do the two buses pass each other?
     iii Roughly how far apart are the buses at 10:40?

9 Real-life graphs

Band 3 questions

6 To film a scene for an action movie, a remote-controlled car is driven off a 50-metre-high cliff.

The height of the car can be calculated using this equation:

$$y = 50 - \frac{1}{5}x^2$$

where y is the car's height and x is its horizontal distance from the cliff.

x and y are both measured in metres.

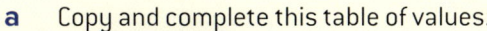

a Copy and complete this table of values.

x (metres)	0	2.5	5	7.5	10	12.5	15
y (metres)	50				30		

b Copy and complete this graph using the data in your table.

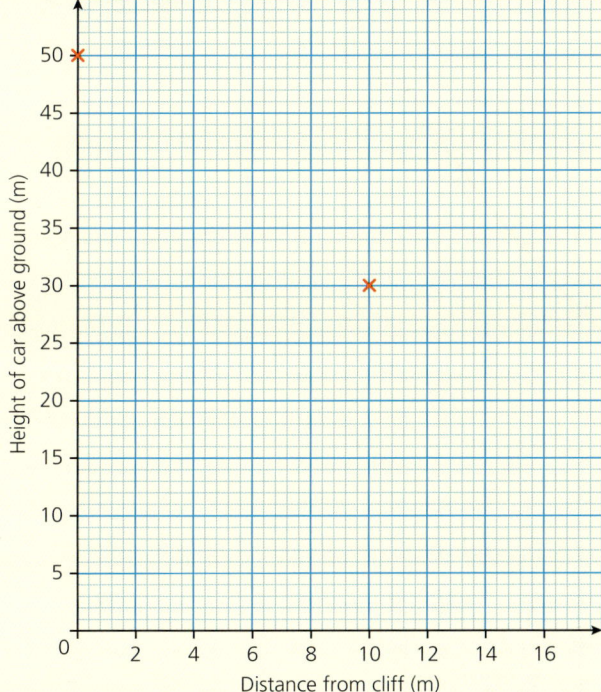

Join the points on your graph with a smooth curve.

Extend the curve so that it touches the x-axis.

Answer these questions using the graph.

c How high above the ground is the car when it is 8 metres from the cliff?

d How far from the cliff is the car when it is at a height of 10 metres?

Is your answer exact or an approximate reading from the graph?

e Roughly how far from the cliff does the car land?

195

7 Wyn was born on 14th May 2013.

His dad kept a record of his weight until he was 7 in the year 2020.

The graph shows Wyn's weight and the average weight of a boy up to the age of 7.

a When Wyn was born, was he below the average weight for a boy, or above?

b Roughly when did this change?

c Roughly what was Wyn's weight when he was 4 years old?

d Roughly how old was Wyn when his weight first went above 20 kg?

e At the age of 7, was Wyn below the average weight of a boy, or above?

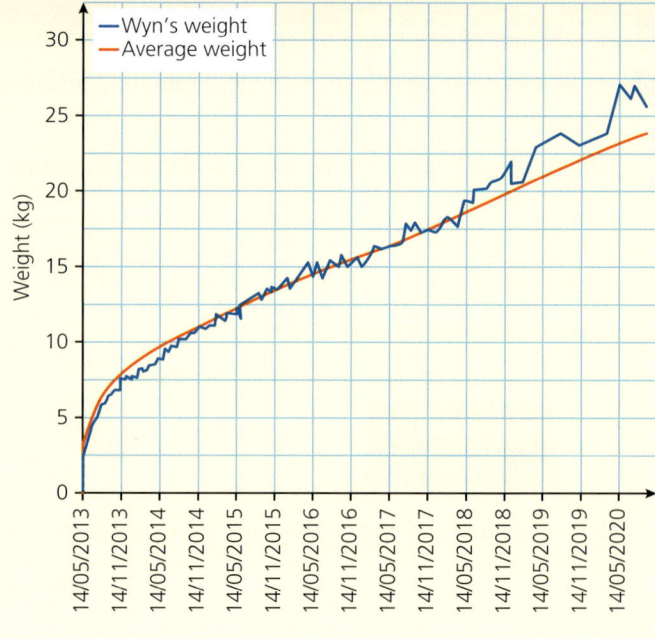

8 Rhys studies the fish population in a large lake.

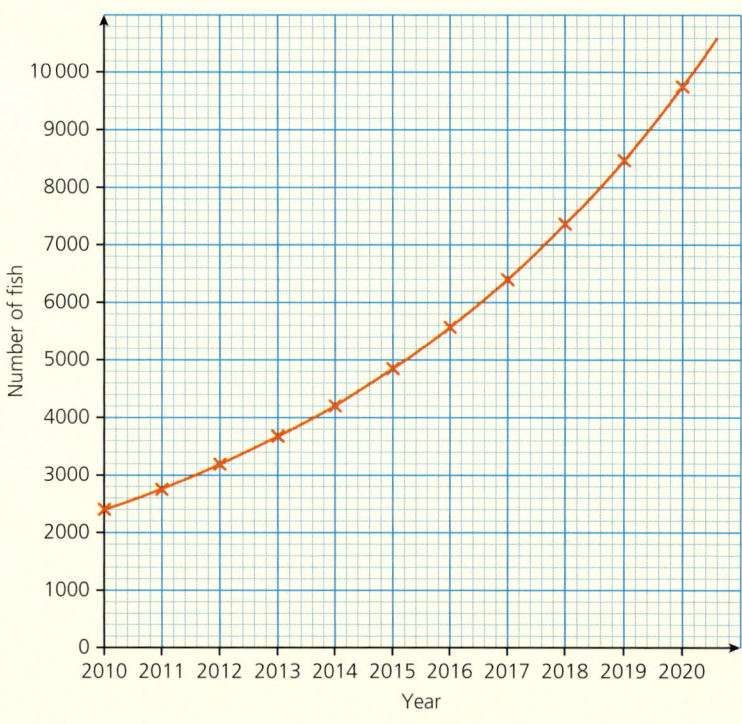

a How many fish are in the lake when Rhys surveys the lake in 2010?

b When roughly is the fish population 5000?

c By roughly what percentage is the population increasing each year?

d Do you think this growth is exponential? Explain your answer.

e Do you think this type of growth could continue for the next 20 years? Explain your answer.

f If this growth continues for one more year, what figure would Rhys expect for the fish population in 2021?

10 Transformations

Coming up...
- Reflection and translation
- Rotation
- Enlargement

Translation code words

+	−	×	÷	/	:	space
)	T	U	V	W	X	Y
(	S	F	G	H	I	Z
?	R	E	start	A	J	1
.	Q	D	C	B	K	2
,	P	O	N	M	L	3
0	9	8	7	6	5	4

a Beginning at the start square, move using these vectors.

$$\begin{pmatrix}-2\\1\end{pmatrix}, \begin{pmatrix}4\\-3\end{pmatrix}, \begin{pmatrix}0\\3\end{pmatrix}, \begin{pmatrix}-1\\-3\end{pmatrix}, \begin{pmatrix}0\\0\end{pmatrix}, \begin{pmatrix}-2\\2\end{pmatrix}, \begin{pmatrix}-1\\0\end{pmatrix}$$

What word do you spell?

b Decode this message.

$$\begin{pmatrix}0\\2\end{pmatrix}, \begin{pmatrix}-1\\-2\end{pmatrix}, \begin{pmatrix}1\\-1\end{pmatrix}, \begin{pmatrix}-2\\3\end{pmatrix}, \begin{pmatrix}1\\-4\end{pmatrix}, \begin{pmatrix}-1\\2\end{pmatrix}, \begin{pmatrix}5\\3\end{pmatrix}, \begin{pmatrix}-3\\-4\end{pmatrix}, \begin{pmatrix}-1\\-1\end{pmatrix}, \begin{pmatrix}0\\1\end{pmatrix}, \begin{pmatrix}0\\1\end{pmatrix}$$

$$\begin{pmatrix}-1\\1\end{pmatrix}, \begin{pmatrix}5\\2\end{pmatrix}, \begin{pmatrix}-2\\-3\end{pmatrix}, \begin{pmatrix}-3\\0\end{pmatrix}, \begin{pmatrix}1\\0\end{pmatrix}, \begin{pmatrix}4\\3\end{pmatrix}, \begin{pmatrix}-4\\-3\end{pmatrix}, \begin{pmatrix}2\\0\end{pmatrix}, \begin{pmatrix}-3\\1\end{pmatrix}, \begin{pmatrix}5\\1\end{pmatrix}, \begin{pmatrix}-6\\-3\end{pmatrix}$$

10.1 Reflection and translation

Skill checker

1. a Draw axes for values of x and y from 0 to 5, then plot these straight lines, labelling each one.
    i $y = 2$
    ii $y = x$
   b At which point do the two lines intersect?

② What are the equations of the two lines shown below?

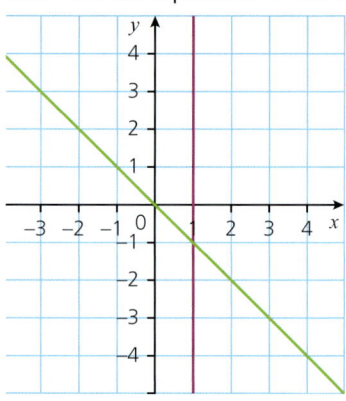

▶ Recap: translations

A **transformation** moves an object according to a rule.

One type of transformation is a **translation**. You learnt about translations in Book 1.

In the diagram object PQR is translated to the image P'Q'R'.

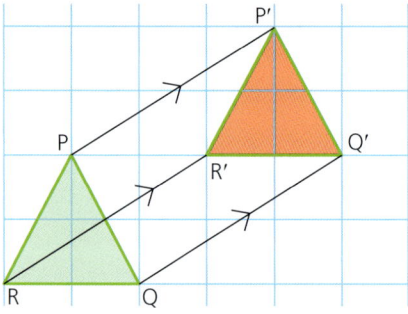

The object slides without turning.

The translation is described as 3 units right and 2 units up.

It can also be written as $\begin{pmatrix} 3 \\ 2 \end{pmatrix}$ ← This way of writing a translation is called a **vector**.

The image and the original object are **congruent**. This means they are the same shape and the same size.

▶ Describing a translation

Worked example

Describe the translation that maps

a A to C   b B to C   c C to A.

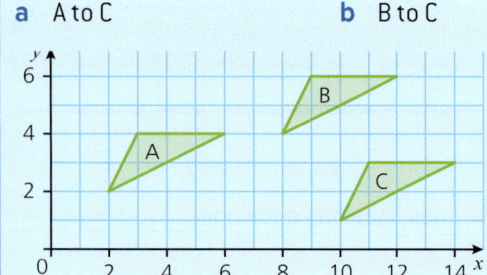

Solution

a The translation mapping A to C is 8 units right and 1 unit down. Using vector notation, this is: $\begin{pmatrix} 8 \\ -1 \end{pmatrix}$.

b The translation from B to C is 2 units right and 3 down, or $\begin{pmatrix} 2 \\ -3 \end{pmatrix}$.

c The translation from C to A is 8 units left and 1 up, so $\begin{pmatrix} -8 \\ 1 \end{pmatrix}$.

Worked example

Translating an object using a vector

a Draw and label an x-axis from -5 to 7 and a y-axis from 0 to 7.
Plot and label these points.
$(1, 0), (3, 1), (4, 3), (5, 1)$
Join the points in order to form a quadrilateral.
Label it A.

b Translate shape A by $\begin{pmatrix} 2 \\ 4 \end{pmatrix}$.
Label the image B.

c Now translate shape B by $\begin{pmatrix} -7 \\ 0 \end{pmatrix}$.
Label the image C.

d Describe the single transformation that maps A to C.

Solution

a–c To translate the shape A using the vector $\begin{pmatrix} 2 \\ 4 \end{pmatrix}$, translate each of the four corners.

The point $(1, 0)$ is moved 2 units to the right and 4 up.
The image point is $(3, 4)$.
Repeat this process with each of the four corners.
Join the four image points and label the shape B.
Then translate each of the corners of shape B 7 units to the left.
Join the four image points and label the shape C.

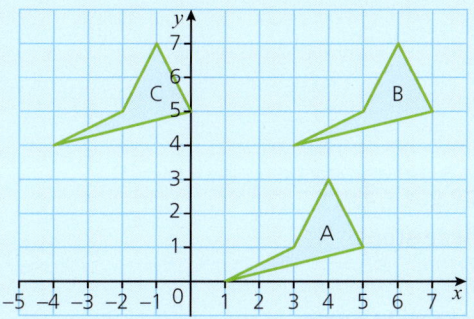

d The transformation A to C is a translation $\begin{pmatrix} -5 \\ 4 \end{pmatrix}$.

▶ Summary: translations

- To translate a shape, translate each of the corners.
  All points on the object move according to the same instructions.
- To describe a translation, start by counting or measuring the units along and up through which a point moves.
- A translation of 4 units to the right and 5 up can be written as $\begin{pmatrix} 4 \\ 5 \end{pmatrix}$.

  Using this notation, right is positive, left is negative, up is positive and down is negative.
- The image after a translation is always congruent to the original object.

Congruent means the same shape and size.

▶ Recap: reflections

In Book 1 you learnt that reflection is a type of transformation.

When an object is reflected in a line, its image is formed on the other side of the line.

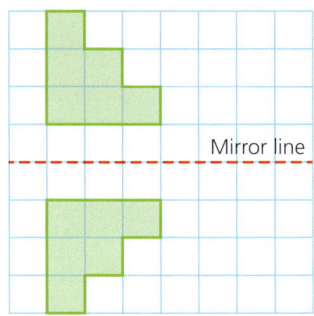

Mirror line

The line is called the **mirror line** or the **line of reflection**.

The image and the original object are **congruent**. This means they are the same shape and the same size.

The line of reflection does not have to be horizontal or vertical – it can be a diagonal line.

Lines that are commonly used for reflection are shown in the diagram below.

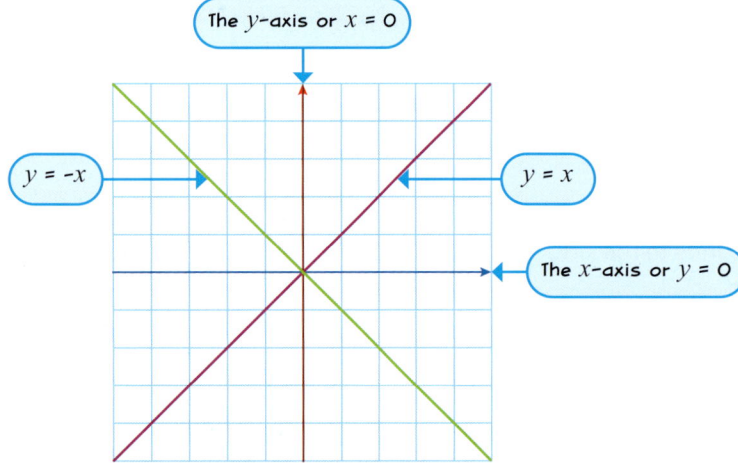

Other commonly used lines of symmetry are vertical and horizontal lines e.g. $x = 3$ or $y = -3$.

It is often obvious where the image of a point should be, but if not, take these steps:

1. Turn your page so that the mirror line is vertical.
2. Draw a line from the point to the mirror line, **meeting the mirror line at a right angle**.
3. Continue the line the same distance on the opposite side of the mirror line, as shown in the example.

10 Transformations

Worked example

Reflect the shape PQRSTU in the mirror line shown.

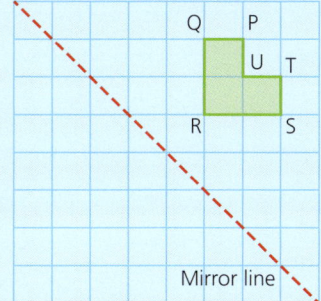

Solution

1. Turn the page to make the mirror line vertical.
2. Draw a line from one vertex of the shape to the mirror line, **meeting the mirror line at a right angle**. This has been done for vertex Q.
3. Continue the line the same distance on the opposite side of the mirror line, as shown.
4. Repeat the reflection for each vertex.
5. Join all the vertices to form the **image** shape, labelled P'Q'R'S'T'U' in the diagram.

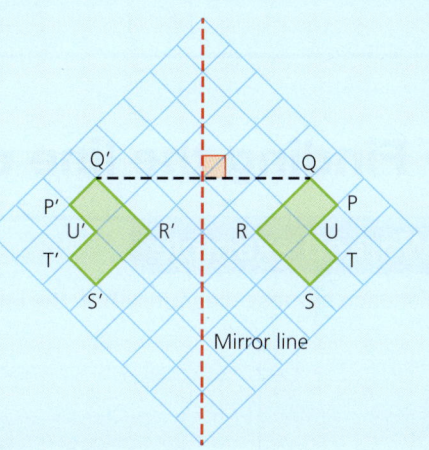

▶ Reflecting a shape on a coordinate grid

A point and its image are always the same distance from the mirror line.

Worked example

a Copy this diagram.

Draw the image of the triangle when it is reflected in the y-axis.

b Are the two triangles congruent?

How do you know?

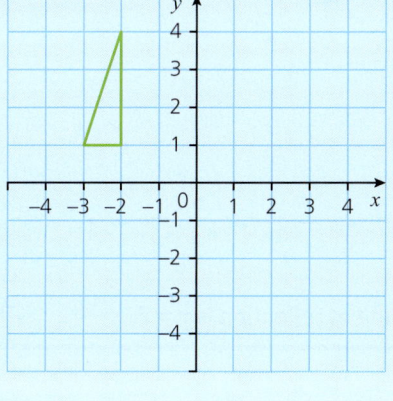

201

Solution

a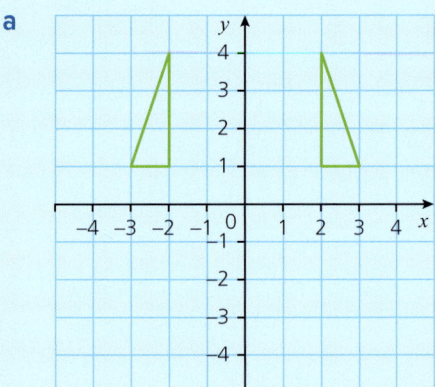

b When an object is reflected, all the lines and angles on the object and its image are the same size.

So, the object and image shapes are congruent.

▶ Finding the line of reflection

Worked example

For each of these diagrams show the line of reflection and give its equation.

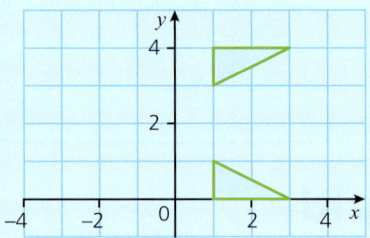 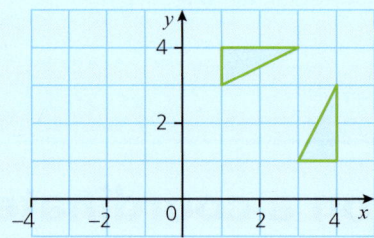 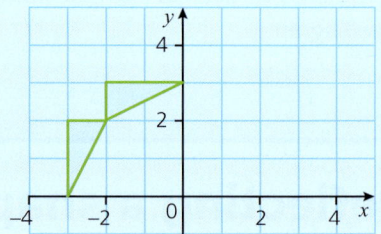

Solution

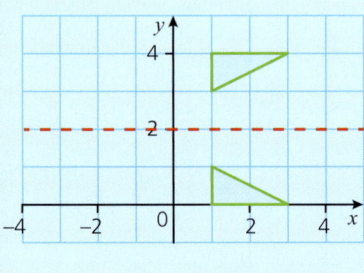 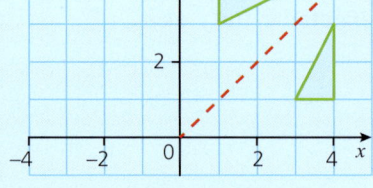 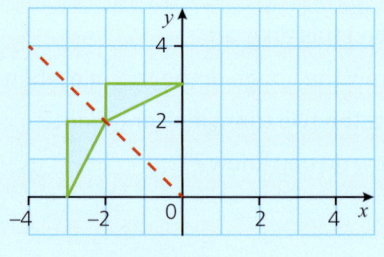

The line of reflection is $y = 2$ The line of reflection is $y = x$ The line of reflection is $y = -x$

To fully describe a reflection you must state the equation of the mirror line.

10 Transformations

Activity

Iwan is playing a computer game.

Shapes appear at the top of the screen.

The aim of the game is to arrange the shapes to completely fill as many horizontal lines as possible.

You can use any translation to move shapes to a final position at the bottom of the screen. Shapes must not overlap.

① This picture shows where the first shape appears (at the top of the screen) and where Iwan places it.
Describe the translation of shape 1.

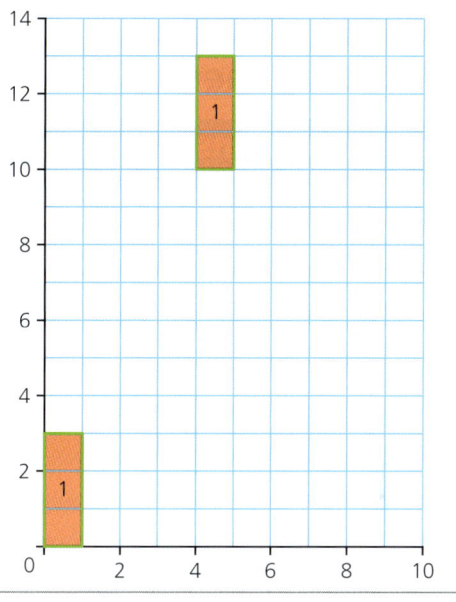

② This picture shows where the second shape appears (at the top of the screen) and where Iwan places it.
Describe the translation of shape 2.

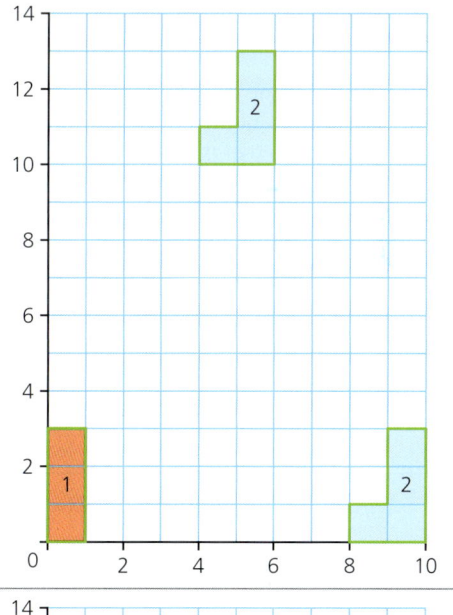

③ This is where the next two shapes appear.
Iwan translates shape 3 by 5 to the right and 10 down.
Copy the diagram in part **2**.
Draw and label the final position of shape 3.
He translates shape 4 by 4 to the left and 10 down.
Draw and label the final position of shape 4 on your diagram.

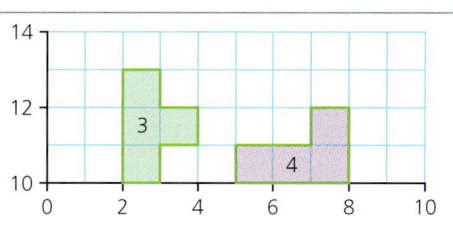

④ Here are the next three shapes.
Decide where to place each one.
  a Draw the final position of each shape on your diagram.
  b Describe the translation of each shape.

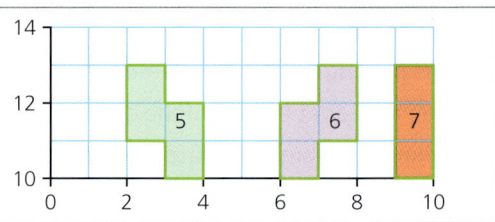

10.1 Now try these

Band 1 questions

① For each diagram write down how many units the blue shape has moved.
Write down if it moved left or right, up or down.

a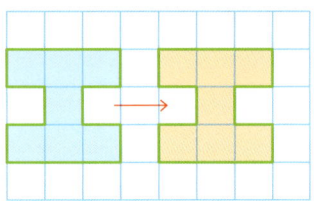

b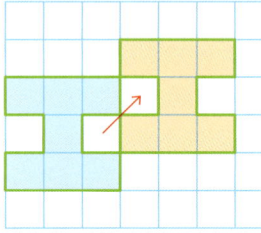

c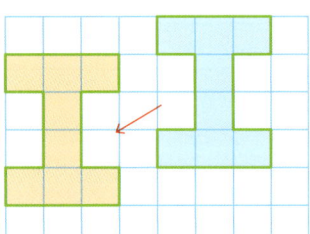

d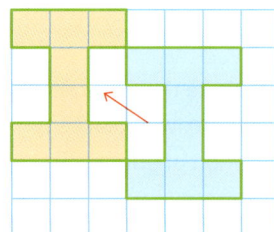

② Copy the diagram.

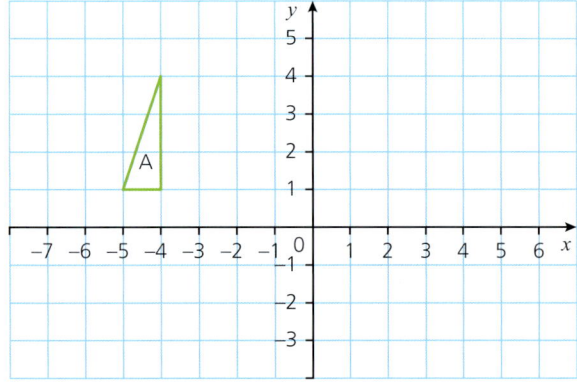

Draw shape A after
  a a translation of 2 to the right and 1 up
  b a translation of 1 to the left and 1 up
  c a translation of 4 down
  d a translation of 6 to the right and 2 down.

3 Copy this diagram, extending the grid as necessary

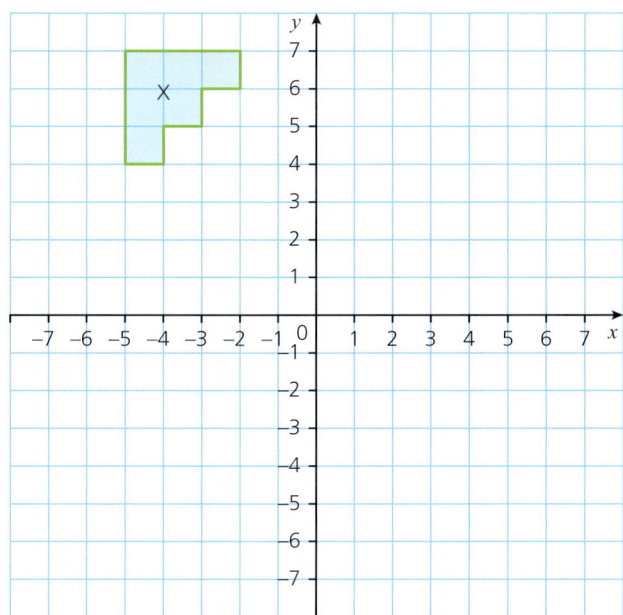

Draw the shapes given by the translations in the table.
Label each of the images.

Translation	Left/right	Up/down
X→Y	3 right	1 up
X→Z	2 right	6 down
X→P	2 left	5 down
Z→Q	4 left	4 down
Z→R	5 right	1 up
R→S	2 left	4 down

4 Look at these shapes.

Copy this table and complete it to describe each translation.

Translation	Translation vector
A→B	$\begin{pmatrix} 8 \\ 1 \end{pmatrix}$
A→C	
A→H	
C→B	
C→G	
C→D	
E→D	
E→F	
G→D	
G→A	
G→H	

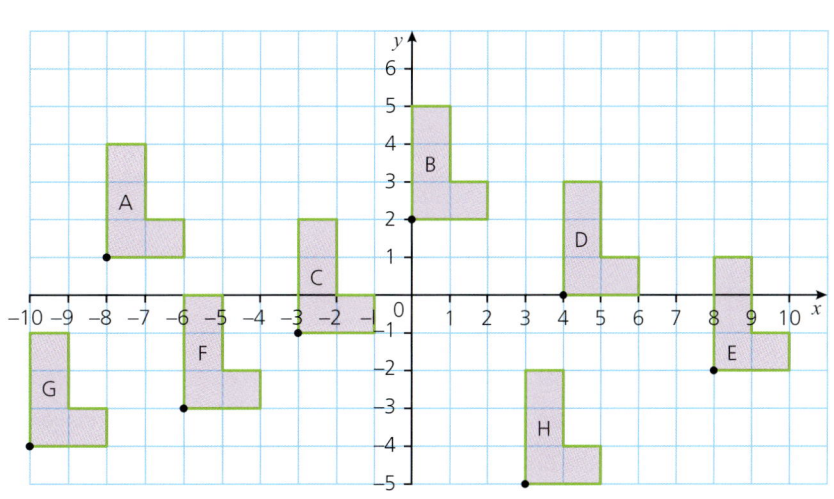

5 a Draw and label x- and y-axes from -7 to 7.
Plot the following points and join them to make a triangle.
$(-6, 3), (-5, 7), (-1, 4)$
Label this triangle P.

b Translate triangle P 5 to the right and 2 down.
Label the new triangle Q.

c What translation will map Q to P?

Band 2 questions

6 a Describe fully the transformation from shape A to shape B.
b Describe fully the transformation from shape A to shape D.
c Describe fully the transformation from shape B to shape C.

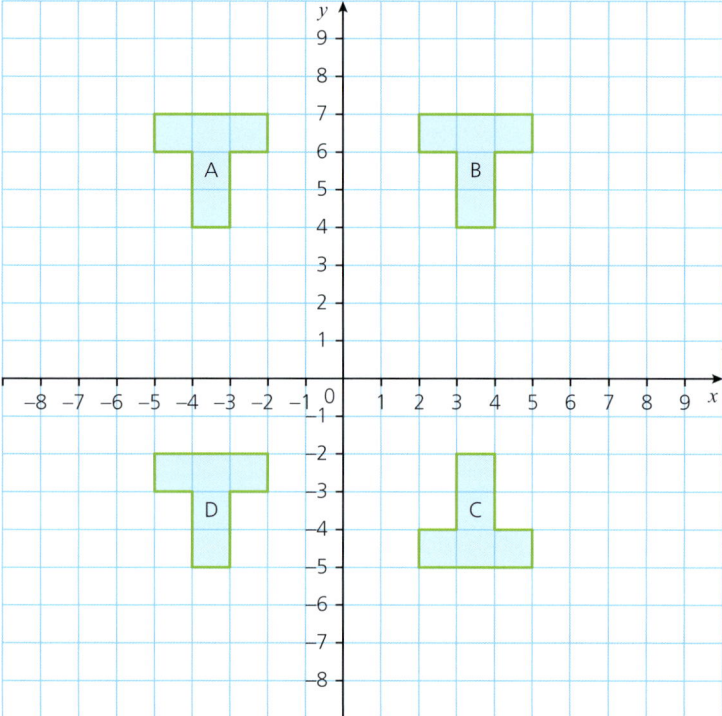

7 Here is a part of a mosaic wall.

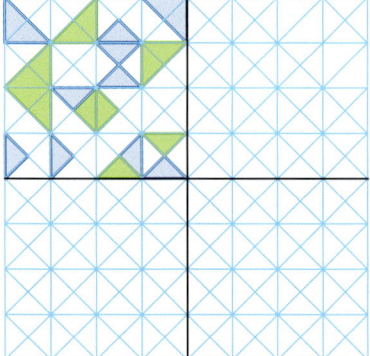

a Copy the diagram. Reflect the coloured pattern in the vertical axis.
b Now reflect the entire pattern in the horizontal axis.

8 **a** Draw x- and y-axes from −7 to 7.

Plot these points and join them to form a triangle.

(−7, −2), (−2, −3), (−5, −6)

Label the triangle A.

b Reflect A in the x-axis and label the image B.

c Reflect B in the y-axis and label the image C.

d Reflect C in the x-axis and label the new image D.

e Reflect D in the y-axis.

What do you notice?

Band 3 questions

9 **a** Copy this diagram and reflect the shape in the y-axis.
What can you say about the image?

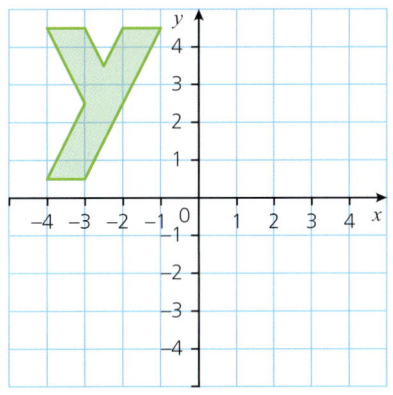

b Copy this diagram and reflect the shape in the y-axis.
What can you say about the image?

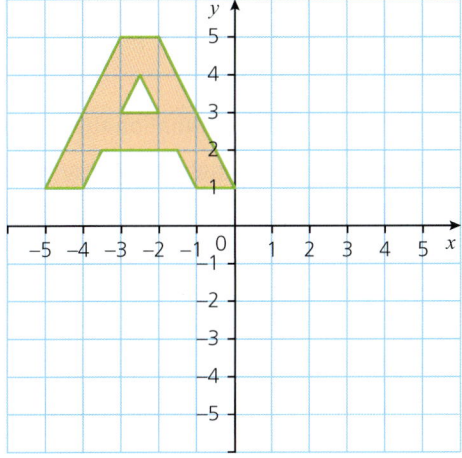

c Why is the letter A the 'right way round' after the reflection, but not the letter Y?

10
a Copy the diagram.
b Reflect the green lines in the y-axis.
c Now reflect both sets of lines in the x-axis.
d Describe the shape you have drawn.

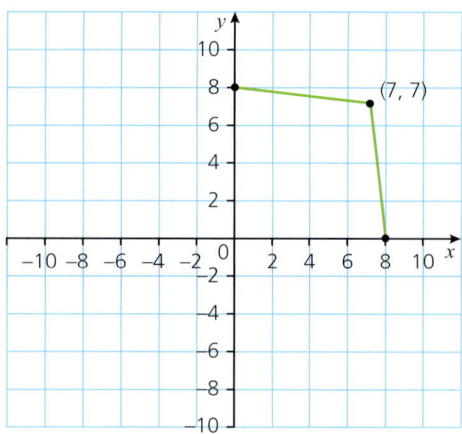

11
a Draw and label axes from −8 to 8.
b Plot the following points and join them to form a triangle.
(1, 2), (3, 7), (6, 1)
Label the triangle A.
c Reflect triangle A in the x-axis and label the image B.
d Write down the coordinates of the vertices of triangle B.
e Translate triangle B by $\begin{pmatrix} -5 \\ -1 \end{pmatrix}$.
Label the image C.
f Reflect triangle C in the x-axis.
Label the image D.
g Describe the transformation that maps D to A.

12
a Write down the translation vector where the object does not move.
b Write down a sequence of three different translations where the shape returns to its original position.

10.2 Rotation

Skill checker

Fill in the blanks in these sentences, then find the words in the word search puzzle below.

The words can appear horizontally, vertically or diagonally. They can appear forwards or backwards.

There are ten words to find. Good luck!

Clues

1. A 90° angle is a _ _ _ _ _ _ R of a _ _ L _ turn.
2. 180° is half a _ _ R _.
3. 270° is _ _ R _ _ quarters of a full turn.
4. 360° is O _ _ full turn.
5. There are several different types of transformation. For example, you can T _ _ _ _ _ _ _ _ , _ _ F _ _ _ _ or R _ _ _ _ _ a shape.
6. You can use a _ _ _ _ _ R to translate a shape.
7. (1, 3) is an example of a _ _ _ _ I.

K	Q	F	Z	T	R	D	U	R
W	T	U	C	P	O	I	N	T
Y	C	L	A	N	T	E	O	F
J	E	L	S	R	A	O	N	E
N	L	I	B	A	T	H	R	G
R	F	T	H	R	E	E	U	P
U	E	V	E	C	T	O	R	L
T	R	A	N	S	L	A	T	E

▶ Introduction to rotations

Another type of transformation is a rotation.

A rotation is performed by turning an object.

A full turn is 360°, so

- a quarter turn is 90° (a right angle)
- a half turn is 180°
- a three-quarters turn is 270°.

In this diagram the orange triangle is the image of the green triangle when it is rotated through 90° clockwise about the origin.

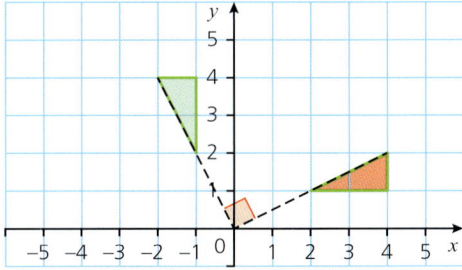

In this diagram the orange flag is the image of the green one when it is rotated through 180° about the point (1, 0).

A rotation of 180° can be clockwise or anticlockwise.

After any rotation, the image shape is always **congruent** to the original shape.

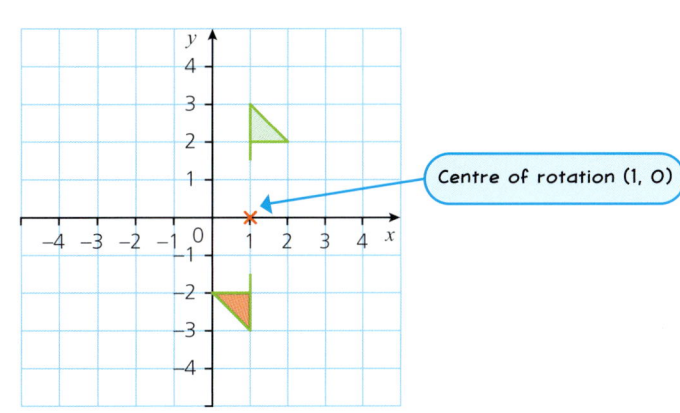

Centre of rotation (1, 0)

▶ Rotating a shape about the origin

Worked example

a Rotate shape A by 90° in a clockwise direction, centre (0, 0). Label the resulting image B.

b Rotate shape A by 180°, centre the origin.
   Label the resulting image C.
   Why does the direction not matter for this rotation?

Solution *Using tracing paper is the easiest way to rotate a shape. Follow these steps.*

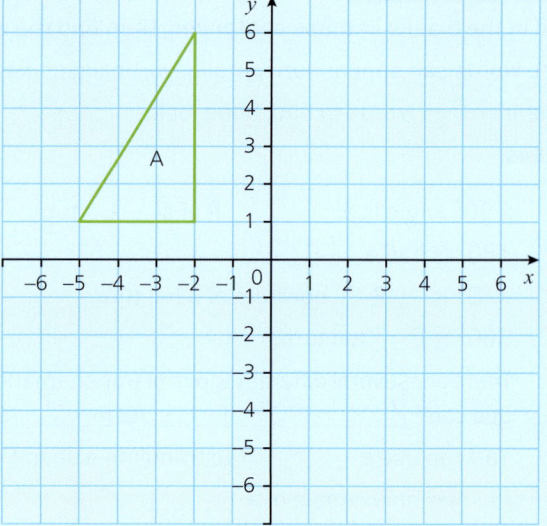

1 Trace the **original** shape AND some of the gridlines onto tracing paper.
2 Without moving the tracing paper, put the point of your pencil on the centre of rotation to fix that point.
3 Rotate the tracing paper. After a 90° or 180° rotation the gridlines on the tracing paper line up with those on the page.
4 Copy the shape back onto the page as the **image**.

a–b

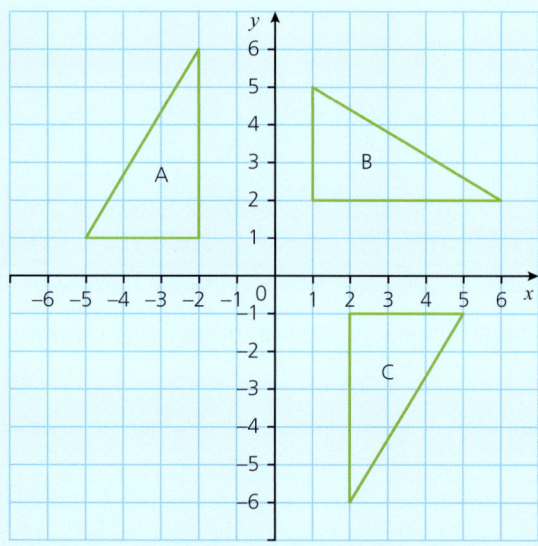

For part **b**, the direction does not matter because a turn of 180° is the same whichever direction you turn.

10 Transformations

▶ Describing a rotation

Worked example

Look at this diagram showing three arrowheads.

a Describe the transformation from shape A to shape B.
b What are the coordinates of the four vertices of shape B?
c Ffion says that shape B is congruent to shape A.
   Geraint says that Ffion is wrong because the shape is upside down.
   Say who is right and explain why.
d Describe the transformation from shape A to shape C.

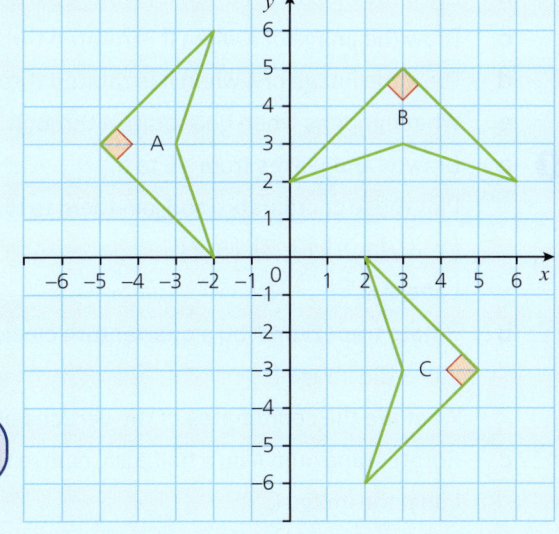

Solution

To describe a rotation you must give the angle, the direction and the centre of rotation.

a This is a rotation through 90° clockwise about (0, 0).
b The coordinates of the vertices of B are: (0, 2), (3, 3), (6, 2) and (3, 5).
c Ffion is right. Congruent shapes have the same size and shape. The **orientation** does not matter.
d This is a rotation through 180° about (0, 0).

Orientation means which way up the shape is.

To describe a rotation of 180° you don't need to give the direction. Both clockwise and anticlockwise are correct.

10.2 Now try these

Band 1 questions

1 Copy this diagram.

a The triangle A is rotated through a quarter turn in a clockwise direction about the origin.
   Draw its image and label it B.
b The triangle A is rotated through a quarter turn in an anticlockwise direction about the origin.
   Draw its image and label it C.
c Describe the rotation that maps triangle B onto triangle C.

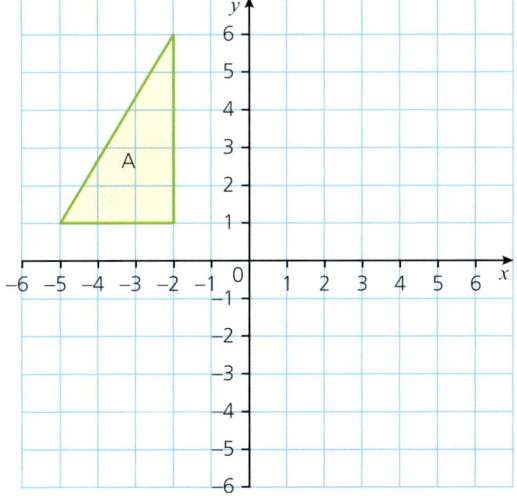

2 a Draw x- and y-axes from −5 to 5.
      Plot the points A(2, 1), B(2, 5) and C(4, 1).
      Join them to form a triangle.
   b Rotate the triangle through 90° clockwise about the origin.
      Label the image A′B′C′ and write down its coordinates.
   c Rotate A′B′C′ by a further 90° clockwise about the origin.
      Label the image A″B″C″ and write down its coordinates.

Remember

You can use tracing paper in all questions on rotations.

211

3 a The vertices of triangle A are (0, 0), (1, 0) and (1, 4).
        Draw the triangle on a grid.
    b Draw the image of A when it is rotated through 90° in a clockwise direction, centre the origin.
    c Draw the image of A when it is rotated through 180°, centre the origin.
    d Draw the image of A when it is rotated through 270° in a clockwise direction, centre the origin.
    e What happens when you rotate A through 360°, centre the origin?

4 a Draw x- and y-axes from −5 to 5.
        Plot these coordinates and join them to make a trapezium.
        (1, 2), (1, 5), (3, 5), (5, 2)
        Label it A.
    b Rotate trapezium A by a quarter turn clockwise, centre the origin.
        Label the image B.
        What are the coordinates of the vertices of B?
    c Rotate trapezium A by a half turn, centre the origin.
        Label the image C.
        What are the coordinates of the vertices of C?
    d What single transformation would map B onto C?

Band 2 questions

5

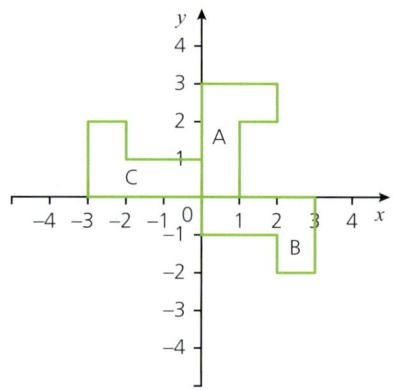

  a Describe each of these rotations.
        i   A to B   **ii**   A to C   **iii**   B to A
  b Draw and label x- and y-axes from −4 to 4.
        Copy shape A onto your diagram.
        Now rotate A through 180° about the origin.
        Label the image D.
  c **i** Is shape D congruent to shape A?   **ii**   Explain your answer.

6 Draw this arrow after each of these transformations.
    Make it four squares long each time.
    a A rotation 90° clockwise, centre the origin.
    b A rotation 270° anticlockwise, centre the origin.
    c A rotation 180° clockwise, centre the origin.
    d A rotation 180° anticlockwise, centre the origin.

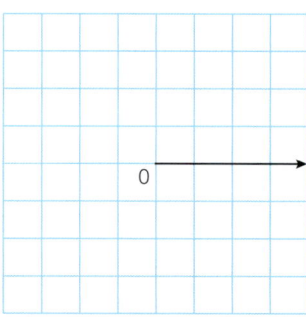

7 This clock has three hands.

Through what angle does each of the hands turn in
- **a** 1 minute
- **b** 30 seconds
- **c** 15 seconds?

Band 3 questions

8 Draw a pair of axes from −6 to 6.
- **a** Plot these points and join them to form a hexagon in the shape of the letter L.
  (0, 0), (−5, 0), (−5, 6), (−3, 6), (−3, 2), (0, 2)
  Label this object A.
- **b** Rotate A by a right angle clockwise, centre the origin.
  Label the image B.
- **c** Rotate B by a right angle clockwise, centre the origin.
  Label the image C.
- **d** What single transformation maps A to C?

9 The diagram shows a plan of a children's roundabout.
N and T show the starting positions of Nia and Tilly.

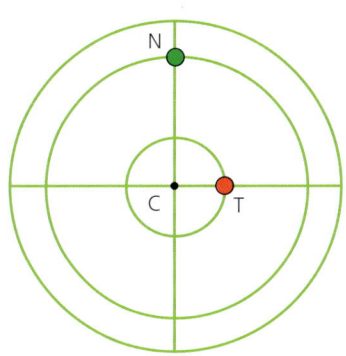

- **a** Make a copy of the diagram.
  On your diagram show the positions of Nia and Tilly after each of these transformations from their starting positions.
  - **i** A clockwise rotation of 90° about C.
    Label them N_1 and T_1.
  - **ii** A clockwise rotation of 270° about C.
    Label them N_2 and T_2.
- **b** Describe two transformations for each of these mappings.
  - **i** N_1 to N_2
  - **ii** T_1 to T_2

10 Look at shapes A, B, C and D in the diagram.
All four shapes are congruent.

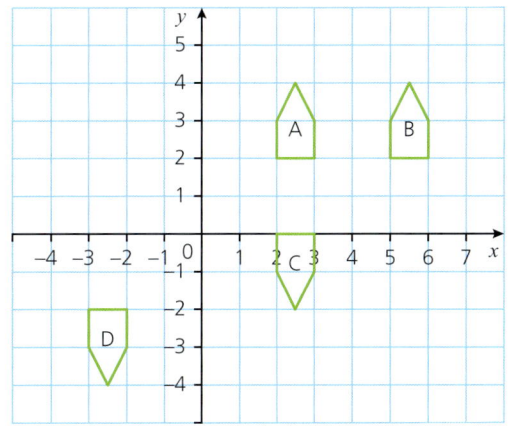

- **a** Describe fully the transformation from C to D.
- **b** Describe fully the reflection that maps A to B.
  Describe fully the translation that maps A to B.
- **c** Describe fully the reflection that maps A to C.
  Describe fully the rotation that maps A to C.
  Is there a translation mapping A to C?
- **d** Can you explain why there is a difference in the types of transformation mapping A to B compared with A to C?
- **e** Describe fully a transformation mapping A to D.

11 Describe fully these transformations.

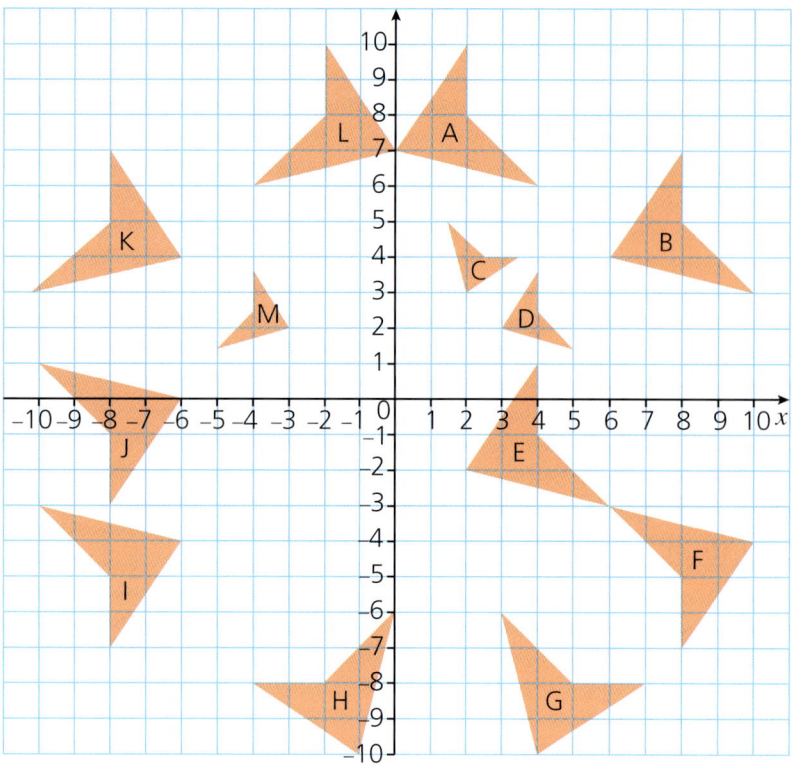

a	K to L	b	A to B	c	K to B	d	I to B	e	K to I
f	C to M	g	F to E	h	F to I	i	A to L	j	H to G
k	C to D	l	J to I	m	F to J				

12 Look at the picture of a spanner turning a bolt.

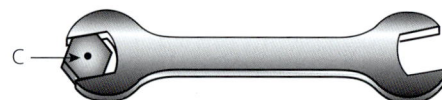

The bolt head is a regular hexagon.

Point C is the centre of rotation.

- **a** The spanner is rotated 60° clockwise about C.

  What can you say about the appearance of the bolt before and after the rotation?
- **b** Write down three other clockwise rotations which do not change the appearance of the bolt.
- **c** Write down two anticlockwise rotations which do not change the appearance of the bolt.
- **d** What is the rotational symmetry of the bolt? ◀ *You learnt about rotational symmetry in Book 1.*

13 A car drives round a racetrack. A to B is a translation of 1 km due East.

- **a** Describe the transformation C to D.
- **b** B to C is a rotation of 180° about Q.

  Describe the transformation D to A.
- **c** Use a combination of transformations to describe the movement of the car from
  - **i** A to C
  - **ii** B to D.

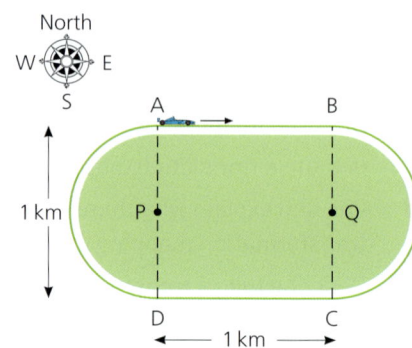

10 Transformations

10.3 Enlargement

Skill checker
True or false?

Are these statements true or false? Explain your answers.

1. If a picture measures 6 cm by 4 cm, it could be enlarged so it would fit in a frame measuring 15 cm by 10 cm.
2. The Elizabeth Tower (Big Ben) is 96 metres high and has a 15-metre square base. A scale model 64 cm high has a square base measuring 10 cm by 10 cm.
3. A scale drawing of a classroom has side lengths that are one tenth of the actual sizes. The area of the scale drawing is also one tenth of the area of the real classroom.
4. A passport photo must be 35 mm wide by 45 mm high. Ben has a photo of himself measuring 7 cm by 8 cm. If he reduces this photo in size, he could use it as a passport photo.
5. A running track has two straight edges and two semicircular ends. The overall length could be doubled if the two straight edges are doubled in length.
6. The bark of a cork tree is used to make bottle corks. A cork tree doubles in height in 5 years. In this time the thickness of the trunk also doubles. Twice as many bottle corks can be made from the bark now.
7. A scale model of a boat is made for the bath. The model's width is $\frac{1}{100}$ of the actual width. The length of the model should be $\frac{1}{100}$ of the actual boat's length.
8. A road is being expanded to take more traffic. The width of the road is doubled, going up from 2 lanes to 4 lanes. The length of the road will also double.

Discussion activity

- Rank the four types of transformation (rotation, reflection, enlargement and translation) in order from easiest to hardest. Why have you put them in this order?
- Three of the four types of transformation (rotation, reflection, enlargement and translation) will always give a shape that is congruent to the original shape. Which is the one transformation that doesn't?

▶ Introduction to enlargements

Enlargement is a transformation that changes the size of an object.

One shape is an **enlargement** of another if all the angles in the shape are the same and the lengths of the sides have all been increased by the same scale factor. ◀ *A shape and its enlargement are called 'similar' shapes.*

In this diagram the pentagon ABCDE is enlarged by a scale factor of 3 to A'B'C'D'E'.

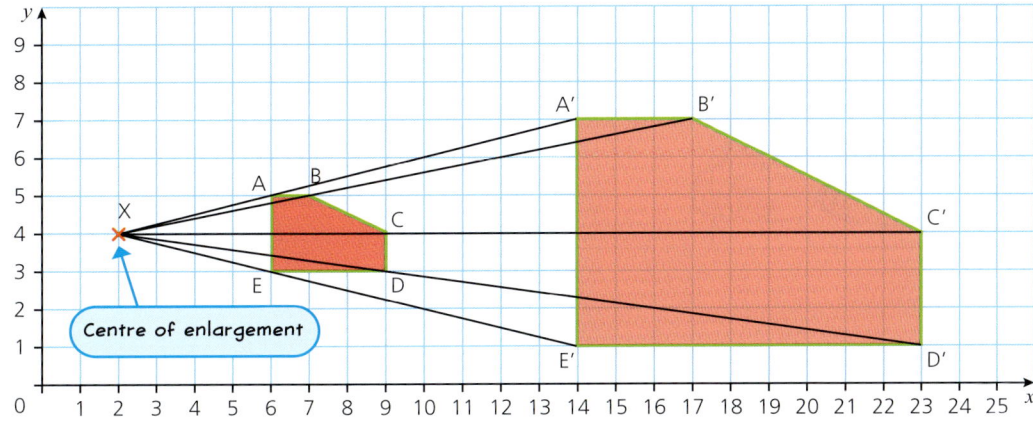

The position of the image depends on the centre of enlargement X.

Because the scale factor of the enlargement is 3
- side A'B' is three times as long as AB
- side E'D' is three times as long as ED, etc.
- the distance XB' is three times the distance XB
- the distance XD' is three times the distance XD, etc.

The term enlargement is also used in situations where the shape is made smaller.

In these cases, the scale factor is between 0 and 1, for example 0.5 or $\frac{1}{3}$.

In the diagram the scale factor for the enlargement of A'B'C'D'E' to ABCDE is $\frac{1}{3}$.

Cross-curricular activity

4 cm

8 cm

In art, enlargement refers to increasing or expanding an image.

Using a grid as a guide is a method used by artists when they want to enlarge an image. They draw a grid (or squares) over the original image and then draw an equal ratio, larger grid on their desired work surface.

The first, smaller drawing has been enlarged using the grid method, and the larger image is also shown.

What is the scale factor of enlargement between the smaller and larger drawings?

In your next Art lesson, select one of your favourite photographs, pictures or paintings, then use the grid method to make an enlarged copy. What scale factor have you used?

▶ Finding the scale factor of an enlargement

Worked example

The diagram shows an enlargement from centre P.
The small rectangle is the original shape; the larger rectangle is the image.
a What is the scale factor of the enlargement?
b What happens if you change the centre of enlargement?

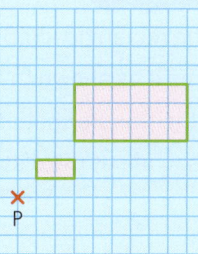

Solution

a The scale factor is 3 because all the sides of the image are three times as long as those of the original shape.
b The image would be in a different place.

▶ Enlarging a shape

Follow these steps to draw an enlargement.

① Choose one of the vertices of the original shape.
② Draw a 'ray' from the centre of enlargement to this vertex.
③ Multiply the length of this ray by the scale factor.
④ Draw the new ray from the centre of enlargement to the image vertex.

> If an enlargement has a scale factor between 0 and 1, the image will be smaller than the original shape.

Worked example

Plot the points A(2, 4), B(4, 4), C(4, 6) and D(2, 6).
Join the points to form a square.
Enlarge ABCD using (0, 5) as the centre of enlargement and a scale factor of 2.
Label the image A'B'C'D'
a What are the coordinates of the points A', B', C' and D'?
b What shape is A'B'C'D'?

Solution

a A'(4, 3), B'(8, 3), C'(8, 7), D'(4, 7)
b A square

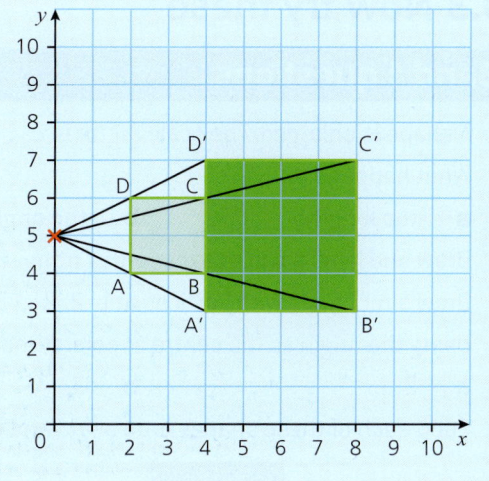

Describing an enlargement

Worked example

Angharad draws the rectangle ABCD.

She then transforms it as shown to A'B'C'D'.

a Describe fully the transformation from ABCD to A'B'C'D'.

b What can you say about the distances AB and A'B'?

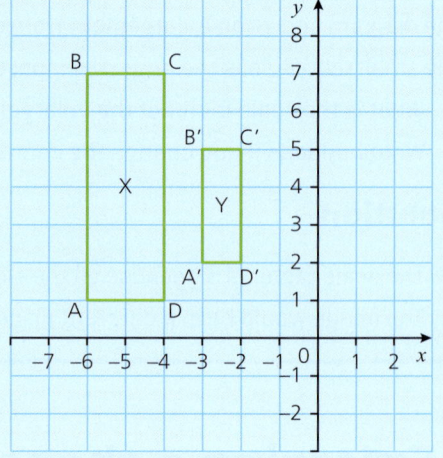

Solution

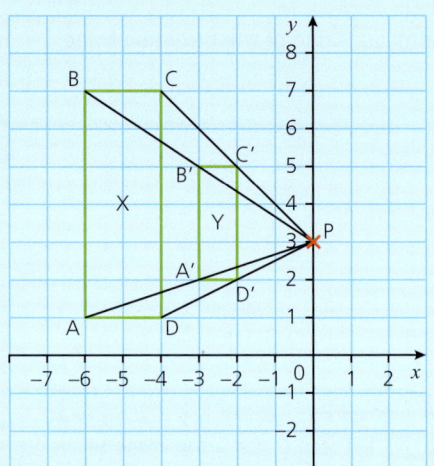

a Join point A with its image A', B with B', etc. Extending these lines, they meet at the point P(0, 3). This is the centre of enlargement.

The length A'B' is half of AB, etc. The scale factor is $\frac{1}{2}$.

This transformation is an enlargement, with a scale factor of $\frac{1}{2}$ and a centre of enlargement (0, 3).

b Since the scale factor is $\frac{1}{2}$, the distance A'B' is half of the distance AB.

10.3 Now try these

Band 1 questions

1 A shape is enlarged. The scale factor is 2.
   What happens to
   a the lengths   b the angles?

2 Draw and label a pair of axes, with x from -7 to 10 and y from -4 to 10.
   Draw a rectangle with vertices at $(-2, 3)$, $(3, 3)$, $(3, -1)$, $(-2, -1)$.
   Using the origin as the centre of enlargement, enlarge the rectangle with a scale factor of
   a 2   b 3.

3 Copy each of these shapes onto a grid and enlarge them by a scale factor of 3.

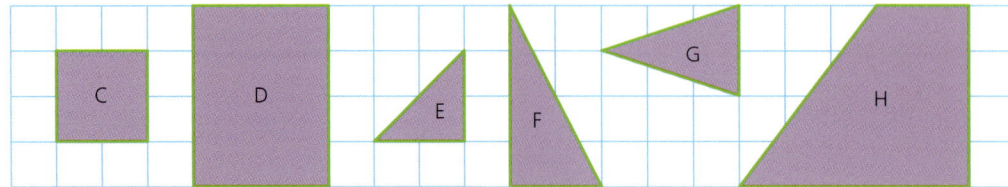

4 Copy this diagram.

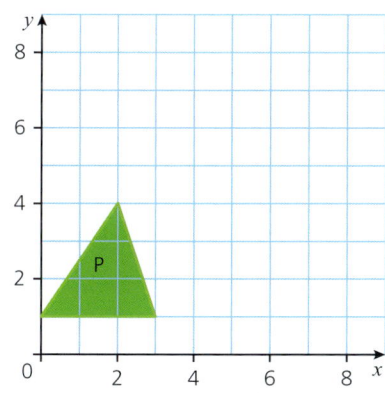

a Enlarge shape P with scale factor 2 and centre of enlargement (0, 2).
   Label the image Q.
b Enlarge Q with scale factor $\frac{1}{2}$ and centre of enlargement (8, 2).
   Label the image R.
c Describe the transformation that maps P to R.

5 Copy each of these diagrams.
  Enlarge each shape by a scale factor of $\frac{1}{2}$ using the origin as the centre of enlargement.

a 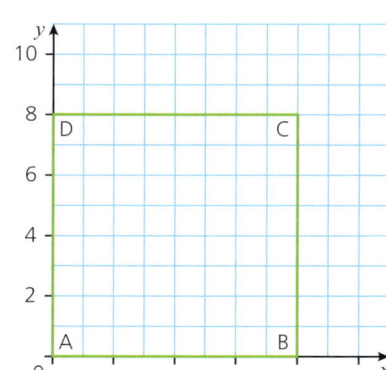   b 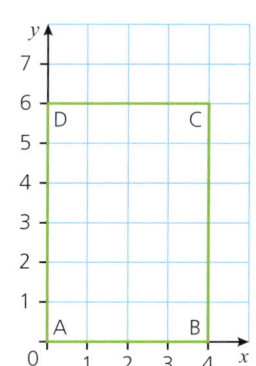   c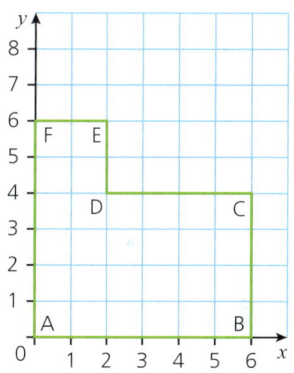

Band 2 questions

6 You will need a protractor for this question.
  a Draw and label an x-axis from 0 to 10 and a y-axis from 0 to 8.
     Plot these points and join them to form a triangle T.
     (6, 8), (6, 4), (10, 6)
  b Measure the angles of triangle T.
  c Enlarge triangle T by scale factor $\frac{1}{2}$ using (1, 8) as the centre of enlargement.
  d Measure the angles of the image.
     What do you notice?

7 Draw an x-axis from 0 to 16 and a y-axis from 0 to 8.
  a Draw the trapezium whose vertices are P(2, 8), Q(8, 8), R(5, 2) and S(2, 2).
  b Using (14, 5) as the centre of enlargement
     i   enlarge PQRS by scale factor $\frac{1}{3}$ and label the image P'Q'R'S'
     ii enlarge PQRS by scale factor $\frac{2}{3}$ and label the image P''Q''R''S''.

c P′Q′R′S′ is an enlargement of P″Q″R″S″.
  i What are the coordinates of the centre of enlargement?
  ii What is the scale factor?

Band 3 questions

8 The enlargement reading on a photocopier is 100% when the copy is to be the same size as the original.

When the reading is 120% then each length is increased by 20%.

a What enlargement reading do you use if you want
  i each length decreased by 10%
  ii a 5 cm line increased to 7 cm
  iii a 6 cm line reduced to 4.5 cm?

b On the original, a rectangle is 20 cm by 12 cm. The enlargement is set to 125%.
  i What is the area of the original rectangle?
  ii What is the area of the enlarged rectangle?
  iii What is the percentage increase in area from the original to the enlargement?

9 Russian dolls are wooden toys in the shape of dolls that fit inside each other.

a There are five Russian dolls in a set. The smallest has a width of 1 cm and a height of 2 cm.

The second doll is twice the size of the first.

The third doll is three times the size of the first, and so on.

Find the width and height of each doll in the set.

Start with the smallest and end with the largest.

Write your answers in a copy of this table.

Width	Height
1 cm	2 cm

b What is the scale factor of the enlargement from the largest doll to the smallest?

10 In the diagram there are six different shapes and their enlargements.

One shape does not pair up with any other.

a Match the shapes to their enlargements.
b Name the shapes.
c Work out the scale factor of enlargement for each pair of shapes.

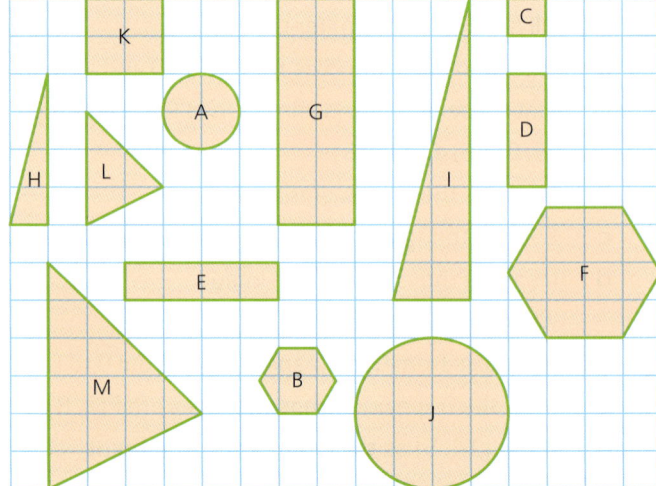

10 Transformations

11 a Copy this diagram.
    b Draw an enlargement of WXYZ, scale factor 2, centre P.
       Label the image W'X'Y'Z'.
    c Draw an enlargement of WXYZ, scale factor 3, centre P.
       Label the image W''X''Y''Z''.
    d What transformation maps W'X'Y'Z' to W''X''Y''Z''?

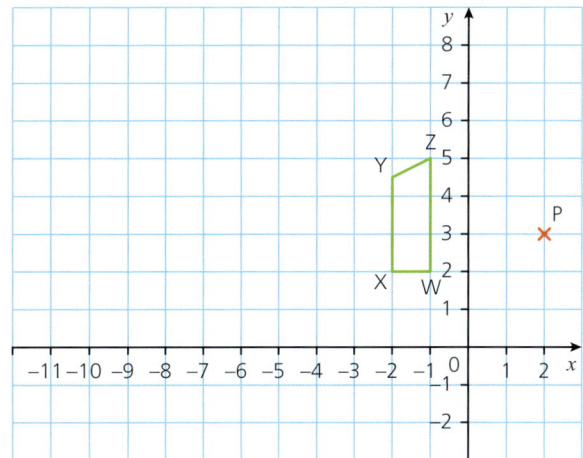

12 Megan draws shape PQRS.
    She transforms it to the image XYZW.

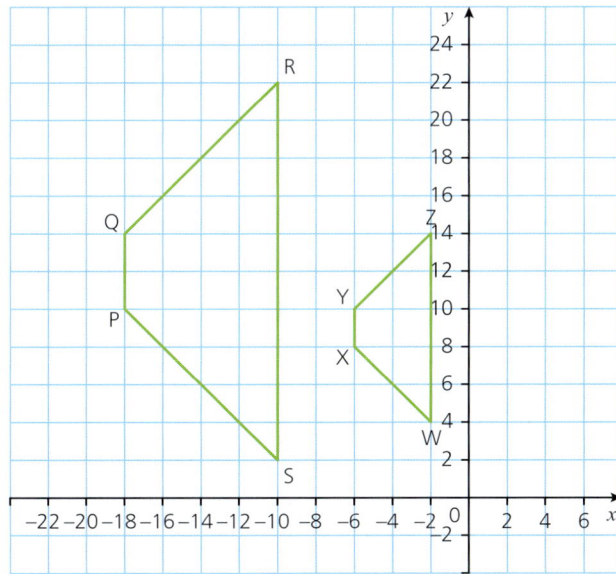

Describe fully the transformation which maps shape PQRS to shape XYZW.

13 P is the centre of enlargement for each of these pairs of shapes.
    Find the scale factor of the enlargement from the larger shape to the smaller in each case.

    a
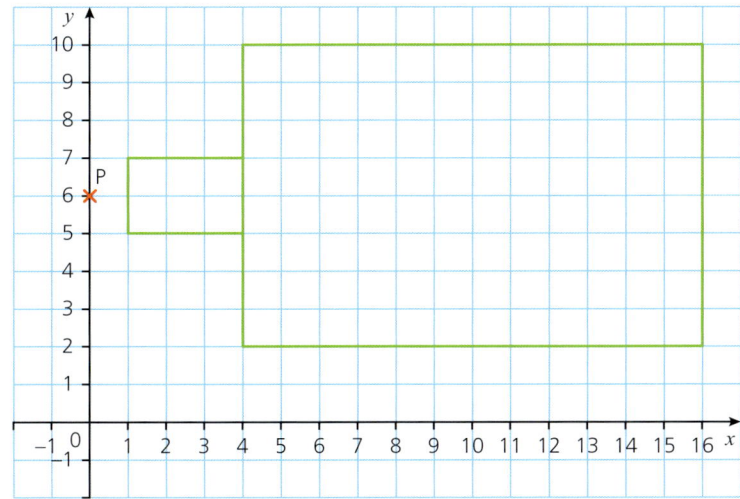

221

b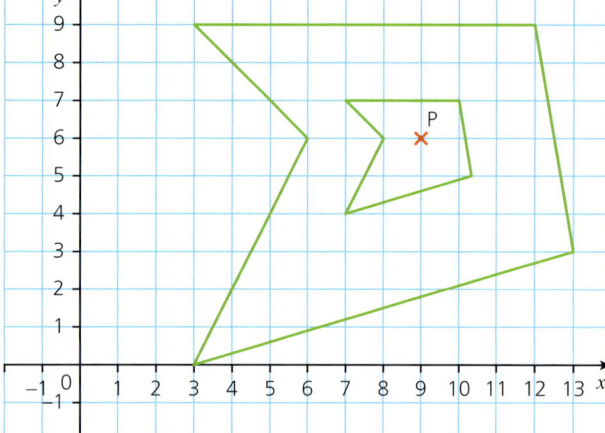

c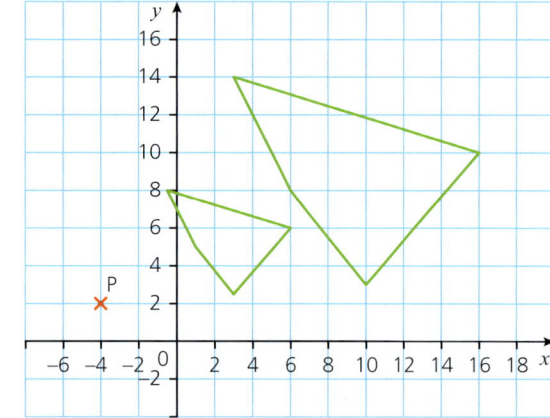

Key words

Here is a list of the key words you met in this chapter.

Congruent Enlargement Reflection Rotation
Transformation Translation Vector

Use the glossary at the back of this book to check any you are unsure about.

Review exercise: transformations

Band 1 questions

1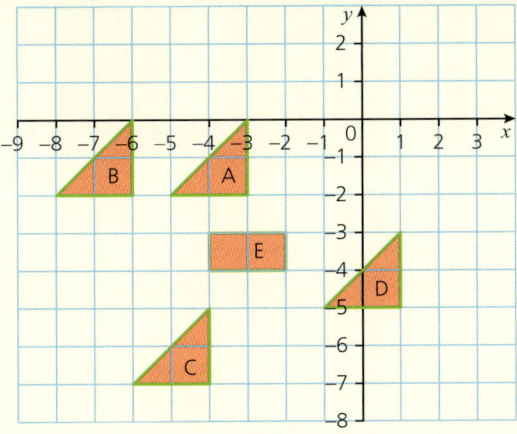

    a Describe each of these translations.
        i A to B   **ii** A to C   **iii** A to D   **iv** B to C   **v** B to D   **vi** C to D

    b Why is A to E impossible?

2 Draw x- and y-axes from -3 to 8.
On the same grid draw these images.

    a Shape A after translation by $\begin{pmatrix} 1 \\ 4 \end{pmatrix}$ **b** Shape B after translation by $\begin{pmatrix} 0 \\ 3 \end{pmatrix}$

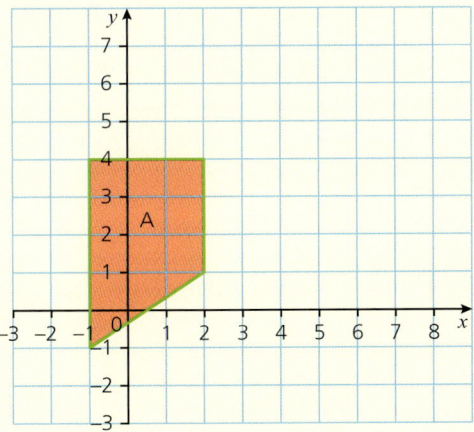
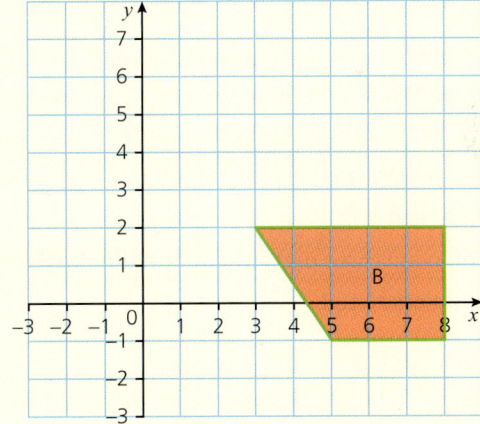

c Shape C after translation by $\begin{pmatrix} -6 \\ 0 \end{pmatrix}$

d Shape D after translation by $\begin{pmatrix} -3 \\ -6 \end{pmatrix}$

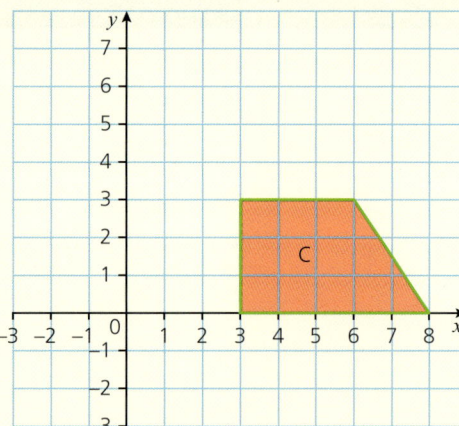
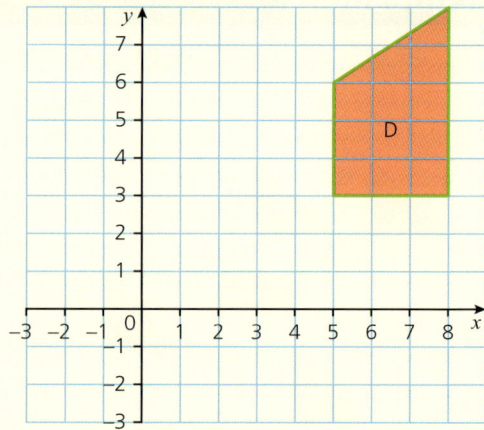

3 The diagram shows **the same word** written in both lower case letters and capital letters.

b	i	?	
B	I	?	

The last letter is missing from the word in both cases.

The short red line is a vertical mirror line.

The longer red line is a horizontal mirror line.

When the word is written in lower case it has vertical symmetry. When it is reflected in the vertical mirror line, it appears the same.

When the word is written in capital letters it has horizontal symmetry. When it is reflected in the horizontal mirror line, it appears the same.

What is the missing letter of the word?

4 a You only need one piece of information to describe a translation.

What is it?

b You need three pieces of information to describe a rotation fully.

What are they?

c   Look at the diagram.

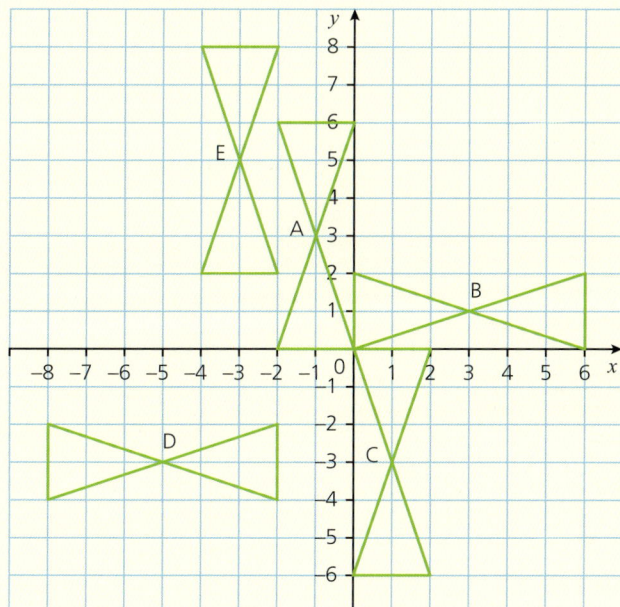

What translation could be used for each of these transformations?
i   A to E   ii   A to C   iii   E to C   iv   B to D

d   What rotation could be used for each of these transformations?
i   A to B   ii   A to C   iii   B to D   iv   C to E

Band 2 questions

5 Copy this flag.

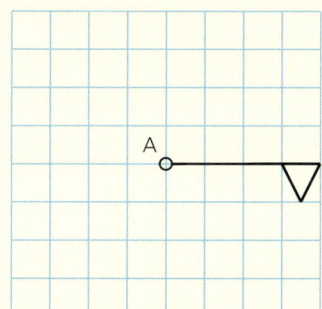

Rotate the flag about point A through
a   a quarter turn anticlockwise   b   a half turn clockwise   c   a three-quarter turn clockwise.

6 The triangle A in the diagram is rotated through 90° anticlockwise with centre of rotation D(−2, 2).

Its image is B.

The triangle B is rotated through 90° clockwise with centre of rotation E(1, 1).

Its image is C.

Find the single transformation that maps A onto C.

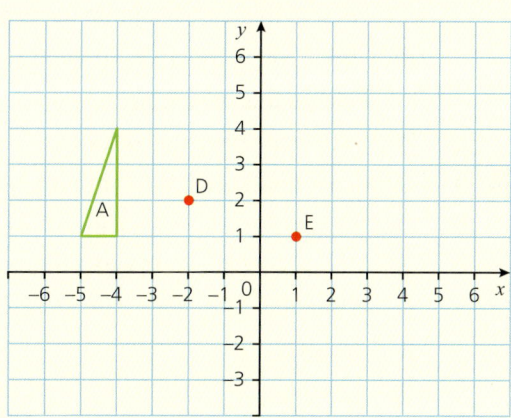

7 P is the centre of enlargement for this pair of shapes.

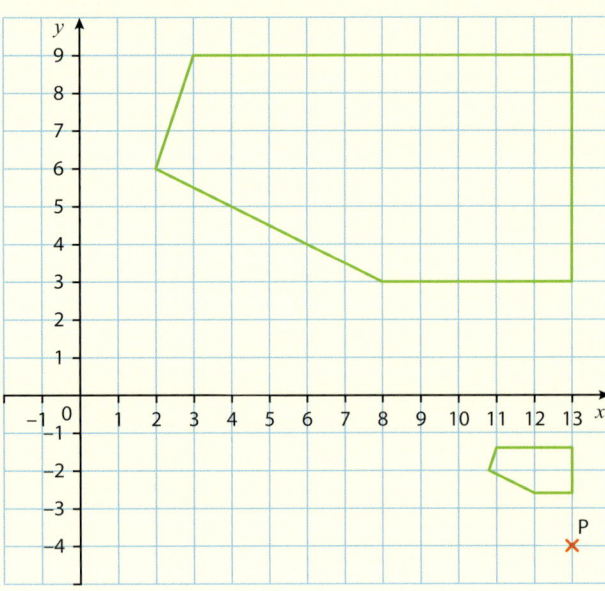

a Find the scale factor of the enlargement from the smaller shape to the larger shape.
b Find the scale factor of the enlargement from the larger shape to the smaller shape.

Band 3 questions

8 Look at the word ZERO, written in capital letters.

It has been reflected in the horizontal line shown.

a The letters E and O look the same as the originals. What property of these letters allows them to look the same after the reflection?

b Write all the numbers from ONE to TEN in capital letters.

Reflecting them in a horizontal line, would any of these words look exactly the same after the reflection?

c The letter T remains unchanged after a reflection in a vertical line. What property must a letter have for this to happen?

d Can you think of any words that would look the same after a reflection in a vertical line?

9 Draw an x-axis from -8 to 14 and a y-axis from 0 to 14.

a Draw these three parallelograms.

  i A$(-4, 11)$, B$(-2, 13)$, C$(0, 12)$, D$(-2, 10)$

   Enlarge this parallelogram using scale factor 2 and centre of enlargement $(-8, 12)$.

  ii E$(-4, 3)$, F$(-4, 6)$, G$(-2, 5)$, H$(-2, 2)$

   Enlarge this parallelogram using scale factor 2 and centre of enlargement $(-8, 2)$.

  iii J$(9, 5)$, K$(9, 8)$, L$(11, 10)$, M$(11, 7)$

   Enlarge this parallelogram using scale factor 2 and centre of enlargement $(14, 8)$.

b What 3D shape have you drawn?

10 a Draw coordinate axes with both the x-axis and y-axis running from -3 to 8.

b Plot these points: A$(-3, 1)$, B$(-3, 3)$, C$(-1, 1)$ and D$(-1, 3)$.

c Join these pairs of points with straight lines.

  i A and B
  ii B and C
  iii C and D.

d What letter have you drawn?

e Translate the entire shape using the vector $\begin{pmatrix} 4 \\ 0 \end{pmatrix}$. Label the image of shape ABCD as A′B′C′D′.

f Rotate the lines A′B′ and C′D′ by 90° clockwise, using a centre of rotation (4, 0).
Label the image of point A′ as A″. Label the images of the other three points B″, C″ and D″.

g Reflect the diagonal line B′C′ in the mirror line $x = 4$.

h Enlarge the shape A″B″C″D″ using a scale factor of 2 and a centre of enlargement (7, −2).
What letter have you drawn?

11 Copy and complete this table. It shows the scale factor for the enlargement of the shape in the first row to the shape in the first column.

		From shape			
		A	B	C	D
To shape	A	1	2	3	4
	B	$\frac{1}{2}$	1	1.5	
	C		$\frac{2}{3}$		
	D				

Consolidation 4: Chapters 8–10

Band 1 questions

1. Write down the gradient and the coordinates of the y-intercept for each of these lines.
   a. $y = 5x + 3$
   b. $y = 7x + 6$
   c. $y = 5 - 10x$
   d. $y = 7x$
   e. $y = -x$
   f. $y = 2x - 1$
   g. $y = 4 + 8x$
   h. $y = 5 - 3x$
   i. $y = 8 + 3x$

2. A line has a gradient of $\frac{2}{3}$ and y-intercept of -2. What is its equation?

3. A ship is sailing past a lighthouse.
   The graph below shows the distance between the ship and the lighthouse at certain times through the day.

   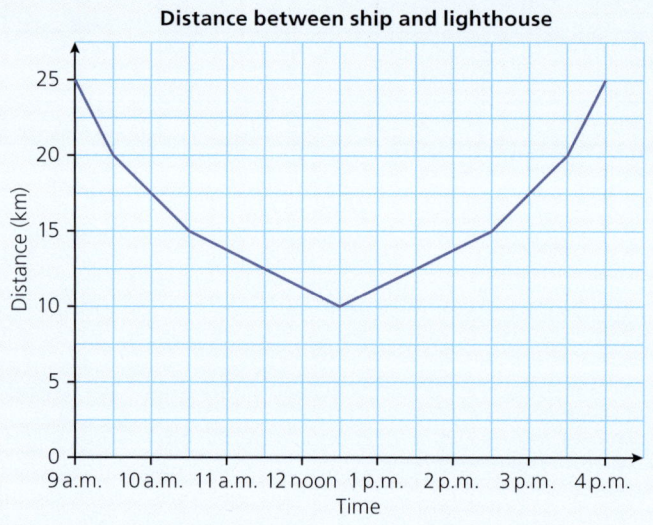

   a. How far is the ship from the lighthouse at 10:30 a.m.?
   b. There are two times during the day when the ship is 20 km from the lighthouse. What are these times?
   c. At what time is the ship closest to the lighthouse?

4. Work out the gradient of the line joining each of these pairs of points.
   a. $(5, -3)$ and $(3, 1)$
   b. $(-4, -7)$ and $(2, 2)$
   c. $(-8, -1)$ and $(7, -6)$
   d. $(-4, -2)$ and $(-9, 4)$

Consolidation 4

5 Bethan goes to school on Monday.

After school, she goes to her friend Bronwen's house, where they spend some time doing their homework together.

The graph below shows Bethan's distance from home at different times of the day.

a   How long does it take Bethan to travel to school?
b   How far from Bethan's home is her school?
c   What time does Bethan arrive at Bronwen's house?
d   How far from home is Bethan at 4 p.m.?
e   How long did Bethan spend doing her homework?
f   On Tuesday Bethan went straight home after school finished at 15:00.
    Her journeys to school and home again both took the same time as her journey to school on Monday.
    Draw her travel graph for Tuesday.

Band 2 questions

6 Look at the diagram.
   a   Describe fully two different transformations that take shape A to shape B.
   b   Translate shape A using the vector $\begin{pmatrix} 2 \\ -4 \end{pmatrix}$.
       Label the image shape C.

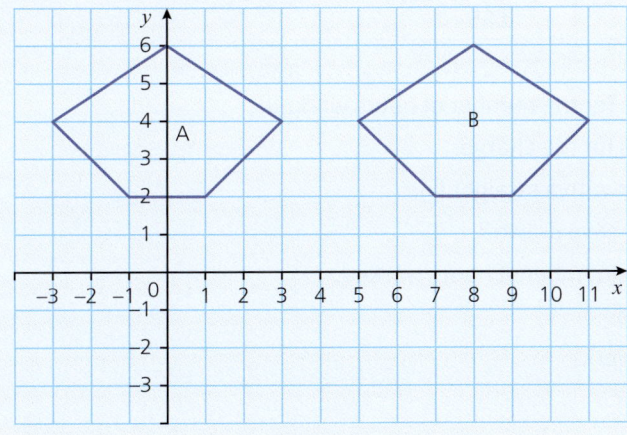

7 Some birds fly a long way to be somewhere warmer for the winter.

The graph below shows the distance some birds have flown on their journey from Norway to Scotland.

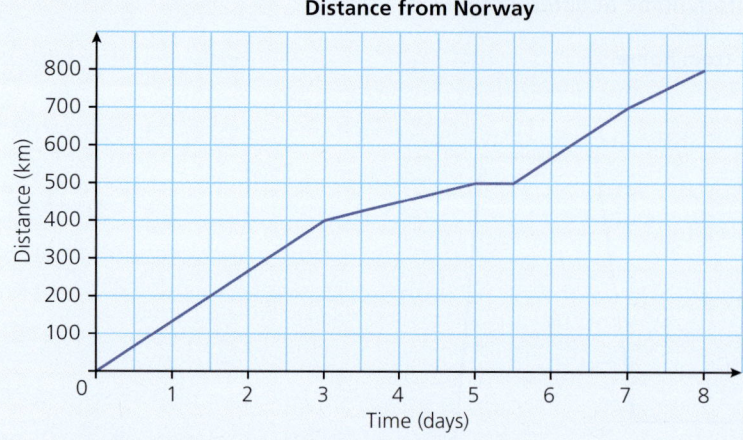

Distance from Norway

- **a** How far had the birds travelled after 3 days?
- **b** On their way to Scotland, the birds stopped at the Shetland Islands.
  How many days did it take them to reach the Shetland Islands?
- **c** How long did the birds rest in the Shetland Islands?
- **d** The birds took 8 days to complete the journey. How far was the entire journey from Norway to Scotland?

8 Look at these equations of straight lines.

- **A** $y = -3x - 4$
- **B** $y = 4 - 3x$
- **C** $y = 3x - 4$
- **D** $y = 4 + 3x$
- **E** $y = 4x + 3$
- **F** $y = 4x - 3$

- **a** Which lines are parallel? There is more than one pair of parallel lines.
- **b** Which pairs of lines have the same y-intercept?

9
- **a** Draw x- and y-axes.
  Use the same scale on each axis and take values from −4 to 4.
- **b** Plot the points A(−4, −1), B(−1, 3), C(2, 2) and D(2, −3).
- **c** Work out the gradient of the lines BC and AD.
- **d** What type of quadrilateral is ABCD?

10 Emyr the electrician charges a £15 callout fee.

For every 30 minutes he works he charges an additional £10.

- **a** Copy and complete this table:

Number of hours worked, n	1	2	3	4	5
Charge, C		£55			

- **b** Draw horizontal and vertical axes.
  Use values from 0 to 5 on the horizontal axis for the number of hours worked.
  Use values from 0 to 120 on the vertical axis for the charge.
  Plot the points from your table and join them with a straight line.
- **c** What is the equation of the straight line you have drawn?
  Use the letters C for the charge and n for the number of hours worked.

Use your graph to answer the following questions:
- **d** How much does Emyr charge for $3\frac{1}{2}$ hours' work?
- **e** How long does Emyr work if he charges £65?

Consolidation 4

Band 3 questions

11 Pinocchio's nose grows when he lies.

If he stops talking, his nose stays the same length.

It shrinks again when he tells the truth.

This is a graph of the length of his nose after he started talking on Monday morning.

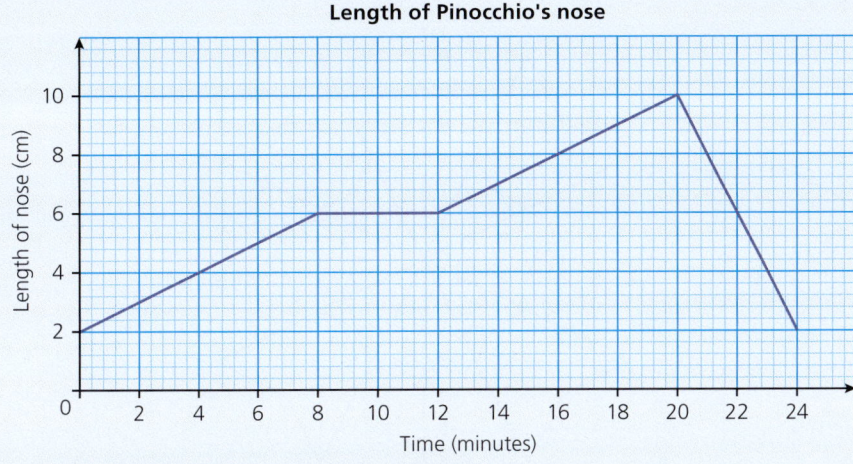

a How long is Pinocchio's nose normally?
b For how long was he telling lies altogether?
c Find the length of Pinocchio's nose after 14 minutes.
d At what times was Pinocchio's nose 8 cm long?
   There is more than one answer.
e For how long was he telling the truth to make his nose shrink back to its normal size?
f What is the name given to a graph of this type, made up of two or more straight line segments?

12 While studying a disease in cattle, scientists found that the number of toxic cells in one cow's bloodstream increased over time, according to this equation:

$N = 14 \times (1.15)^t$

where N is the number of the toxic cells in the bloodstream per litre and t is the time in hours.

A graph of the number of toxic cells against time is shown:

Answer the following questions using the graph.

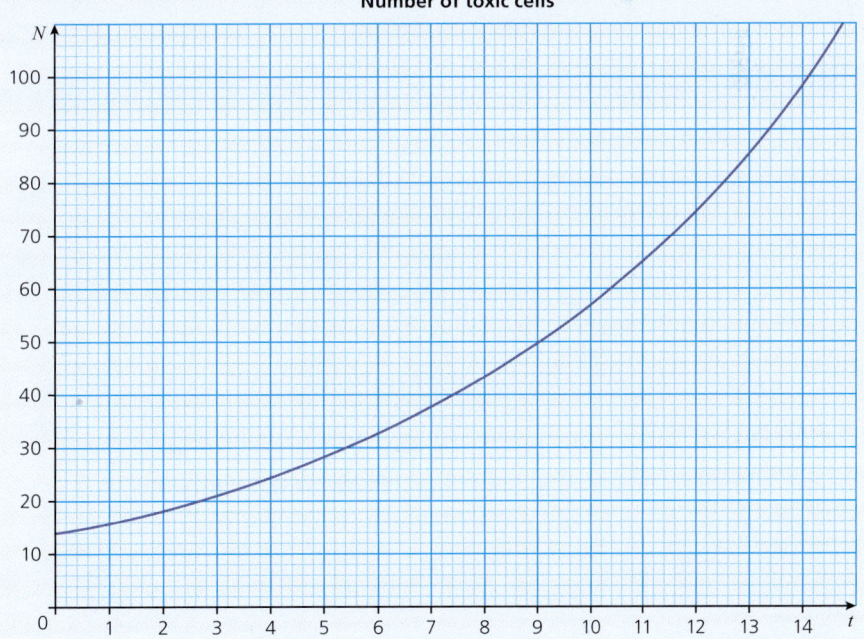

a Find the initial number of toxic cells in this cow's bloodstream.
b Find the number of toxic cells in the bloodstream after 5 hours.
c Find the time it takes for the number of toxic cells in the bloodstream to first exceed 70 per litre.

13 A designer begins by plotting the four points (0, 0), (6, 3), (4, 4) and (3, 6). She joins these points to make a quadrilateral as shown in the first diagram.

She then performs three transformations to make the logo shown in the second diagram: an enlargement, a reflection and a rotation.

Can you describe the three transformations completely?

There is more than one way to answer this.

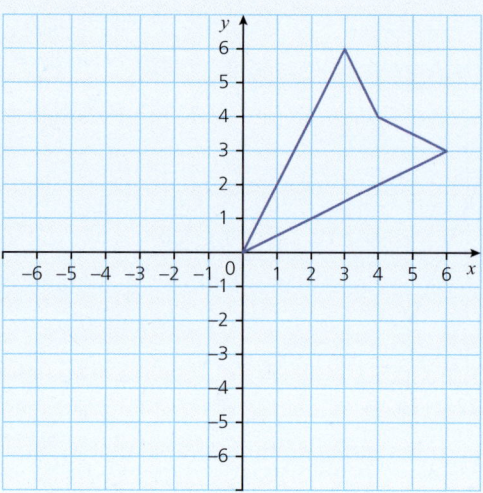
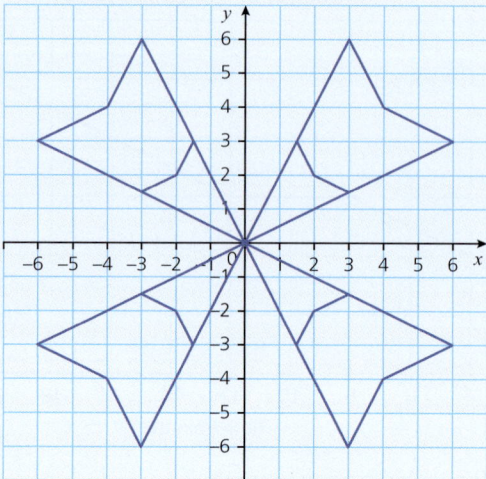

11. Prisms and cylinders

Coming up...

▶ Finding the volume of a prism
▶ Finding the volume of a cylinder
▶ Finding the surface area of prisms and cylinders

Water pouring puzzles

Solve these water pouring puzzles.

You are by a river and you have these two jugs.

5 litres   3 litres

How can you use the jugs to measure out exactly 4 litres?

You are by a river and you have these two jugs.

9 litres   4 litres

How can you use the jugs to measure out exactly 6 litres?

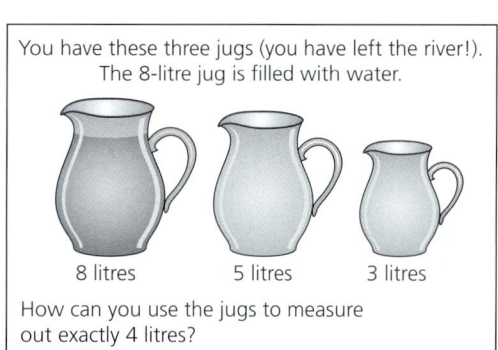

You have these three jugs (you have left the river!). The 8-litre jug is filled with water.

8 litres   5 litres   3 litres

How can you use the jugs to measure out exactly 4 litres?

11.1 Volume of a prism

Skill checker

① **a** Match each shape with the appropriate area formula.

| Triangle | Rectangle | Parallelogram | Square | Trapezium |

lw $\frac{1}{2}(a+b)h$ $\frac{1}{2}bh$ s^2 bh

b Draw a diagram of each shape and mark on your diagram the sides used in the formula.

c Which two formulas could be used for more than one shape?

② Work out the area of each of these shapes.

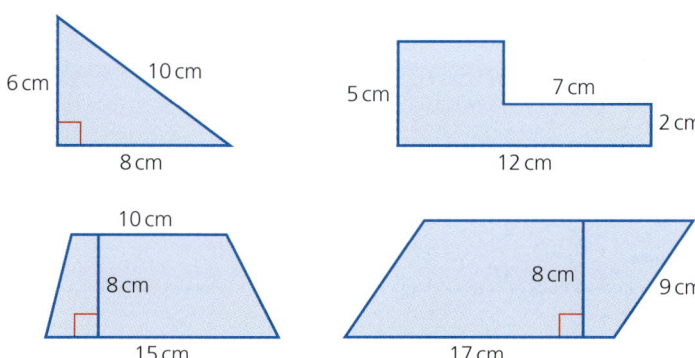

233

Volume of cubes and cuboids

Remember how to find the volume of a cube and a cuboid.

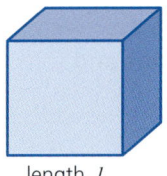

length, l

Volume of a cube = side length^3
Volume of cube = l^3

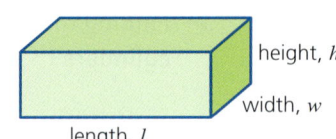

Volume of a cuboid = length × width × height
Volume of a cuboid = lwh

Worked example

The volume of the cube is the same as the volume of the cuboid. Work out the height of the cuboid.

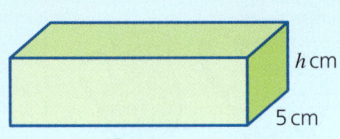
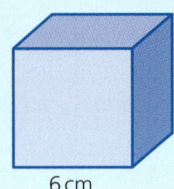

Solution

Volume of cuboid = $12 \times 5 \times h = 60h \text{ cm}^3$

Volume of cube = $6^3 = 216 \text{ cm}^3$

Volume of cuboid = Volume of cube

$$60h = 216$$
$$h = \frac{216}{60}$$
$$h = 3.6$$

So the height of the cuboid is 3.6 cm.

Discussion activity

Cuboid A and Cuboid B are shown below.

The sides of Cuboid B are three times longer than the sides of Cuboid A.

The volume of Cuboid A is 16 cm^3.

Medwyn says that the volume of Cuboid B will be three times the volume of Cuboid A.

Kate says that the volume of Cuboid B will be 3^3 times the volume of Cuboid A.

Who is correct, Medwyn or Kate?

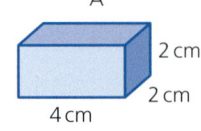
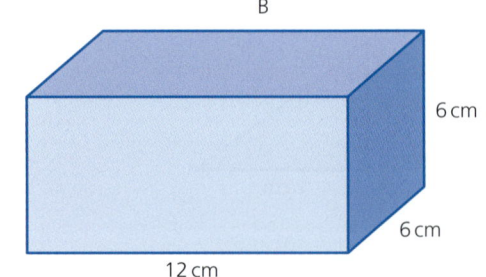

Communication using symbols

11 Prisms and cylinders

▶ Volume of a prism

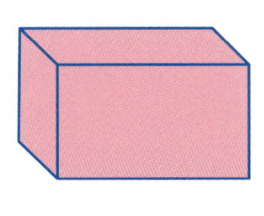
Cuboid

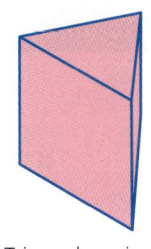
Triangular prism

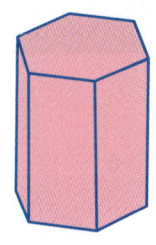
Hexagonal prism

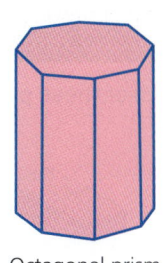
Octagonal prism

> **Remember**
> A prism is a three-dimensional shape with the same cross-section all along its length.

The volume of a prism is measured in cubic units such as cubic centimetres (cm^3) or cubic metres (m^3).

Volume = area of cross-section × length

Or height.

Worked example

Work out the volume of each of these prisms.

a
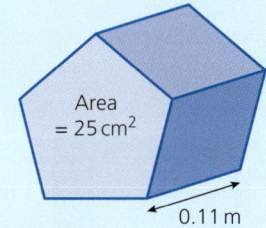
Area = 25 cm²
0.11 m

b
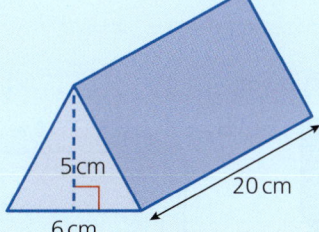
5 cm
6 cm
20 cm

Solution

a Area of cross-section = 25 cm²
   Length = 0.11 m = 11 cm   *Take care! Make sure you don't mix your units!*
   Volume = area of cross-section × length
          = 25 × 11
   Volume = 275 cm³   *Don't forget your units.*

b Area of cross-section = $\frac{1}{2}$ × 6 × 5 = 15 cm²   *Area of triangle = $\frac{1}{2}$ × base × height*
   Volume = area of cross-section × length
          = 15 × 20
   Volume = 300 cm³

11.1 Now try these

Band 1 questions

1 Work out the volume of each of these prisms.
Each cube is 1 cm³.

a 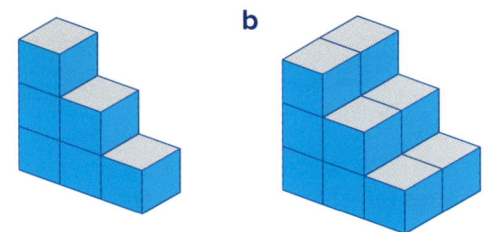 b  c  d

e  f 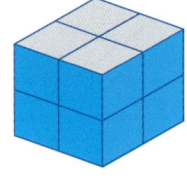 g

2 Calculate the volume of each of these cuboids.

a 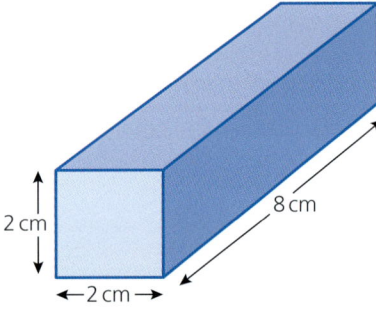 b 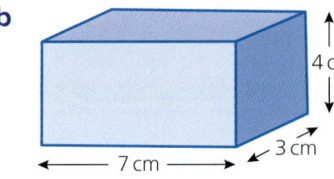 c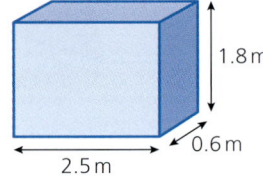

3 Calculate the volume of these cubes.

a 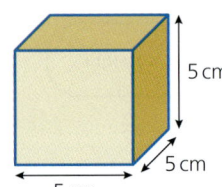 b 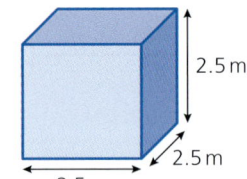 c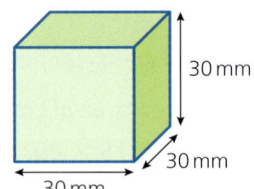

4 Match each of these objects with a suitable volume shown below.
   a   A shoe box
   b   A classroom
   c   A matchbox
   d   A large swimming pool

   2500 m³   1500 mm³   9000 cm³   50 m³

11 Prisms and cylinders

Band 2 questions

5 Work out the volume of these prisms. *Watch out! Check the units carefully.*

a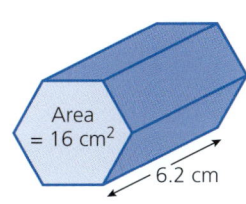

b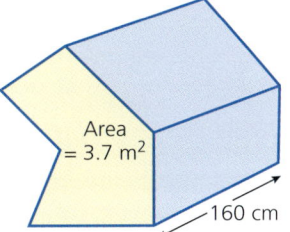

c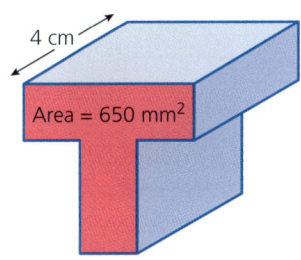

6 Work out the volume of each of these triangular prisms.

a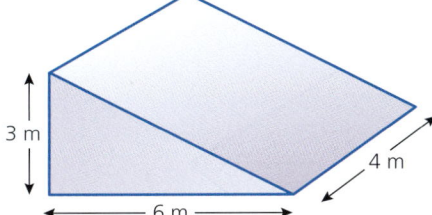

b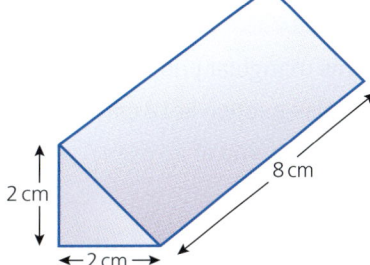

c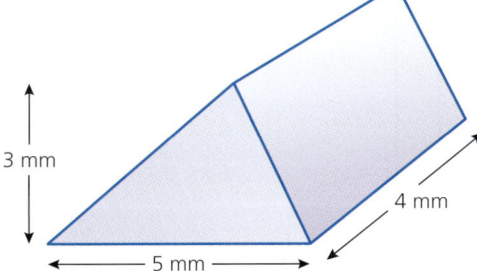

7 For each of these prisms
  i state the name of the shape of the end-face
  ii find the volume of the prism.

a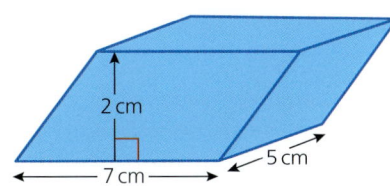

b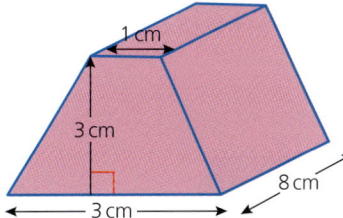

8 Calculate the volume of this prism.

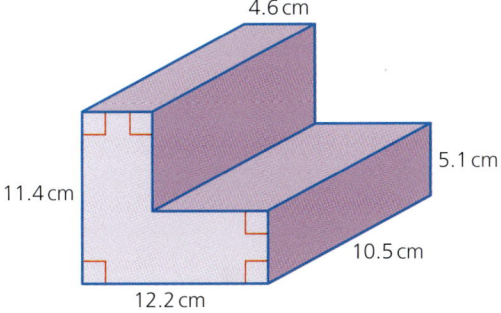

Give your answer correct to the nearest cubic centimetre.

9 The volume of a cube is 343 cm³.
   What is the length of one side of the cube?

10 The volume of this cuboid is 240 m³.

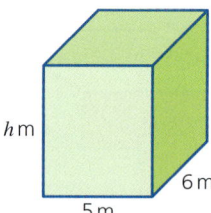

   Work out the height of the cuboid.

Band 3 questions

11 The volume of this triangular prism is 96 cm³.
   The area of the cross-section is 12 cm².
   Calculate the value of x and of y.

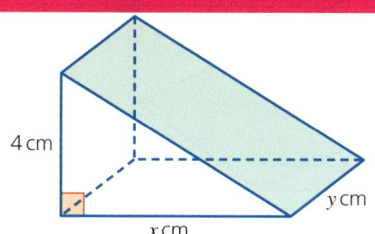

12 A carton of orange juice is in the shape of a cuboid.
   The orange juice is 3 cm below the top of the carton.
   Ted lies the carton flat on the side that says 'ORANGE'.
   What is the depth of the orange juice now?

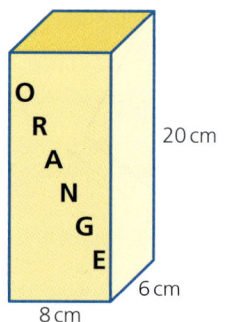

13 A silver bar is in the shape of a prism with a trapezium as its cross-section.
   The bar is melted down and used to make a triangular prism of the same volume.
   Calculate the height of the triangular prism.

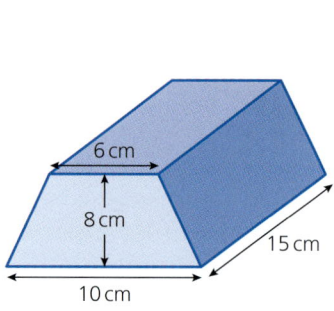 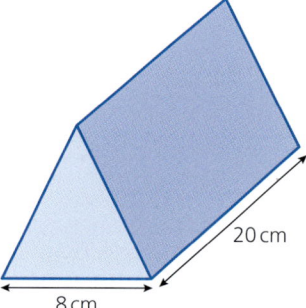

14 Calculate the volume of this triangular prism.

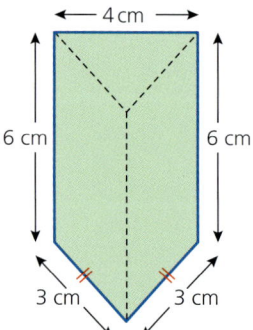

> **Note**
> You'll need to use Pythagoras' theorem.

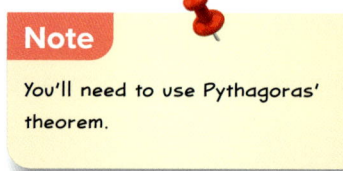

11.2 Volume of a cylinder

Skill checker

Give your answers correct to 3 significant figures.

① Write down the formula for the area of a circle.

② Work out the area of each of these circles.

a 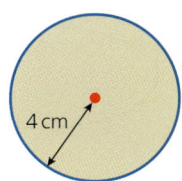 b 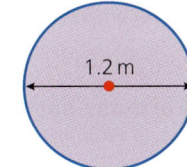 c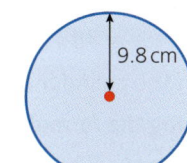

③ Calculate the radius of this circle.

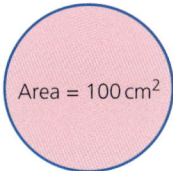

Activity

Consider an enterprise activity in which you are catering for a large group of people, serving portions of pizza.
Pizza Pitsa sells two sizes of pizza: one with a radius of 7 inches, and another with a radius of 12 inches.
Stef says that two 7-inch pizzas will give him more pizza than one 12-inch pizza.
Is Stef correct?
How many 7-inch pizzas are needed so that Stef and his six friends get at least 100 cm² of pizza each?

▶ Find the volume of a cylinder

A cylinder is like a prism with a circular base.
The area of its cross-section is πr^2.
So its volume is $\pi r^2 \times h$.

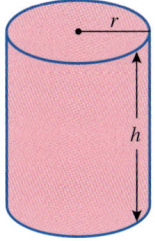

Volume of cylinder $= \pi r^2 h$

Note

In Science lessons, you often calculate the volume of solids, liquids and gases.

The unit of measurement used in Science for volume is millilitres (ml), where 1 ml = 1 cm³.

A measuring cylinder can be used to measure the volume of a liquid.

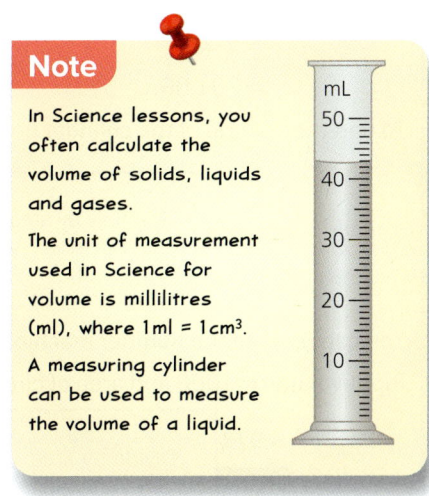

Worked example

Work out the volume of this cylinder.

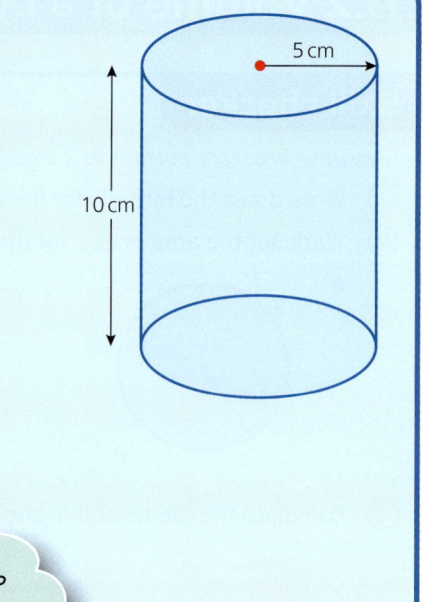

Solution

Method 1: Using the formula, $V = \pi r^2 h$

Volume of cylinder $= \pi r^2 h$

$= \pi \times 5^2 \times 10$

$= 785.398...$

The volume of the cylinder is 785 cm³ (to 3 significant figures).

Method 2: Using the formula, volume of prism = area of cross-section × height

Area of circle $= \pi r^2$

$= \pi \times 5^2$

$= 78.539...$ ← Store this number in your calculator – don't round it!

Volume of cylinder = area of cross-section × height

$= 78.539... \times 10$

$= 785.398...$

The volume of the cylinder is 785 cm³ (to 3 significant figures).

Can you see that these two methods are really the same?

Sometimes you'll be given the volume and you'll need to work back to find the length or height of the cylinder.

Worked example

A cylinder has a diameter of 16 cm and a volume of 1000 cm³.
Work out the length of the cylinder.

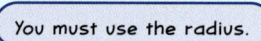

Solution

Diameter = 16 cm so radius = 8 cm ← You must use the radius.

Method 1: Using a reverse number machine

Start by working out the area of the circular cross-section:

Area of circle $= \pi \times r^2$

$= \pi \times 8^2$

$= 201.06...$

To find the volume of the cylinder you can input the length into this number machine.

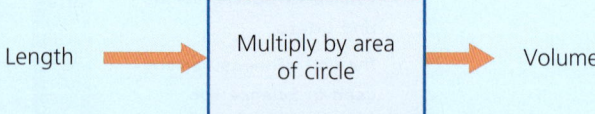

You can reverse the number machine to find the length given the volume.

Input volume = 1000 and area of circle = 201.06...

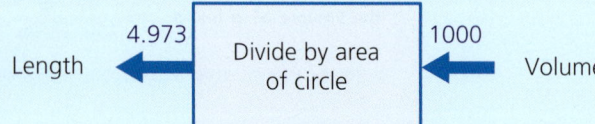

So the length = 4.97 cm (to 3 s.f.).

Method 2: Solving an equation

$\pi r^2 \times l = $ volume of cylinder

$\pi \times 8^2 \times l = 1000$

$201.06 \times l = 1000$

$\div 201.06...$ $\div 201.06...$

$l = 4.973$

So the length is 4.97 cm (to 3 s.f.).

You may also need to work back from the volume of a cylinder to find the radius.

Worked example

A cylinder has a height of 12 cm and a volume of 600 cm³.

Work out the radius of the cylinder.

Solution

Method 1: Using a reverse number machine

Here is a number machine to work out the volume of a cylinder.

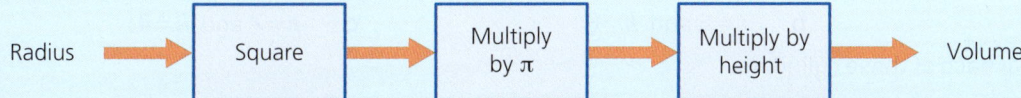

Here is the reverse number machine.

Input volume = 600 and height = 12.

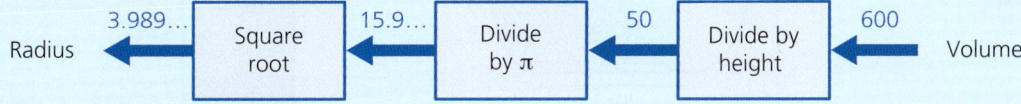

So the radius is 3.99 cm (to 3 s.f.).

Method 2: Solving an equation

$\pi r^2 \times h = $ volume of cylinder

$\pi \times r^2 \times 12 = 600$

$12\pi \times r^2 = 600$

$\div 12\pi$ 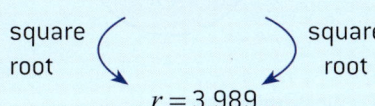 $\div 12\pi$

$r^2 = 15.9...$

square root   square root

$r = 3.989...$

So the radius is 3.99 cm (to 3 s.f.).

11.2 Now try these

Give your answers correct to 3 significant figures.

Band 1 questions

1 Look at this number machine.

Radius → Square → Multiply by π → Multiply by height → Volume

Use the number machine to work out the volume when
- **a** radius = 5 cm and height = 8 cm
- **b** radius = 7 cm and height = 15 cm
- **c** radius = 2 m and height = 3 m
- **d** radius = 10 mm and height = 18 mm.

2 Work out the value of πr^2 when
- **a** $r = 6$
- **b** $r = 8$
- **c** $r = 10$.

3 Work out the value of these.
- **a** $\pi \times 4^2 \times 10$
- **b** $\pi \times 12^2 \times 15$
- **c** $\pi \times 3^2 \times 4$

4 Work out the value of $\pi r^2 h$ when
- **a** $r = 3$ and $h = 5$
- **b** $r = 6$ and $h = 6$
- **c** $r = 7$ and $h = 4$.

5 Work out the volume of each of these cylinders.

a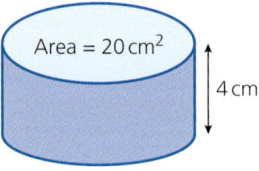
Area = 20 cm², 4 cm

b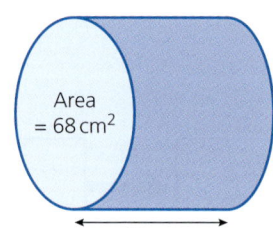
Area = 68 cm², 20 cm

c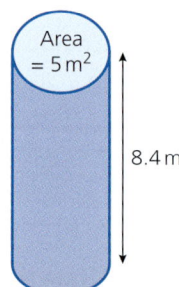
Area = 5 m², 8.4 m

Band 2 questions

6 a Work out the area of each of these circles.

i 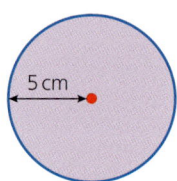 5 cm

ii 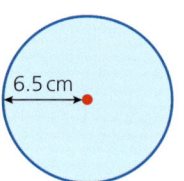 6.5 cm

iii 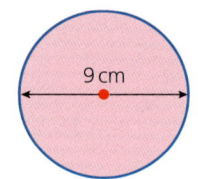 9 cm

b Use your answers to part **a** to help you work out the volume of each of these cylinders.

i 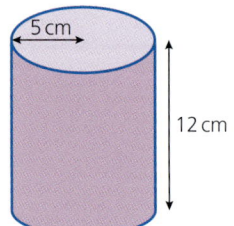 5 cm, 12 cm

ii 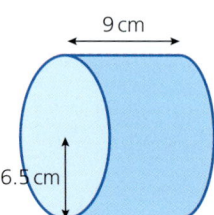 9 cm, 6.5 cm

iii 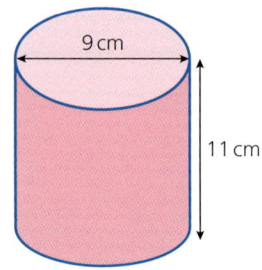 9 cm, 11 cm

11 Prisms and cylinders

7 Osian is buying topsoil to fill a flower bed in his garden.
He is going to fill a rectangular flower bed of area 50 m² to a depth of 0.15 m.
What volume of topsoil does he need?

8 Work out the volume of each of these cylinders.

a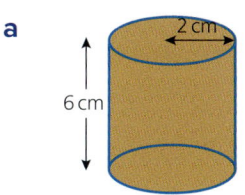

b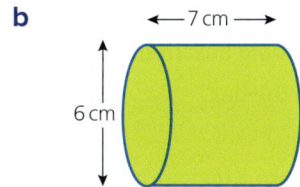

9 A circular lawn has a path round it.

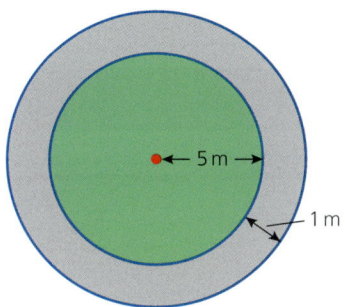

a Find the area of
  i the lawn
  ii the lawn and path
  iii the path.

The path is covered in gravel 5 cm thick.

b Find the volume of the gravel.

10 A cylinder has base area 16 cm² and volume 80 cm³.
What is its height?

Band 3 questions

11 A large jug is in the shape of a cylinder.
It is filled with lemonade to a depth of 33 cm.
How many of these cylindrical glasses can it fill?

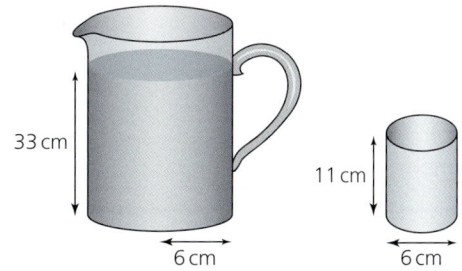

12 Work out the missing length on each of these cylinders.

a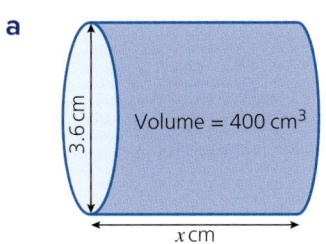

b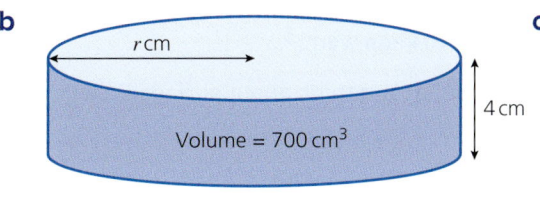

c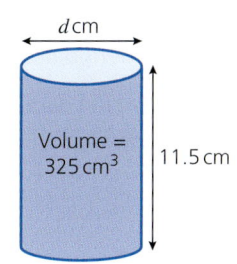

Strategic competence

⑬ A cylinder is cut in half along its length.

Work out the volume of this shape.

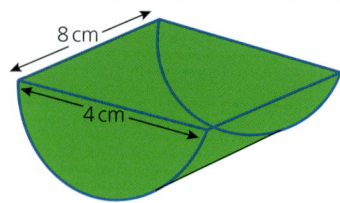

⑭ A cylindrical candle has diameter 6 cm and length 15 cm.

Carwyn packs it into a cuboid box in which it just fits.

Find the volume of air in the box.

⑮ Anna is a scientist.

She pours 50 ml of acid into a measuring cylinder of diameter 3 cm.

Find the depth of the acid in the cylinder.

11.3 Surface area

Skill checker

Give your answers correct to 3 significant figures.

① Write down the formula for the circumference of a circle.

② Work out the circumference of each of these circles.

a 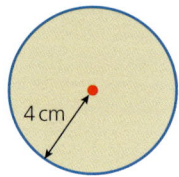   b 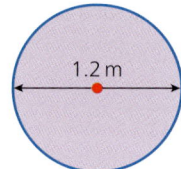   c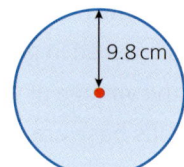

③ For each prism in parts a–c below write down

    i   how many faces each prism has

    ii what shape the faces are.

  a cube

  b triangular prism

  c hexagonal prism

Cross-curricular activity

In your Geography and Science lessons, investigate how animals keep warm in cold climates. While doing so, consider this question:

▶ Why do penguins huddle together to keep warm?

You could model each penguin as a prism and think about surface area to help you answer.

▶ Finding the surface area of a prism

The **surface area** of a prism is the total area of all the faces.

Worked example

Calculate the surface area of this cuboid:

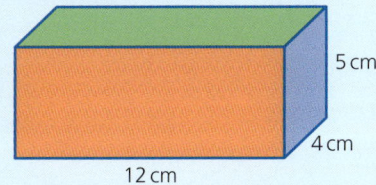

Solution

A cuboid has 6 faces.

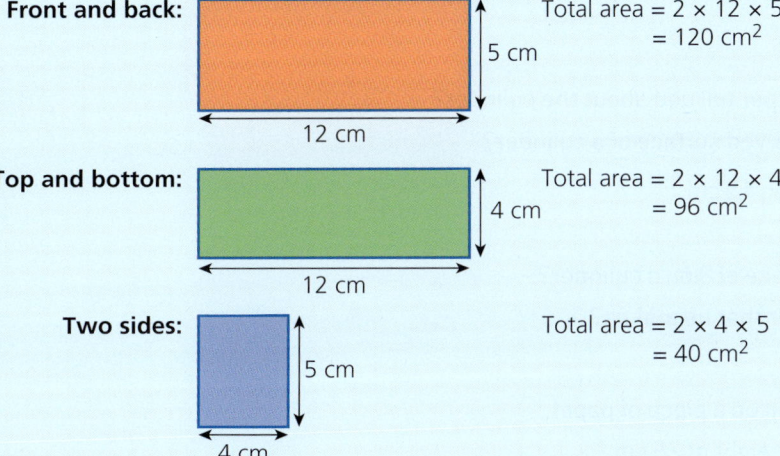

Front and back: Total area = 2 × 12 × 5 = 120 cm²

Top and bottom: Total area = 2 × 12 × 4 = 96 cm²

Two sides: Total area = 2 × 4 × 5 = 40 cm²

Surface area = 120 cm² + 96 cm² + 40 cm²
= 256 cm²

Worked example

Calculate the surface area of this prism.

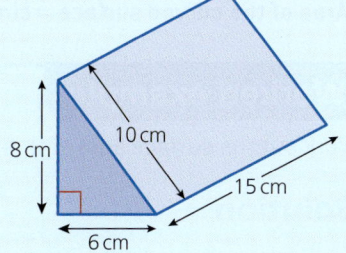

Solution

Work out the area of each face individually.

There are five faces altogether.

Area of one triangular face = $\frac{1}{2}$ × base × height
= $\frac{1}{2}$ × 6 × 8
= 24 cm²

The two triangular faces are the same.

Area of other triangular face = 24 cm²
Area of rectangular base = 6 × 15 = 90 cm²
Area of rectangular side face = 8 × 15 = 120 cm²
Area of sloping rectangular face = 10 × 15 = 150 cm²
Total surface area = 24 cm² + 24 cm² + 90 cm² + 120 cm² + 150 cm²
= 408 cm²

Curriculum for Wales Mastering Mathematics: Book 3

▶ Finding the surface area of a cylinder

Activity

Take a piece of paper.

Roll up your paper to make a hollow cylinder like this:

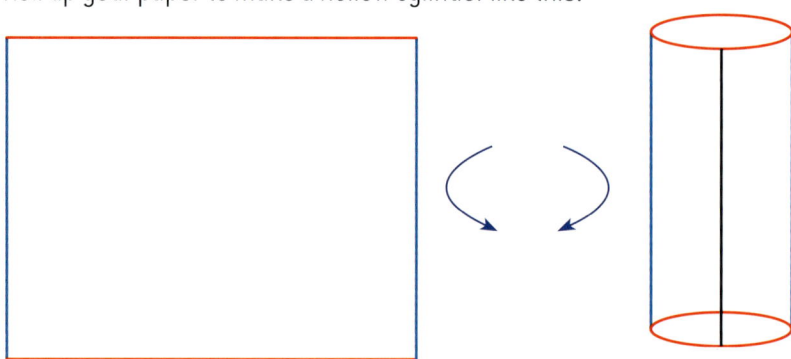

a   What do the width and length of your paper tell you about the cylinder?

How can you work out the area of the curved surface of a cylinder?

b   Siôn has a piece of paper that is 20 cm by 15 cm.

He rolls it up to make a cylinder.

 i   What is the area of the curved surface of Siôn's cylinder?

 ii What is the diameter of the cylinder that he makes?

Find two different answers.

c   Malika makes a hollow cylinder by rolling up a piece of paper.

Her cylinder has a radius of 6 cm and a height of 25 cm.

 i   What size is her piece of paper?

 ii What is the area of the curved surface of Malika's cylinder?

In the activity you found that the curved surface of a cylinder can be unrolled to make a rectangle.

Area of the curved surface = circumference of the cylinder × height

Worked example

Work out the surface area of this cylinder.

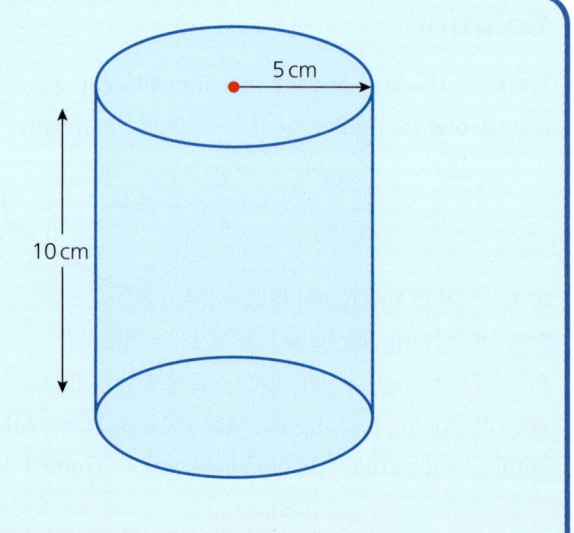

Solution

Area of top circle = πr^2

$= \pi \times 5^2$

$= 78.53... \text{ cm}^2$ ← *Don't round yet!*

Area of bottom circle = $78.53... \text{ cm}^2$

Diameter = $2 \times 5 \text{ cm} = 10 \text{ cm}$

Circumference = πd

$= \pi \times 10$

$= 31.41... \text{ cm}$

Communication using symbols

Area of curved surface = circumference × height
$$= 31.415... \times 10$$
$$= 314.15... \text{ cm}^2$$
Total surface area = curved surface + top circle + bottom circle
$$= 314.15... \text{ cm}^2 + 78.53... \text{ cm}^2 + 78.53... \text{ cm}^2$$
$$= 471.2... \text{ cm}^2$$
$$= 471 \text{ cm}^2 \text{ (to 3 s.f.)}$$

Maths in context

Archimedes, a Greek mathematician, discovered that the surface area of a sphere is equal to the surface area of a cylinder when two conditions are met.

The radius of the cylinder is equal to
- the radius of the sphere **and**
- the height of the cylinder.

Can you find the formula for the surface area of a sphere?

11.3 Now try these

Give your answers correct to 3 significant figures.

Band 1 questions

1 Look at this cube.

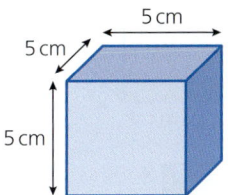

    a How many faces does the cube have?
    b Work out the area of one face.
    c Work out the surface area of the cube.

2 Work out the surface area of these cubes.

  a 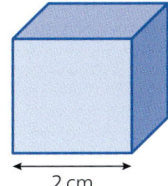   **b** 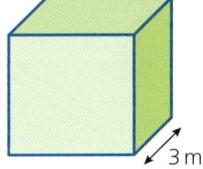   **c**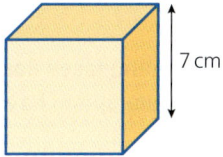

3 Look at this cuboid.
    a How many faces does the cuboid have?
    b Work out the area of the face marked
        i A   **ii** B   **iii** C.
    c Work out the total surface area of the cuboid.

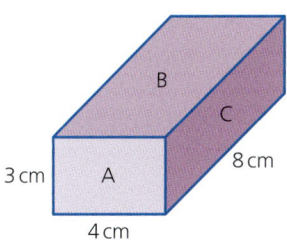

4. Here is the net of a cuboid.
   Each rectangle makes a face of the cuboid.
   a What is the area of the bottom face B?
   b What is the area of the top face?
   c What does the area of the net tell you?
   Peter folds up the net to make a cuboid.
   d What is the volume of the cuboid?

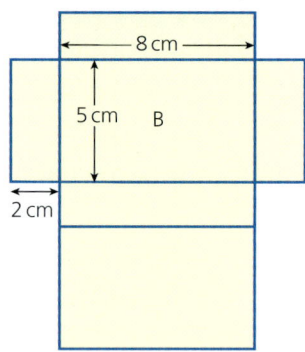

Band 2 questions

5. Find the surface area of each of these cuboids.
   a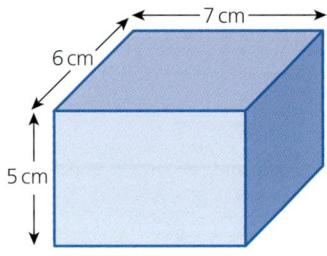
   b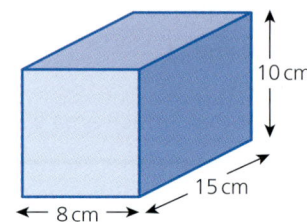

6. Find the volume and surface area of each of these cuboids.
   Give your answers in suitable units.
   a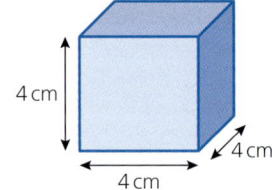
   b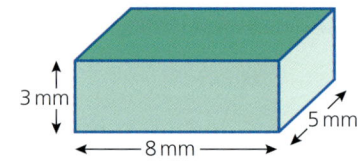
   c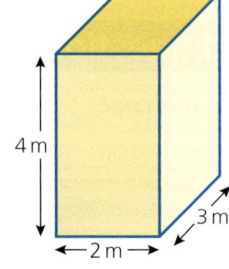

7. The diagram shows Nerys's water tank.
   It is full of water.
   a Work out the volume of water in the tank.
   b Given that $1000 \, cm^3 = 1$ litre, write down the volume in litres.
   The tank is made of metal. It has no top.
   c Work out the area of metal used to make the tank.

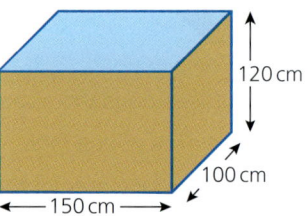

8. a How many faces does this triangular prism have?
   b Work out the total area of the two triangular end faces.
   c Calculate the surface area of the prism.

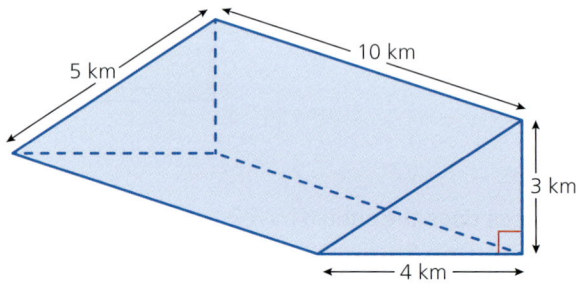

11 Prisms and cylinders

9 Calculate the surface area of these triangular prisms.

a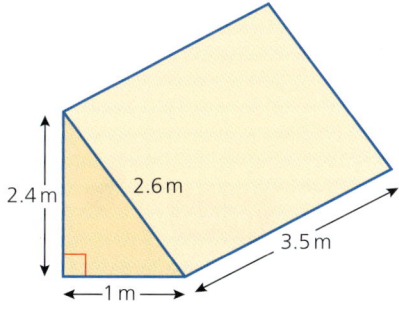

b

10 Calculate the surface area of each of these wooden letters.

a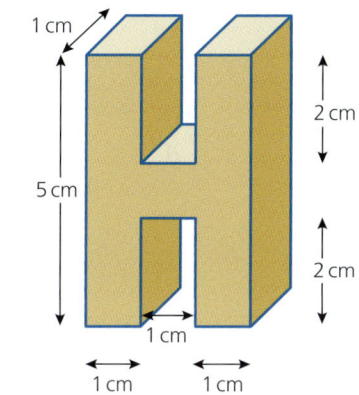

b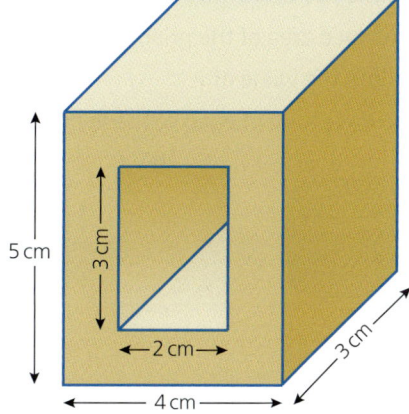

11 A cube has a surface area of 486 cm². Calculate the volume of the cube.

Band 3 questions

12 A cube has a surface area of x cm² and a volume of x cm³.
Work out the value of x.

13 The surface area of this cuboid is 258 cm².

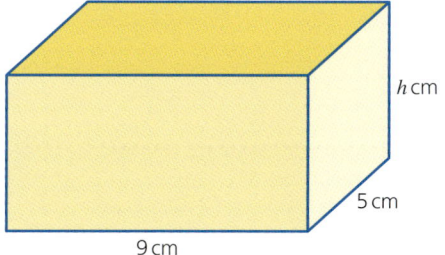

Calculate the height of the cuboid.

⑭ Look at this triangular prism.

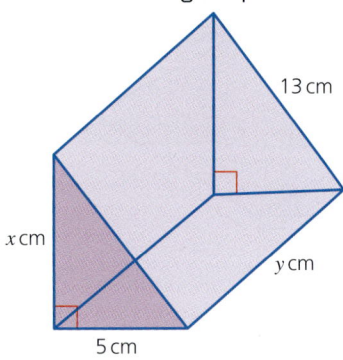

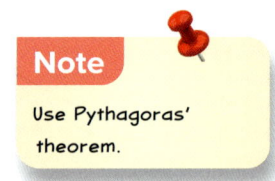

Note

Use Pythagoras' theorem.

a Calculate the value of x.
b The surface area of the prism is 360 cm².
   Calculate the value of y.

⑮ Find the surface area of each of these cylinders.

a 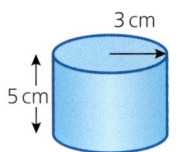   b 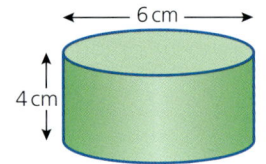   c 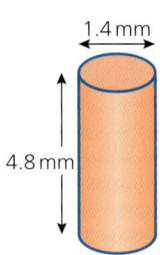   d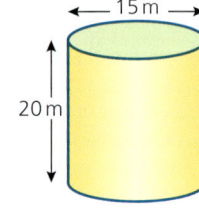

⑯ Owen is painting the outside of a ship's funnel.
The funnel is an open cylinder of diameter 4 m and height 20 m.
A tin of paint covers 8 m².
How many tins of paint does Owen need?

⑰ Aled draws a net of a cylinder on paper.
The sheet of paper measures 20 cm by 15 cm.
He cuts out the net and makes a cylinder.
Its radius is 2 cm and its height is 10 cm.
Show that the net will fit on the paper.

Key words

Here is a list of the key words you met in this chapter.

Area	Circle	Circumference	Cube
Cuboid	Cylinder	Diameter	Prism
Radius	Surface area	Volume	

Use the glossary at the back of this book to check any you are unsure about.

11 Prisms and cylinders

Review exercise: prisms and cylinders

Give your answers correct to 3 significant figures.

Band 1 questions

1 Work out the volume of each of these cuboids.

a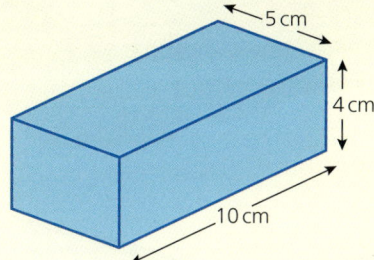

b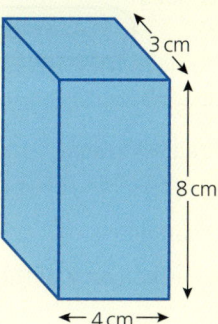

c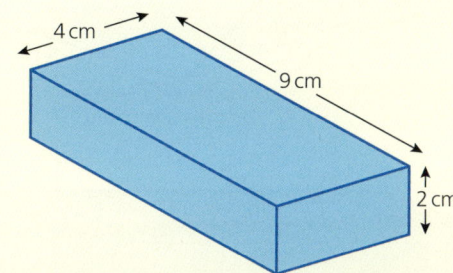

2 Work out the surface area of each of these cuboids.

a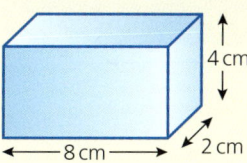

b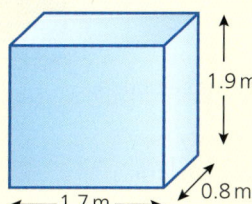

c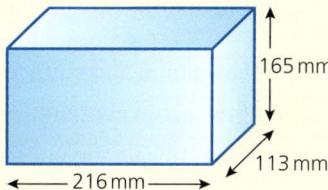

3 Look at Llinos's new chicken shed.

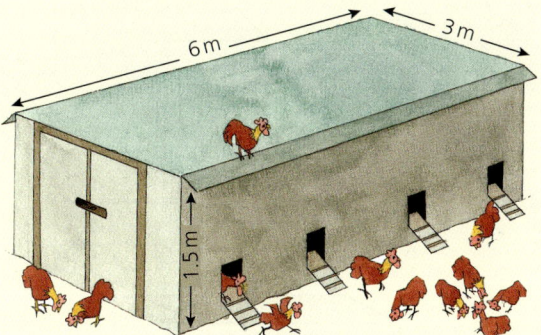

a What is its volume?

Each chicken needs 0.5 m³ of space in the shed.

b How many chickens can Llinos have?

Llinos paints the outside of the shed (but not the roof) brown.

One tin of paint covers 10 m².

c How many tins of paint does she buy?

Curriculum for Wales Mastering Mathematics: Book 3

Band 2 questions

4 A new outdoor swimming pool is planned.
This is one of the designs.

  a What is the volume of the paddling section?
  b What is the volume of the swimming section?
  c How many litres of water does the pool hold altogether? 1 litre = 1000 cm³

The bottom and inside walls of the entire pool are to be covered with tiles.

  d Find the area to be covered with tiles.

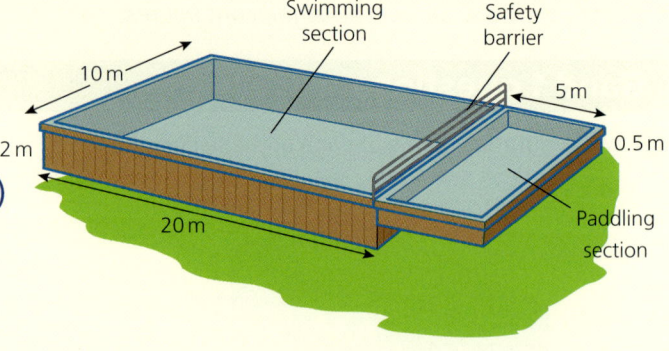

5 Look at this triangular prism.
The end face is an isosceles triangle.
Work out the volume and the surface area of the prism.

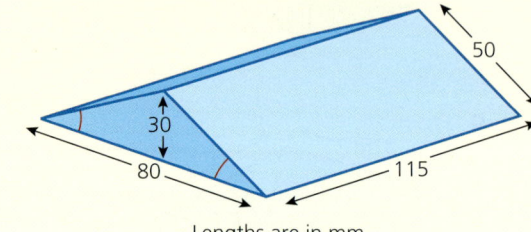

Lengths are in mm

6 A cuboid has length l cm, width w cm and height h cm.
Using l, w and h, write down the formula for
  a its volume
  b its surface area.
State the unit for each formula.

7 Find the volume and surface area of each of these solids.
All lengths are in centimetres.
Remember to give the units for each of your answers.

  a 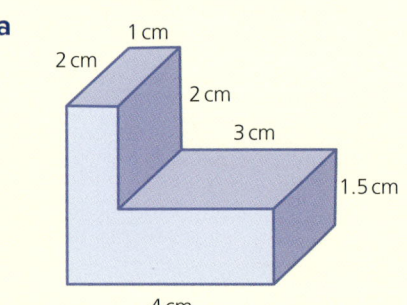   b 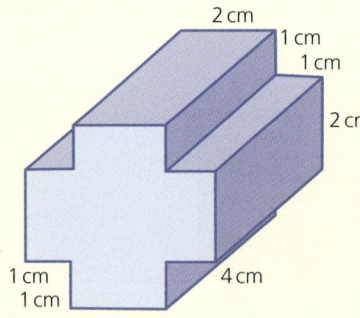   c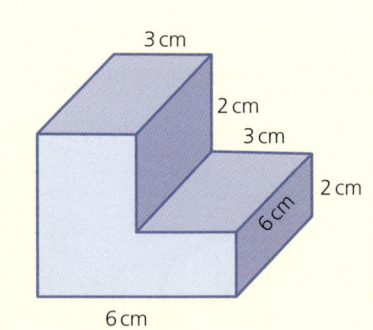

8 A cube of side 5 cm is removed from a cuboid shaped block.
Calculate the volume and surface area of the remaining shape.

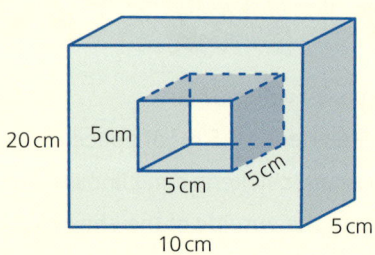

11 Prisms and cylinders

Band 3 questions

9 Calculate
  i the volume
  ii the surface area of each of these cylinders.
  a
  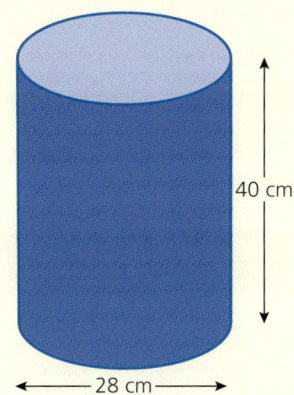
  b

10 a Calculate the volume of this triangular prism.
  b Calculate the surface area of the prism.

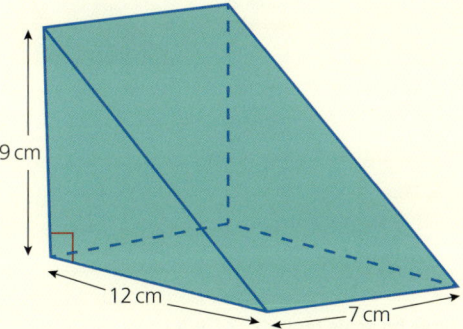

11 The surface area of this cuboid is 446 cm². Calculate the length of the cuboid.

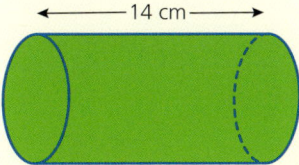

12 The volume of this cylinder is 396 cm³.
  a Calculate the radius of the cylinder.
  b Calculate the surface area of the cylinder.

13 A fence is made from 25 cylindrical posts each with a diameter of 6 cm and a height of 92 cm.
  a Find the volume of wood used to make the fence posts.
  b The curved surface and the top of each post are painted. Find the area of the fence that is painted.

14 Lowri is making a wedding cake with three cylindrical layers.
  The radii are 6 cm, 10 cm and 20 cm.
  The heights of the layers are 5 cm, 6 cm and 8 cm, respectively.
  a Find the total volume of cake she needs to make.
  She covers the top and sides of each layer with icing 3 mm thick.
  b Find the volume of icing she needs to make.

253

12 Trigonometry

Coming up...

- Recognising similar shapes
- Solving problems with similar triangles
- Labelling the sides in right-angled triangles
- Using trigonometry to find
  - the length of an unknown side
  - the size of an unknown angle in a right-angled triangle
- Using trigonometry to solve problems

Enlargements

Draw a shape on squared paper.

Enlarge your shape using a scale factor of 2.

How many times does your original shape fit inside the enlarged shape?

Investigate further using different starting shapes and different scale factors.

Here are two shapes to get you started.

Try some of your own shapes.

Is it always possible to fit your original shape inside the enlarged shape without cutting it up?

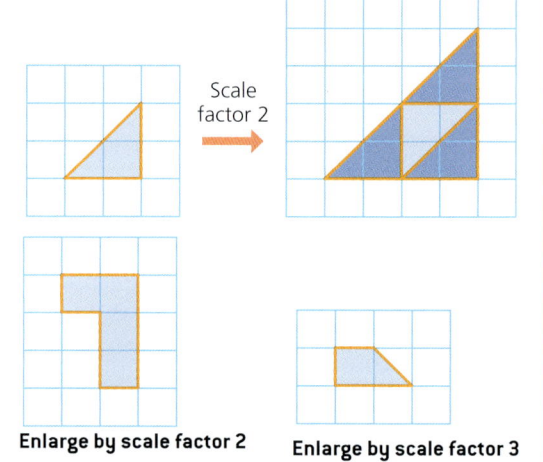

Enlarge by scale factor 2   Enlarge by scale factor 3

12.1 Similarity

Skill checker

Match the rectangles that are enlargements of each other.

Which rectangle is the odd one out?

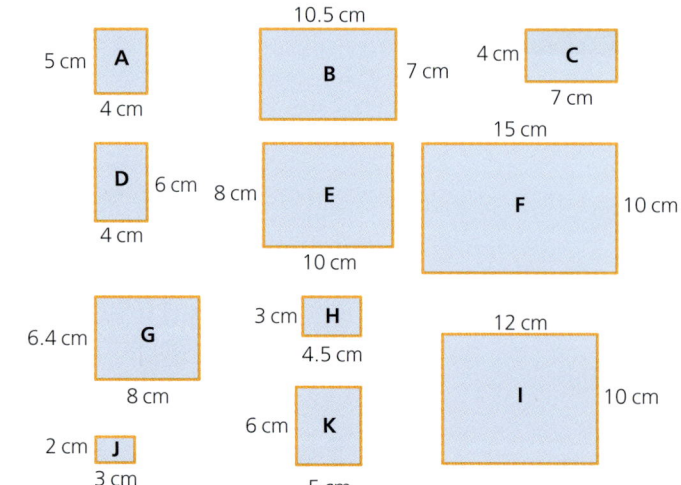

12 Trigonometry

> **Discussion activity**
>
> Discuss whether each of the statements below is true or false:
> - All equilateral triangles with the same area are congruent.
> - All rectangles with the same area are congruent.
> - All squares with the same perimeter are congruent.
> - Circles with the same radius are congruent.

When one shape is an enlargement of another shape the two shapes are said to be **similar**.

Similar shapes have the **same** angles but are different sizes.

Worked example

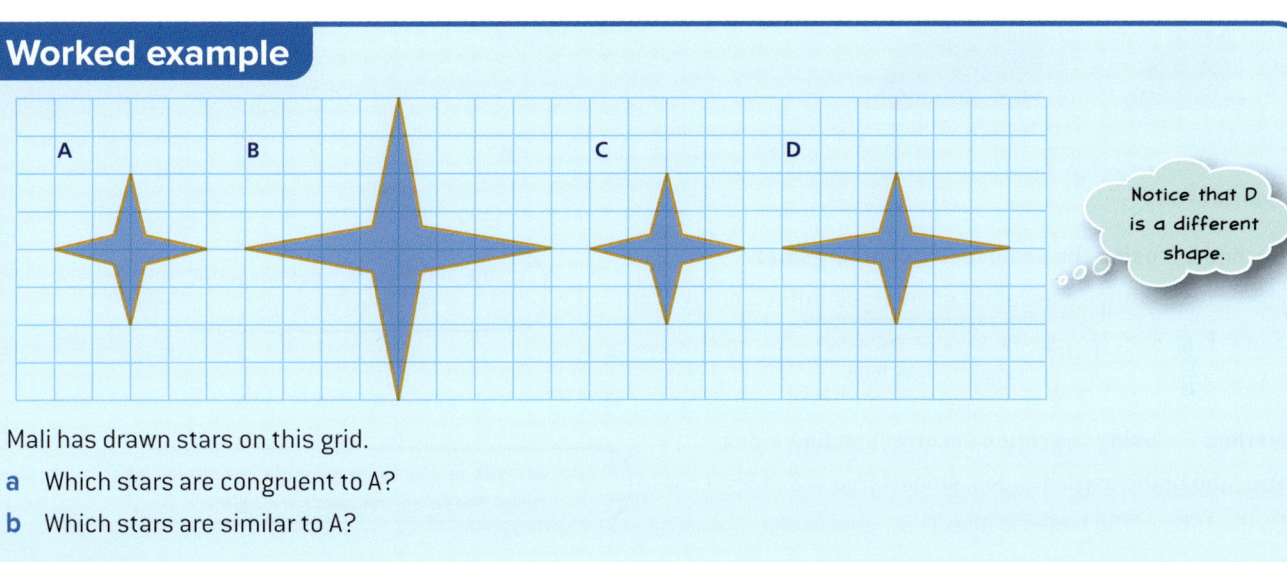

Mali has drawn stars on this grid.

a Which stars are congruent to A?

b Which stars are similar to A?

Solution

a Congruent shapes are the same shape and size.

   Only A and C are congruent.

b Similar shapes are the same shape but they are usually different sizes.

   A, B and C are all similar.

Notice that D is a different shape.

Watch out! All congruent shapes are similar but similar shapes are rarely congruent.

Look at these parallelograms.

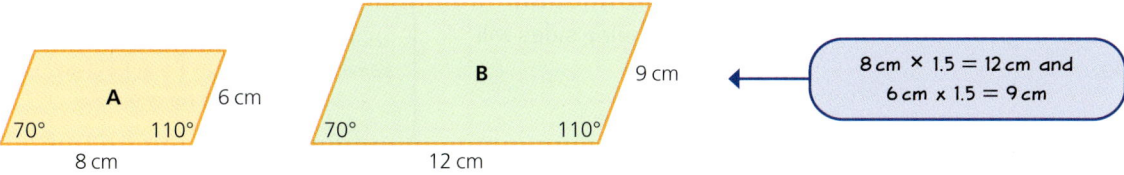

8 cm × 1.5 = 12 cm and 6 cm × 1.5 = 9 cm

All the lengths in B are 1.5 times longer than those in A but the corresponding angles remain the same. Parallelograms A and B are similar.

They are the same shape but different sizes.

The lengths of the sides in similar shapes are all in the same **ratio**. For example:

Parallelogram A   base : slant height = 8 : 6
$$= 4 : 3$$

Parallelogram B   base : slant height = 12 : 9
$$= 4 : 3$$

When two shapes are similar you can use the scale factor of enlargement or the ratios of their sides to work out missing lengths.

Worked example

These triangles are similar.
What is the value of x?

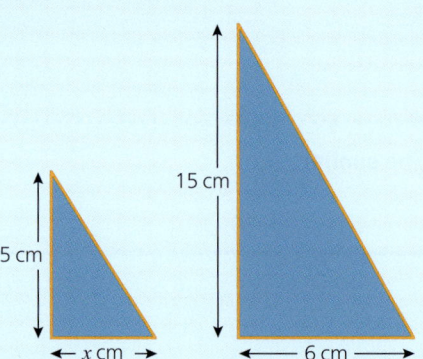

Solution

Method 1 – using the scale factor of enlargement

First work out the scale factor of enlargement: $15 \text{ cm} \div 5 \text{ cm} = 3$

So $x = 6 \div 3$
$ = 2$

> The sides are 3 times longer in the larger triangle.

> So the base of the small triangle is 2 cm long.

Method 2 – using the ratios of corresponding sides

The ratio of the vertical sides, small : large is $5 : 15 = 1 : 3$.
So the ratio of the bases is also $1 : 3$.
For the bases: $3x = 6$
$ x = 2$

> So the sides of the large triangle are three times those of the small triangle.

Method 3 – using the ratios of sides

In the large triangle the ratio of the sides, base : height is $6 : 15 = 1 : 2.5$.
For the small triangle: $2.5x = 5$
$ x = 2$

> So the height of each triangle is 2.5 times the base.

You need to be able to recognise when two triangles are similar. When two triangles have the **same angles** they are similar and corresponding sides will be in the same ratio.

> Watch out! This is only true for triangles. Other shapes may have the same angles but not be enlargements of each other. For example:
> □ and ▭ have the same angles but they are not similar.

Worked example

Look at this diagram.
CD is parallel to BE.

a Show that triangles ABE and ACD are similar.

b Work out the length of
  i BE
  ii BC.

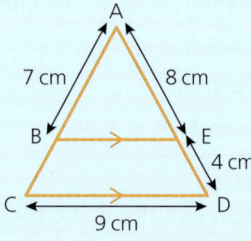

Solution

a

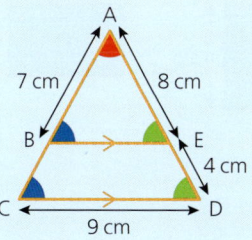

As CD and BE are parallel, the angles at B and C are corresponding angles so they are equal.

The angles at D and E are also corresponding angles so they are equal.

The angle at A is common to both triangles.

Since triangles ABE and ACD have the same angles, the triangles are similar.

b i It often helps to redraw the two triangles separately.

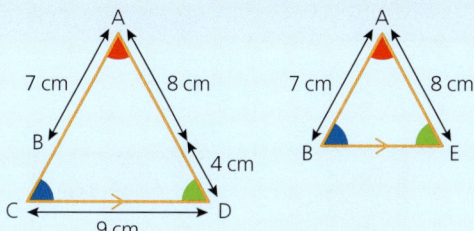

AD = 8 cm + 4 cm = 12 cm and AE = 8 cm

The scale factor is 12 ÷ 8 = 1.5 ← *So triangle ACD is 1.5 times larger than triangle ABE.*

1.5 × BE = CD

So BE = 9 cm ÷ 1.5 = 6 cm

ii You can't find BC directly, but you can work out AC.

1.5 × AB = AC

AC = 1.5 × 7 = 10.5

BC = 10.5 cm − 7 cm = 3.5 cm

12.1 Now try these

Band 1 questions

1 Choose the correct words from the box to complete the sentences below.

| sides | angles | similar | congruent | rotations | enlargements | ratio | reflections |

_____ shapes have all _____ the same and all sides in the same _____.

They are _____ of each other.

2 Steffan writes down two statements about similar shapes.

> A   All squares are similar.
>
> B   All rectangles are similar.

Are Steffan's statements correct? Explain your answers fully.

3 **a** Which rectangles are similar to rectangle A?

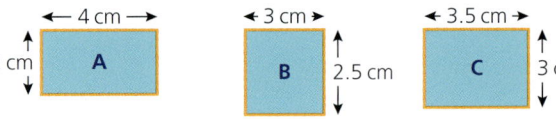
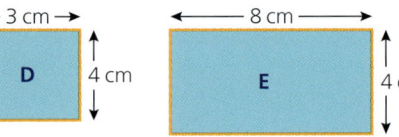
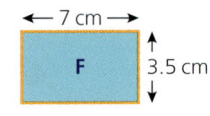

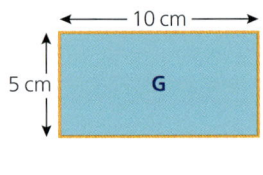
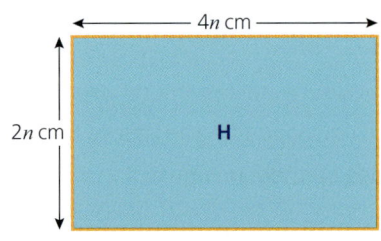

b What is the scale factor for an enlargement mapping A onto each of those rectangles?

Band 2 questions

4 The triangles in each of these pairs are similar.
Find the sides marked x.

a

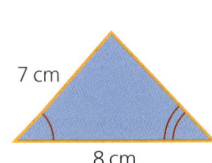

b

5 Triangle ABC is enlarged to give triangle XYZ.
The scale factor is 4.

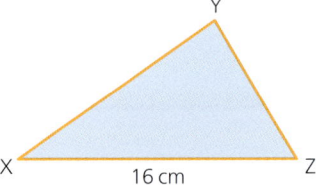

a Find the length
   i XY
   ii YZ
   iii AC.

b What is the ratio of the perimeter of ABC to the perimeter of XYZ?

6 A photograph is 10 cm by 15 cm.
  a The diagrams show some possible enlargements.
Find the missing measurements.

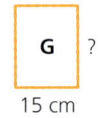

b What is the ratio, the width of E : the width of F?
c Calculate the areas of the photographs E, F and G.

7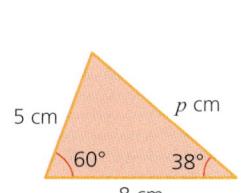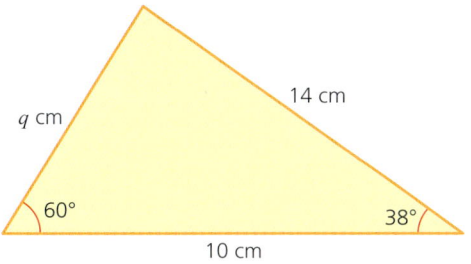

   a Show that these two triangles are similar.

   b Find the value of p and the value of q.

8 You need two sheets of A4 paper, a ruler and a pair of scissors.

Copy this table and complete it with your results.

Size of paper	Longer side (mm)	Shorter side (mm)	Longer ÷ shorter
A4			
A5			
A6			

Measure the sides of a sheet of A4 paper.

Work out the ratio of longer side ÷ shorter side.

Cut the second sheet of A4 paper in half as shown in the diagram.

Each half is now an A5 sheet.

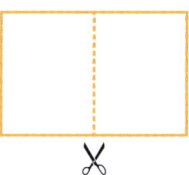

Measure the sides and work out the ratio longer side ÷ shorter side for an A5 sheet.

Cut one of the A5 sheets in half (see diagram) to get two A6 sheets.

Measure the sides and work out the ratio longer side ÷ shorter side for an A6 sheet.

What do you notice about the three ratios that you have worked out?

Band 3 questions

9 These rectangles are similar.

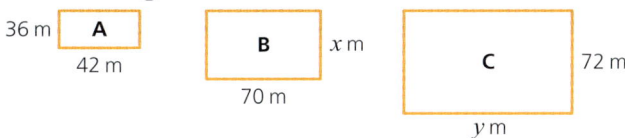

   a Find the lengths marked x and y.

The rectangles represent gardens.

   b **i** Find their perimeters.

      ii Find the ratio of their perimeters.

   c **i** Find the areas of the gardens.

      ii Find the ratio of the areas.

   d Fauzia does scale drawings of all three gardens with a scale of 1 : 2000.

      What are the measurements of her scale drawings?

10 The diagram shows triangle ABC and triangle DEC.

   a Show that triangles ABC and DEC are similar.

   b Find the length of

      i DE   **ii** BE.

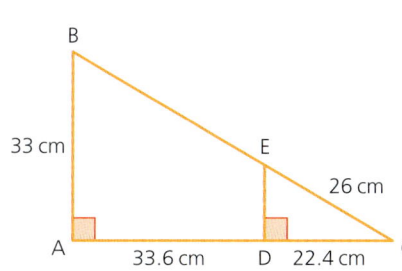

11. The diagram shows triangle ABC and triangle ADE.
    BC and DE are parallel.

    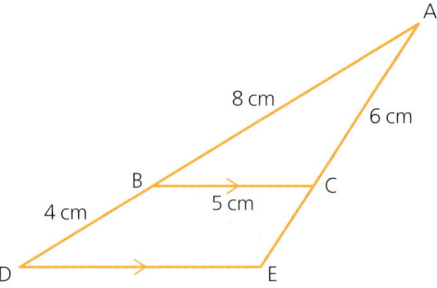

    a   Show that triangles ABC and ADE are similar.
    b   Find the length of
        i   DE   ii   EC.

12. The diagram shows triangle ABC and triangle CDE.

    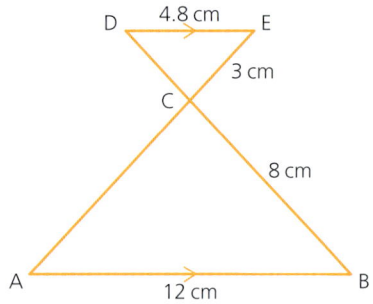

    a   Show that triangles ABC and DEC are similar.
    b   Find the length of
        i   DC   ii   AE.

12.2 Finding a missing side

Skill checker

a   Draw these two triangles accurately.
b   Measure the other two angles.
    What do you notice about the angles?
c   Measure the other side.
    What do you notice about the sides?
d   Work out
    i $\dfrac{\text{base}}{\text{height}}$ ii $\dfrac{\text{base}}{\text{hypotenuse}}$ iii $\dfrac{\text{height}}{\text{hypotenuse}}$
    for each triangle.
    What do you notice?

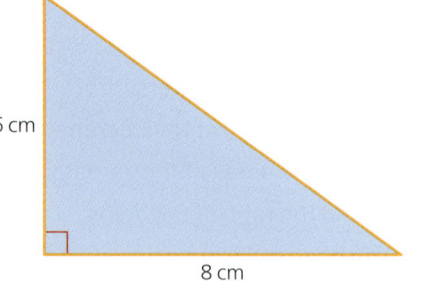

12 Trigonometry

▶ Labelling right-angled triangles

The Greek letter theta, θ, is often used to show an unknown angle. The three sides in a right-angled triangle have special names:

- The **longest side** is the **hypotenuse**.

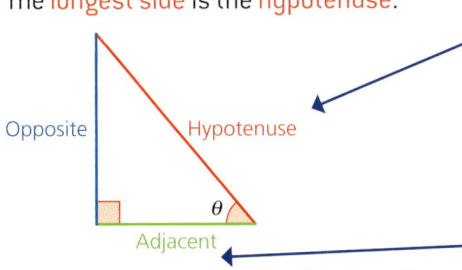

The hypotenuse is fixed – it is the longest side and it is opposite the right-angle.

Watch out! The opposite and adjacent swap around depending on which of the acute (small) angles is θ.

The **adjacent side** has the right-angle at one end of it and θ at the other.

- The **side next to** the angle θ is the **adjacent**.
- The **side opposite** the angle θ is the **opposite**.

Adjacent means 'next to'.

Activity

Label the hypotenuse, opposite and adjacent sides in these right-angled triangles.

① 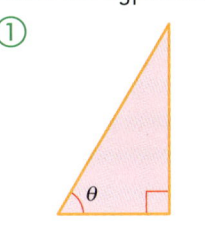   ②    ③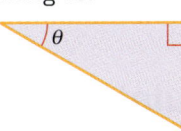

④ 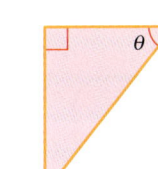   ⑤ (pink triangle)   ⑥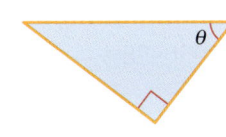

You might find it helps to turn the triangle round before you label it.

▶ Trigonometry

Trigonometry is the study of right-angled triangles and their side lengths and angles. In the next activity you'll be investigating similar right-angled triangles.

Activity

① Accurately draw a right-angled triangle with an angle of 30°.

Accurately measure the lengths of the opposite and the hypotenuse.

Work out the ratio $\dfrac{\text{opposite}}{\text{hypotenuse}}$

Repeat with at least two other right-angled triangles with an angle of 30°.

What do you notice?

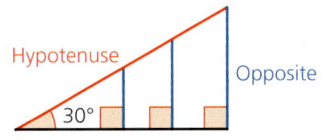

② Work out the sides and angle marked with letters.

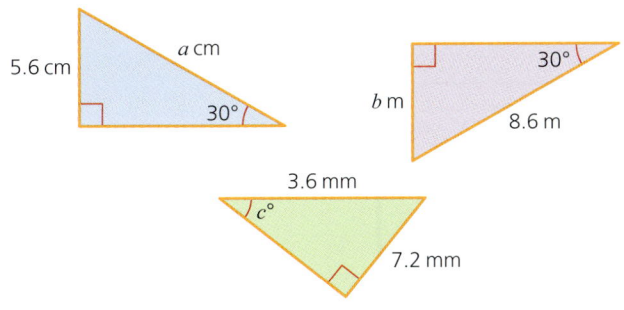

Hint

All these triangles are similar to the triangles you drew in part 1.

261

▶ The sine ratio

In the activity you found that when a right-angled triangle has one angle of 30°, the ratio $\frac{\text{opposite}}{\text{hypotenuse}}$ was always equal to $\frac{1}{2}$. This is because all right-angled triangles with one angle of 30° are **similar**.

> **Remember**
> Similar triangles have sides in the same ratio.

If you repeated the activity using a right-angled triangle with an angle of 40°, you would find that the ratio $\frac{\text{opposite}}{\text{hypotenuse}}$ is approximately 0.64.

In fact, the ratio $\frac{\text{opposite}}{\text{hypotenuse}}$ always has the same value for a given angle θ.

The ratio $\frac{\text{opposite}}{\text{hypotenuse}}$ is called the **sine of the angle θ**.

You write $\sin \theta = \frac{\text{opposite}}{\text{hypotenuse}}$

Say 'sine theta is equal to opposite over hypotenuse'.

So $\sin 30° = \frac{1}{2}$ and $\sin 40° = 0.64\ldots$

> **Hint**
> Check that you get an answer of 0.5: if you don't then you need to make sure that your calculator is in degrees mode.
> Watch out! Your calculator may automatically open a bracket when you press sin, so you need to close it!

Using your calculator

You can use the 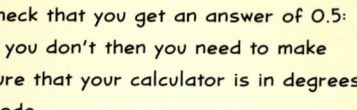 button to find the value of $\sin \theta$ for any angle θ.

To use your calculator to find $\sin 30°$ press .

You can use the sine ratio to find the length of either the opposite or the hypotenuse in a right-angled triangle.

If you know the opposite and θ.

If you know the hypotenuse and θ.

Use your calculator to find the value of
a $\sin 90°$ b $\sin 60°$
c $\sin 20°$.

Worked example

Work out the value of x.

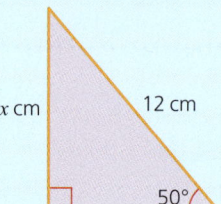

Solution

The two sides are the **opposite**, x, and the **hypotenuse**, 12 cm.

You know that $\theta = 50°$.

Use $\sin \theta = \frac{\text{opposite}}{\text{hypotenuse}}$

Substituting gives: $\sin 50° = \frac{x}{12}$

Rearrange: $12 \sin 50° = x$

You don't need to write a multiplication symbol between 12 and $\sin 50°$.

Use your calculator: $x = 12 \sin 50°$
$= 9.192\ldots$

So the missing side is 9.19 cm (to 3 s.f.).

12 Trigonometry

> **Activity**
>
> You can use this formula triangle to help you rearrange the formula.
>
> Use $\sin\theta = \dfrac{\text{opposite}}{\text{hypotenuse}}$ to work out the value of x and of y.
>
> a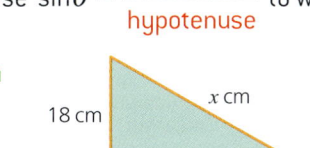
>
> b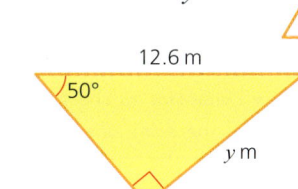

▶ The cosine ratio

The ratio $\dfrac{\text{adjacent}}{\text{hypotenuse}}$ is called the **cosine** of the angle θ.

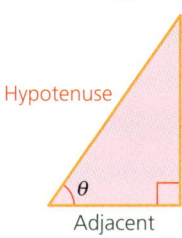

You write it as $\cos\theta = \dfrac{\text{adjacent}}{\text{hypotenuse}}$

You can use the **cos** button on your calculator to find the value of $\cos\theta$ for any angle θ.

For example, $\cos 60° = \dfrac{1}{2}$ ← Use your calculator to check that you get the same answer.

Worked example

Work out the value of x.

Solution

Label the sides of the triangle.

Using $\cos\theta = \dfrac{\text{adjacent}}{\text{hypotenuse}}$ gives:

$\cos 40° = \dfrac{x}{15}$

$x = 15 \cos 40°$

$ = 11.49...$

$ = 11.5$ (to 3 s.f.)

So the missing side is 11.5 cm (to 3 s.f.).

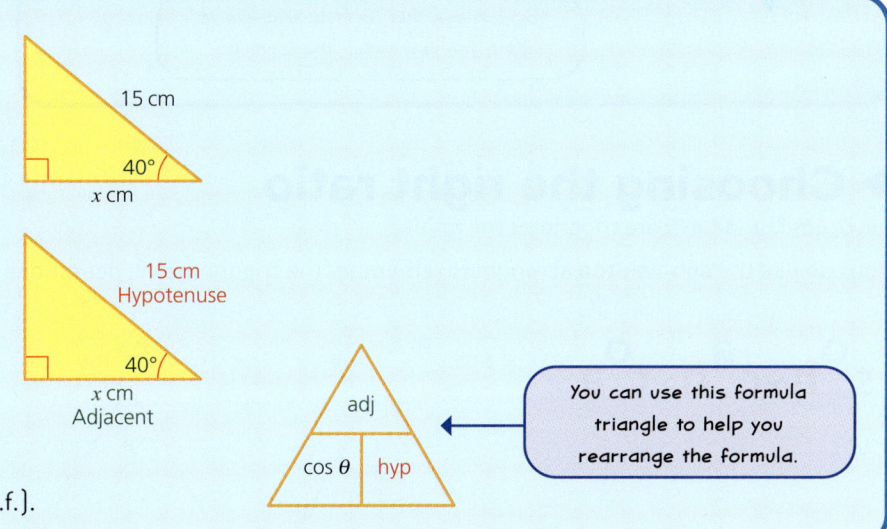

You can use this formula triangle to help you rearrange the formula.

Communication using symbols

263

Curriculum for Wales Mastering Mathematics: Book 3

▶ The tangent ratio

The ratio $\frac{\text{opposite}}{\text{adjacent}}$ is called the **tangent** of the angle θ.

You write it as $\tan\theta = \frac{\text{opposite}}{\text{adjacent}}$

You can use the `tan` button on your calculator to find the value of $\tan\theta$ for any angle θ.

For example, $\tan 50° = 1.191...$ ← Use your calculator to check that you get the same answer.

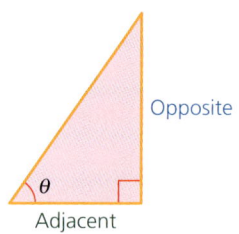

Worked example

Work out the value of x.

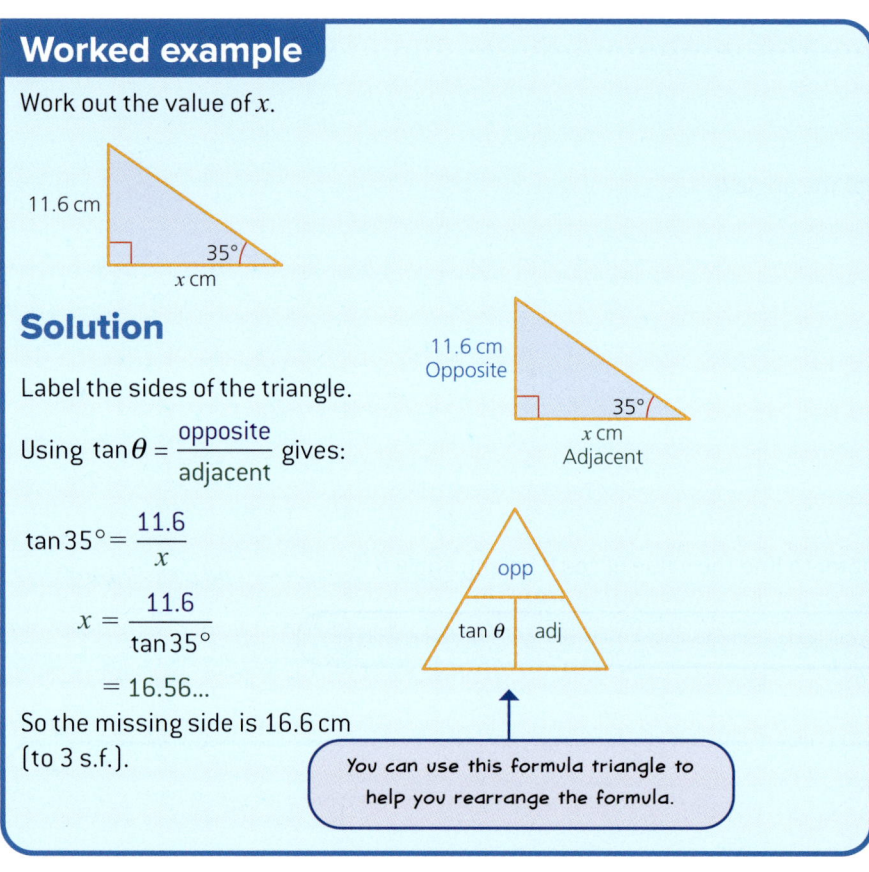

Solution

Label the sides of the triangle.

Using $\tan\theta = \frac{\text{opposite}}{\text{adjacent}}$ gives:

$\tan 35° = \frac{11.6}{x}$

$x = \frac{11.6}{\tan 35°}$

$= 16.56...$

So the missing side is 16.6 cm (to 3 s.f.).

You can use this formula triangle to help you rearrange the formula.

▶ Choosing the right ratio

$\sin\theta$, $\cos\theta$ and $\tan\theta$ are trigonometric ratios.

You can use the **soh-cah-toa** acronym to remember the trigonometric definitions and the formula triangles.

$\sin\theta = \frac{\text{opp}}{\text{hyp}}$ $\cos\theta = \frac{\text{adj}}{\text{hyp}}$ $\tan\theta = \frac{\text{opp}}{\text{adj}}$

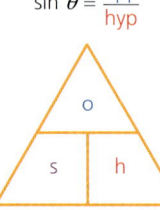 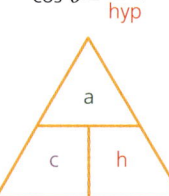 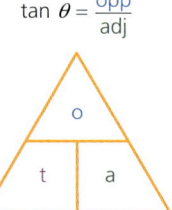

You usually need to decide which ratio is the correct one to use.
- Label the side you are given and the side you want.
- Choose the ratio that connects these two sides

12 Trigonometry

Worked example

Work out the value of x.

Solution

Label the sides:

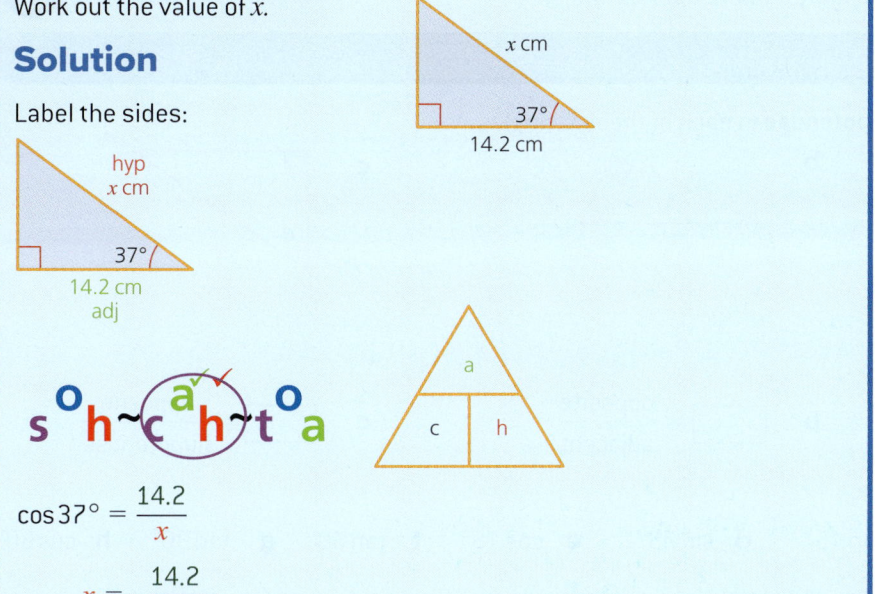

$$\cos 37° = \frac{14.2}{x}$$

$$x = \frac{14.2}{\cos 37°}$$

$$= 17.78...$$

So the missing side is 17.8 cm (to 3 s.f.).

Discussion activity

Pythagoras' theorem and trigonometry can both be used to find the size of an unknown side in a right-angled triangle. For each of the triangles below, discuss whether you could use
- Pythagoras' theorem
- trigonometry
- both Pythagoras' theorem and trigonometry
- neither

to calculate the size of the unknown side.

A

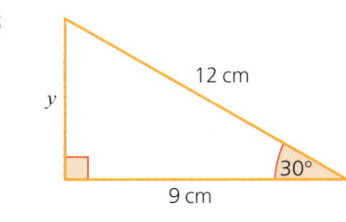

B

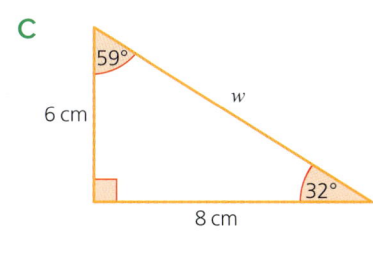

C

Cross-curricular activity

In Design Technology, accuracy is vital to the quality of a finished product. Trigonometry can be used to ensure that the correct angles and side lengths are used in designs. For example, when designing this iron bracket, if the vertical side was too short, the bracket would not be able to bear the required weight.

Ask your Design Technology teacher where else trigonometry is used in their subject area.

Curriculum for Wales Mastering Mathematics: Book 3

12.2 Now try these

Give your answers correct to 3 significant figures where appropriate.

Band 1 questions

1. Label the **opposite**, **adjacent** and **hypotenuse** in each of these triangles.

   a 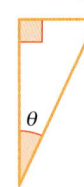   b    c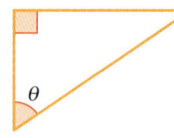

2. Complete the trigonometric ratios.

   a $\boxed{} = \dfrac{\text{adjacent}}{\text{hypotenuse}}$ b $\boxed{} = \dfrac{\text{opposite}}{\text{adjacent}}$ c $\boxed{} = \dfrac{\text{opposite}}{\text{hypotenuse}}$

3. Use your calculator to work out these.

   a $\cos 45°$ b $\sin 45°$ c $\tan 45°$ d $\sin 75°$ e $\cos 15°$ f $\tan 60°$ g $\tan 30°$ h $\cos 60°$

4. Use your calculator to work out these.

   a $2 \times \cos 25°$ b $5.4 \times \sin 30°$ c $2 \tan 45°$ d $7 \sin 32°$ e $8 \div \cos 60°$ f $12 \div \tan 50°$

   g $\dfrac{12}{\sin 30°}$ h $\dfrac{11.7}{\tan 80°}$

Band 2 questions

5. Use the ratio for $\sin \theta$ to find the value of x and of y.

   a 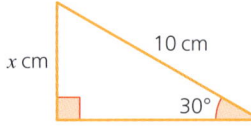   b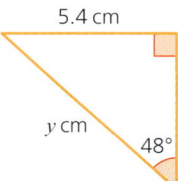

6. Use the ratio for $\cos \theta$ to find the value of x and of y.

   a 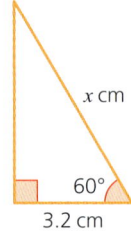   b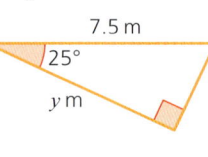

7. Use the ratio for $\tan \theta$ to find the value of x and of y.

   a    b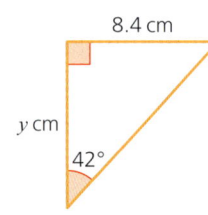

12 Trigonometry

8 Look at these right-angled triangles.

a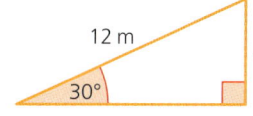
b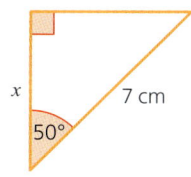
c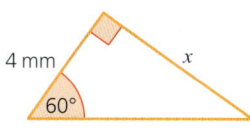

For each triangle:
i   make a sketch and label the sides opposite, adjacent and hypotenuse
ii write down (using opposite, adjacent or hypotenuse) which side you know and which side you want to find
iii write down the trigonometric ratio that involves these two sides
iv use the correct trigonometric ratio and your calculator to find the side length that you want.

Band 3 questions

9 Find the length of each side marked with a letter.

a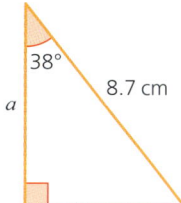
b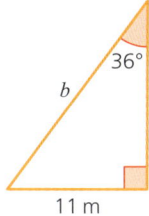
c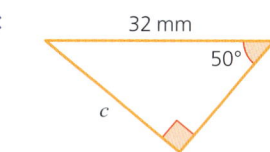
d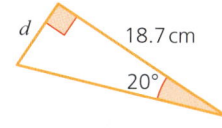

10 Find the area of triangle ABC.

11 The diagram shows a rectangular field ABCD.
AC = 25 m and angle CAD = 50°.
A farmer wants to build a fence around the field.
He needs to put a 2.5 metre gate on each of sides AD and BC.
Fencing costs £45 per 2 metres. Gates cost £225 each.
How much does it cost the farmer to fence and gate the field?

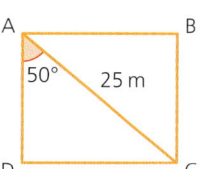

12 Work out the length of AD.

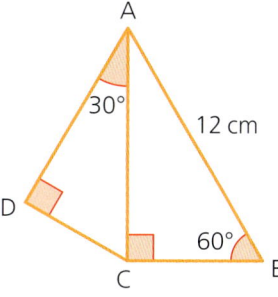

12.3 Finding a missing angle

Skill checker

1. Label the **opposite**, **adjacent** and **hypotenuse** in the triangle.

2. These two triangles are similar.
   a. Work out the value of x and of y.
   b. Write down the value of
      i. $\cos \theta$
      ii. $\tan \theta$.

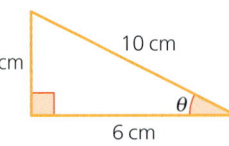 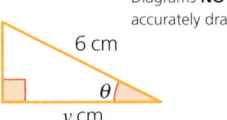

Diagrams **NOT** accurately drawn

▶ Using trigonometry to find an angle

Look at this triangle.
You can see that

$\sin \theta = \dfrac{\text{opposite}}{\text{hypotenuse}}$

$= \dfrac{8}{10}$

$= 0.8$

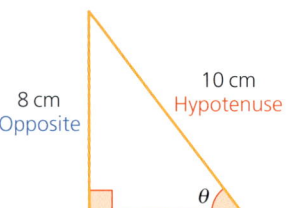

Your calculator uses the **inverse trigonometric functions** $\sin^{-1}$, $\cos^{-1}$ and $\tan^{-1}$ to 'undo' sin, cos and tan.

Sometimes these are written as arcsin, arccos and arctan.

Your calculator can work out what angle has a sine of 0.8.

$\sin \theta = 0.8$

$\theta = \sin^{-1}(0.8)$ ← Say inverse sin for $\sin^{-1}$.

$= 53.1°$ (to 1 d.p.)

You need to use the shift or 2nd F button on your calculator for $\sin^{-1} \theta$, so you need (for example) these keys:

[shift][sin][0][.][8][=]

Activity

1. Use Pythagoras' theorem to work out the adjacent side.

2. Copy and complete the following:

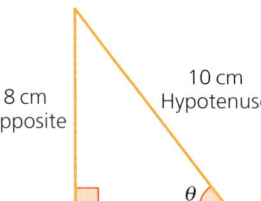

a. $\cos \theta = \dfrac{\text{adj}}{\text{hyp}}$

$= \dfrac{\square}{\square}$

$= \square$

So $\theta = \cos^{-1} \square$

$= \square$ (to 1 d.p.)

b. $\tan \theta = \dfrac{\text{opp}}{\text{adj}}$

$= \dfrac{\square}{\square}$

$= \square$

So $\theta = \tan^{-1} \dfrac{\square}{\square}$

$= \square$ (to 1 d.p.)

What do you notice?

12 Trigonometry

Worked example

a Look at the diagram. Which ratio would you use to find θ?

b Find θ.

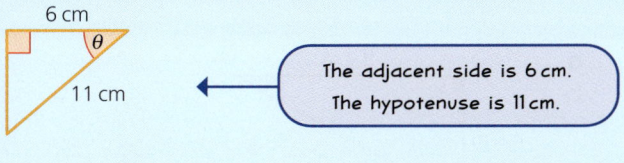

Solution

a You know the adjacent and the hypotenuse so use $\cos\theta = \dfrac{\text{adjacent}}{\text{hypotenuse}}$

The adjacent side is 6 cm.
The hypotenuse is 11 cm.

b $\cos\theta = \dfrac{6}{11}$

So $\theta = \cos^{-1}\left(\dfrac{6}{11}\right)$

Use your calculator to work this out.

$= 56.9°$ (to 1 d.p.)

Angles are usually given correct to 1 decimal place and side lengths to 3 significant figures.

Make sure you know how to use your calculator correctly, for example make sure your calculator is in 'degrees' mode.

12.3 Now try these

Unless exact, give all angles correct to 1 decimal place.

Band 1 questions

1 ABC is a right-angled triangle with a right-angle at B.

The angle BAC is θ.

Find the value of θ when

   a $\cos\theta = 0.5$ **b** $\sin\theta = 0.5$ **c** $\tan\theta = 1$.

2 Use your calculator to find the angles for which

   a $\sin x = 0.2$ **b** $\cos y = 0.45$ **c** $\tan z = 0.8$.

3 Here is a right-angled triangle.

Write down the value of

   a $\cos\theta$ **b** $\sin\theta$ **c** $\tan\theta$.

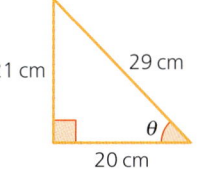

Band 2 questions

4 Use the ratio for $\sin\theta$ to find the angle θ in each triangle.

a

b

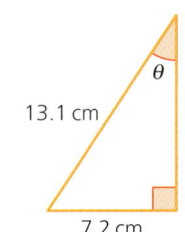

5 Use the ratio for cos θ to find the angle θ in each triangle.

a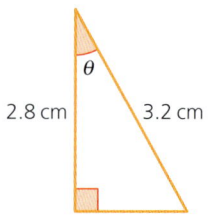

b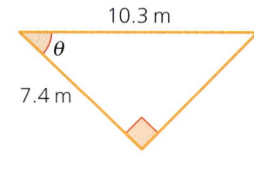

6 Use the ratio for tan θ to find the angle θ in each triangle.

a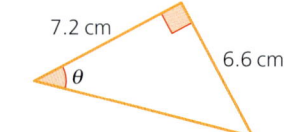

b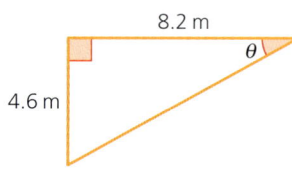

Band 3 questions

7 Find the angle θ in each of these triangles.

a 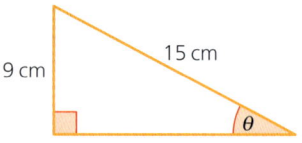   b 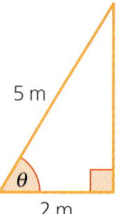   c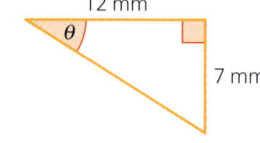

8 Find the angle θ in each of these triangles.

a    b 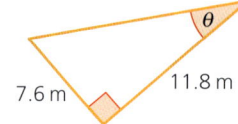   c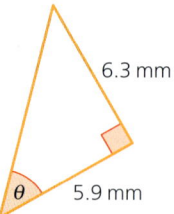

9 In triangle PQR

PQ = 8 mm, PR = 1.2 cm, ∠PQR = 90° and ∠QPR = $x°$.

Work out the value of x.

10 The area of triangle ABC is 24 cm².

Calculate the angle ABC.

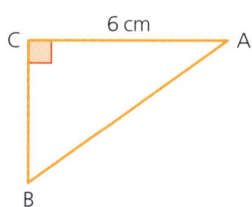

11 Efa writes down some statements about a right-angled triangle.

Decide whether each of Efa's statements is 'always true', 'sometimes true' or 'never true'.

Give a reason for each of your answers.

A cos θ is less than 1

B sin θ could be greater than 1

C tan θ could be greater than 1

D tan θ could equal 1

12.4 Solving problems

Skill checker

1. Write down Pythagoras' theorem.
2. LMN is a right-angled triangle.

   ∠LNM = 90°, ∠NLM = 40° and LN = 7.2 cm

   a. Sketch triangle LMN.
   b. Calculate the length of LM.
   c. Use Pythagoras' theorem to calculate the length of MN.
3. PQR is a right-angled triangle.

   ∠PQR = 90°, PR = 4.5 cm, and PQ = 3.6 cm

   a. Sketch triangle PQR.
   b. Use Pythagoras' theorem to calculate the length of QR.
   c. Calculate the angle RPQ.

▶ Using trigonometry to solve problems

Use these tips to help you use trigonometry and Pythagoras' theorem to solve problems.

- Draw a diagram.
- Look for 'hidden' right-angled triangles.

Note

▶ Make a plan. Sometimes you have to find a different length or angle *before* you can find the one you want.

▶ Remember to use Pythagoras' theorem to find the third side of a right-angled triangle when you know the other two sides.

Angles of elevation and depression

When you look up at an object, the *angle of elevation* is the angle between the horizontal and your line of sight.

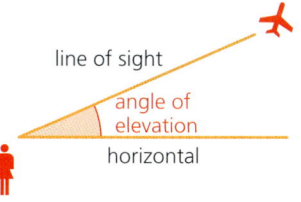

When you look down at an object, the *angle of depression* is the angle between the horizontal and your line of sight.

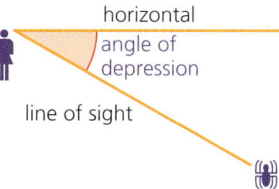

Worked example

A surveyor wants to work out the height of a tree.

She stands 30 m away from the base of the tree on level horizontal ground.

She measures the angle of elevation from her measuring instrument to the top of the tree.

The angle of elevation is 25°.

Her instrument is 1.6 m above the ground.

Calculate the height of the tree.

Solution

Draw a diagram to help you.

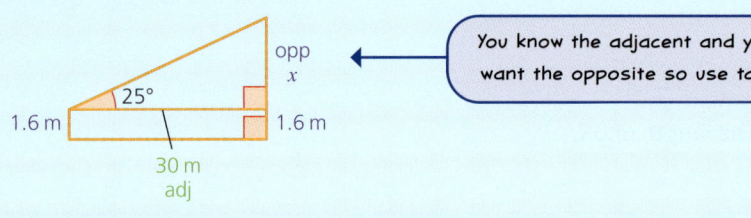

You know the adjacent and you want the opposite so use tan.

$$\tan 25° = \frac{x}{30}$$

$$x = 30 \tan 25°$$

$$= 13.98...$$

$$= 14.0 \text{ (to 3 s.f.)}$$

So the height of the tree is 14.0 m + 1.6 m = 15.6 m (to 3 s.f.).

12.4 Now try these

Unless the answers are exact, round any lengths to 3 significant figures and any angles to 1 decimal place.

Band 1 questions

1. A bird is flying above a rabbit in a field.

   The bird looks down at the rabbit.

   The diagram shows the position of the bird at B and the rabbit at R.

   a   The angle of elevation is angle ☐

   b   The angle of depression is angle ☐

   c   The angle of elevation is the same as angle ☐

   d   RB is called the line of ☐

2 Calculate the value of x and of y.

a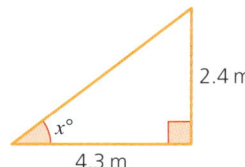

b

3 Calculate the lengths of the sides marked with a letter.

a 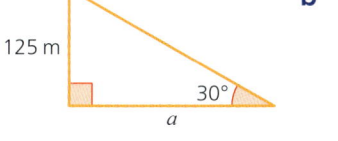 b 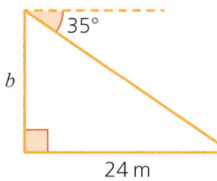 c 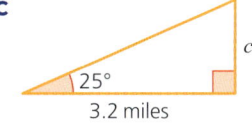 d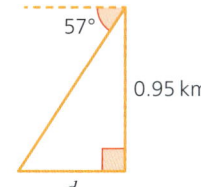

4 Siôn is a window cleaner. His ladder is 6 m long.

The angle between the ladder and the ground is 70°.

Find the height that his ladder reaches up the wall.

5 A rectangular field is 160 m long and 120 m wide.

A straight path along one diagonal cuts across it.

a Find the angle the path makes with the longer side.

b Work out the length of the path.

Band 2 questions

6 A girl flies a kite at the end of a 25 m long string.

The angle of elevation from the girl's hand to the kite is 75°.

Calculate the height of the kite above the girl's hand.

7 Gwyn is in a small boat.

It is 1500 m away from the bottom of a cliff.

The height of the cliff is 230 m.

Find the angle of elevation of the top of the cliff from the boat.

8 A mouse is 5 m away from a 6 m high wall.

The mouse looks up at the wall and sees a cat sitting on the wall.

a Calculate the angle of elevation from the mouse to the cat.

b Write down the angle of depression from the cat to the mouse.

Band 3 questions

9 A ship is 8 km from port on a bearing of 036°.

How far north and how far east is the ship from the port?

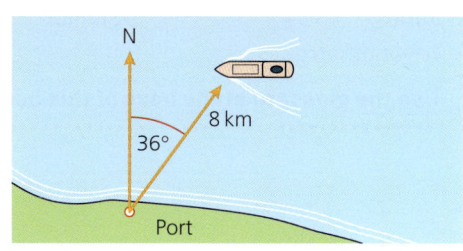

10 An owl sits on a branch at O, part way up a tall tree.

There is a mouse on the ground at point M, 35 m away from the foot of the tree.

The angle of depression from the owl to the mouse is 25°.

The angle of elevation from the mouse to the top of the tree is 40°.

Work out the distance between the owl and the top of the tree.

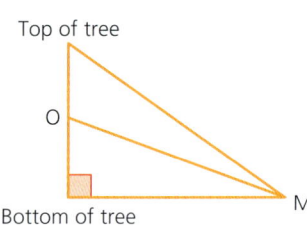

11 Cerys cycles from Afanbach to Brynpandy along a straight road.

Brynpandy is 4 km north and 3 km east of Afanbach.

   a Find the bearing of her journey.

   b How far does she cycle?

   c What is the return bearing?

12 Look at this diagram.

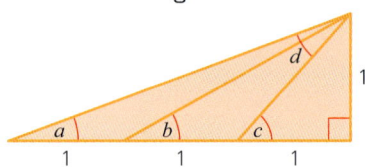

   a Use trigonometry to find the value of $a + b + c$.

   b Use your answer to part **a** to prove that $b = d$.

13 The diagram shows a form of spiral.

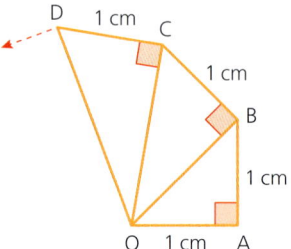

It continues with more triangles.

   a Work out the lengths of

      i OB   **ii** OC   **iii** OD.

Use square root signs in your answers when you need to.

   b How many triangles are needed to complete

      i 180° of the spiral   **ii** 360° of the spiral?

Key words

Here is a list of the key words you met in this chapter.

Adjacent	Angle of depression	Angle of elevation	Congruent
Corresponding	Cosine (cos)	Hypotenuse	Pythagoras' theorem
Ratio	Similar	Sine (sin)	Tangent (tan)
Trigonometry			

Use the glossary at the back of this book to check any you are unsure about.

12 Trigonometry

Review exercise: trigonometry

Unless the answers are exact, round any lengths to 3 significant figures and any angles to 1 decimal place.

Band 1 questions

1 Which **two** of the following are correct?

A opposite × sin θ = hypotenuse

B cos θ × hypotenuse = adjacent

C tan θ = $\dfrac{\text{adjacent}}{\text{opposite}}$

D hypotenuse = $\dfrac{\text{opposite}}{\sin \theta}$

E opposite × tan θ = adjacent

2 a Construct a triangle with sides of length 6 cm, 12 cm and 7.5 cm.

b Construct a triangle with sides of length 4 cm, 8 cm and 5 cm.

c Measure the angles in each triangle.

What do you notice? Why is this?

3 Group these triangles into sets of similar triangles.

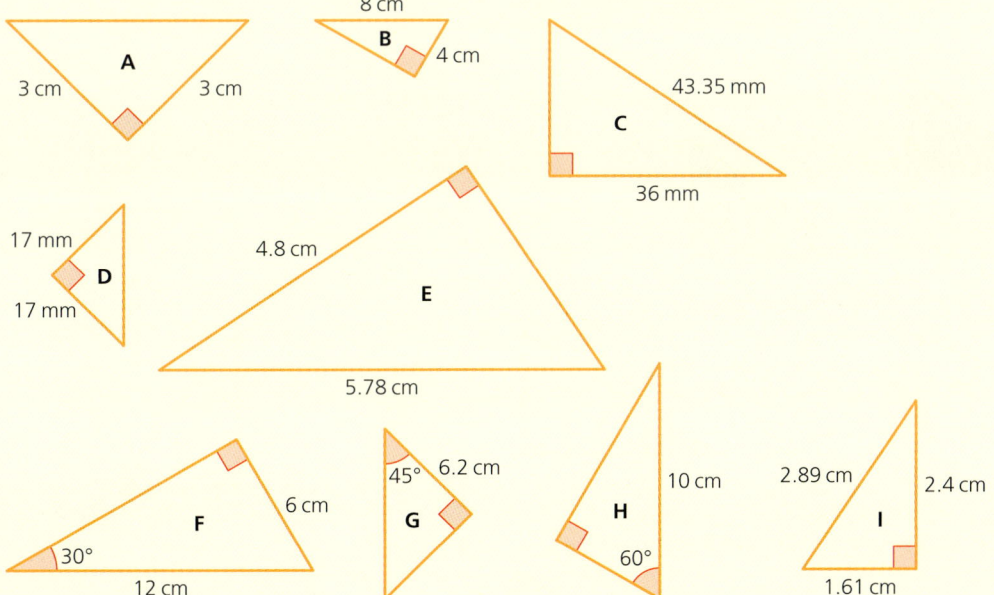

4 Look at the triangle LMN.

Angle LNM is a right angle.

a Find the length of the side LN without using trigonometry.

b Write sin θ, cos θ and tan θ as fractions.

c Now use your calculator to find θ three times, using the three trigonometric ratios.

Check that you get the same answer each time.

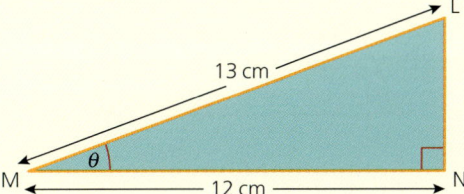

Band 2 questions

5 Triangle ABC is similar to triangle PQR.

Angle ACB = Angle QPR

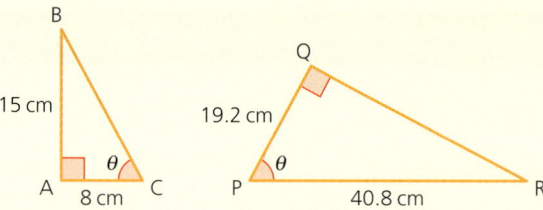

Find the length of

a QR

b BC.

6 The triangles in each of these pairs are similar.

Find the sides marked x.

a

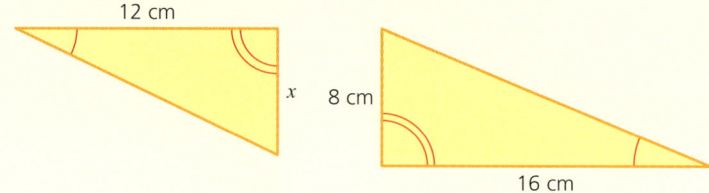

b

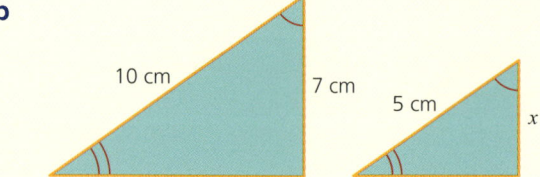

7 For each of these triangles, find the side labelled x.

a 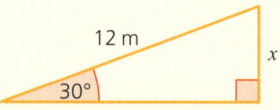   b 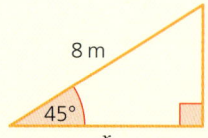   c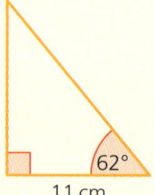

8 Find the angle labelled with a letter in each of these triangles.

a 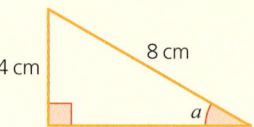   b 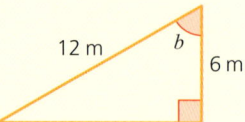   c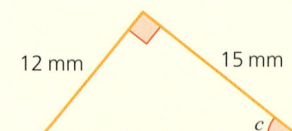

d 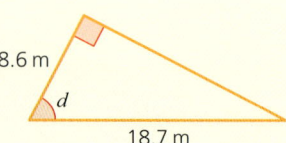   e 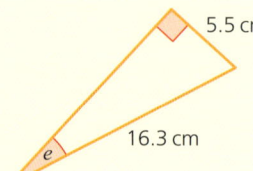   f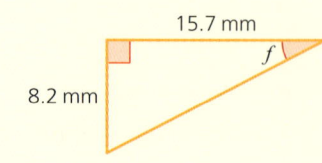

12 Trigonometry

9 Triangle ABC is isosceles.

Angle BAC = 130°

BC = 12 cm

Work out

a the height of triangle ABC

b the area of triangle ABC.

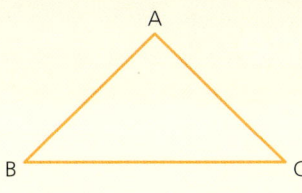

10 A 3 metre long ladder leans against a vertical wall.

The ladder is on level horizontal ground and the foot of the ladder is 90 cm away from the wall.

a Calculate the angle between the top of the ladder and the wall.

A window on the wall is 2.8 metres above the ground.

b i Does the top of the ladder reach the window?

Show all your working clearly.

ii Show how you can use a different method to answer this question.

Band 3 questions

11 Say whether each of these statements is true or false.

a The green triangle on the right shows you that $\sin 60° $ can be $\frac{5}{5}=1$.

b $\sin 90° = 1$

c $\tan 0° = 0$

d Whatever the value of θ, $\sin \theta$ cannot be greater than 1.

e Different triangles give different values for $\tan \theta$.

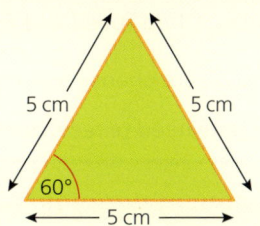

12 A ship is 1.2 km north-east of a lighthouse when the captain radios for help.

A lifeboat is 3.7 km south-east of the lighthouse.

a How far is the lifeboat away from the ship?

b What is the bearing of the ship from the lifeboat?

Give your answer correct to the nearest degree.

13 The diagram shows a trapezium.

The trapezium has one line of symmetry.

Calculate the area of the trapezium.

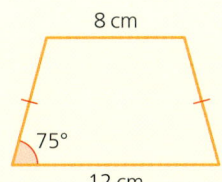

14 PB is a vertical flagpole.

It is tethered by ropes AP and CP.

A, B and C are on level horizontal ground.

AP = 6.3 m, CP = 4.8 m and ∠PAB = 40°.

Calculate

a the height of the flagpole

b the angle BPC.

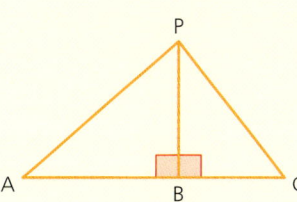

15 The diagram shows the quadrilateral ABCD.

∠BCA = 47°, ∠CAD = 25°, ∠ABC = ∠ACD = 90° and BC = 5.2 m.

Calculate the perimeter of ABCD.

Give your answer correct to the nearest centimetre.

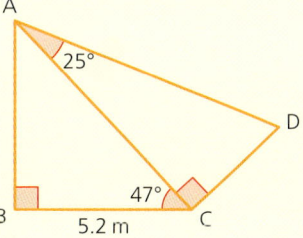

13 Real-life finance

Coming up...
- Income and expenditure
- Budgets
- Foreign currency exchange
- Exchange rates
- Taxation
- VAT

Pocket money puzzle

Kate's parents decide to give her pocket money each week.

They give her the following options:

① a one-off payment of £10 each week

② £2 on Monday, £2.50 on Tuesday, £3 on Wednesday, and so on, increasing by 50p per day (Monday to Friday)

③ 50p on Monday, £1 on Tuesday, £2 on Wednesday, and so on, doubling each day (Monday to Friday).

Which option should Kate choose and why?

How much would Kate need to receive on Monday to receive £10 in total with Option 2?

13.1 Income, expenditure and budgets

Skill checker

Complete each of the calculations below.

① £4.39 + 78p = £_____

② £8.14 − _____ p = £7.89

③ 5 × 29p = £_____

④ 6 × _____ p = £2.88

⑤ £800 increased by 26% is £_____.

⑥ £300 decreased by 14% is £_____.

⑦ $\frac{4}{5}$ of £65 is £_____.

Conceptual understanding

Income is money that a person receives. For adults, most income is employment income – this is the money that they are paid for working. For children, most income comes from pocket money or gifts on special occasions, such as Christmas and birthdays.

Expenditure is the money that a person spends.

It is sensible to have a **budget** to help keep your finances in check. Budgeting is the process of managing your money, and it looks at the balance between your income and expenditure. Creating a budget helps to ensure that you have enough money to cover the necessities and do the things which you would like to do.

Note

Money left over for saving or spending is found by calculating:

Income (money coming in) − Expenditure (money going out)

Profit = Income − Expenditure

Note

In each year, there are
- 12 months
- 52 weeks.

13 Real-life finance

Cross-curricular activity

Make a list of the income you have each month and what your planned expenditure is.

This will help ensure that you have money to spend on the things which you need.

Discussion activity

Does increasing a business's income always increase its profits? Discuss with a partner.

13.1 Now try these

Band 1 questions

1. Lewys is paid £800 each month.
   How much is his income each year?

2. Ceri is paid £100 each week.
   How much is her income each year?

3. Meic receives £10 pocket money each week.
   He receives £80 for his birthday and £120 for Christmas.
   How much is Meic's income over the year?

4. Bethan spends £2.50 each week on her favourite magazine.
   How much does she spend on magazines over the year?

5. Seren buys a new game each month for £5.99.
   What is Seren's expenditure on games each year?

6. Dewi receives his income from pocket money and his birthday.
   He receives £30 pocket money each month.
   His income for a year is £450.
   How much money did he receive for his birthday?

7. Mali is saving to buy a new computer.
   The computer costs £516.
   She plans to save money each month for one year.
   How much money will she have to save each month to buy the computer in one year's time?

8. Owain wants to buy a new pair of rugby boots.
   He is planning on saving £4 each week to pay for the boots.
   The rugby boots Owain would like to buy cost £50.
   How many weeks will it be until Owain can afford to buy them?

9. Ffion buys a bus pass every week of the year.
   She currently spends £3 each week.
   A bus company offers a bus pass for £12 a month.
   Should Ffion switch to paying £12 a month? Explain your reasoning.

Band 2 questions

Fluency

10 Steve gets £10 pocket money each week.
He likes to go to the cinema and buy a magazine each week, and he saves the remaining money.
A cinema ticket costs £4.99 and Steve's magazine costs £2.49.
How much money does Steve save each week?

11 A butcher has weekly costs of £253.
One week, his sales are £408.
What is the butcher's profit for that week?

12 Sally likes to spend her pocket money on magazines, milkshakes and going to the cinema.
She has £10 pocket money each week.
She puts £2 from this money into her savings account each week.
Her favourite magazine costs £1.99 and a cinema ticket costs £4.99.
How much money does Sally have left to spend on milkshakes?

Strategic competence

13 It costs Ben the Barber £200 each week to run his barber shop.
In an average week, he cuts 25 customers' hair.
It costs £9 to have your hair cut at Ben's barber shop.
What is Ben's profit each week?

14 Kate receives money from her family on birthdays and at Christmas.
She is given £80 on her birthday and £20 at Christmas.
Kate would like to subscribe to a television streaming service which costs £8.50 a month.
The streaming service has a minimum subscription of 12 months.
Can Kate afford the subscription using just her birthday and Christmas money?

Logical reasoning

15 Pedr is a personal trainer.
Next month, he has 22 customers booked who will each pay him £15.
Out of this money, Pedr has to pay for the use of the gym for his customers.
He has two options.
Pedr can either pay a £6 fee to the gym for each of his customers, or he can pay a flat monthly rate of £105.
Which of the two options should Pedr choose? Explain your reasoning.

16 Beca's weekly income is £400.
She saves $\frac{2}{5}$ of this money towards buying a car.
Beca likes a car which costs £3000.
How many weeks will it be before Beca can afford the car?

Strategic competence

17 The pie chart below shows what Elfyn does with his money each month.
Elfyn spends £40 each month on petrol.
How much money does Elfyn save each month and what is his total income?

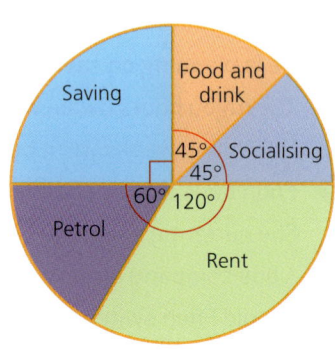

Band 3 questions

18 Lowri gets £10 pocket money each week from her parents.

She gets £5 pocket money each fortnight from her grandparents.

Lowri spends £8 per week going to the gym and £4 per week going to the cinema, then saves the rest.

How much would Lowri have saved after 10 weeks?

19 When Oliver was aged 7, he was given £30 from his uncle, £20 from his aunty and £25 from his grandparents for Christmas.

Oliver's 8th birthday is on 31st December.

On his 8th birthday, he is given £20 from his uncle, £20 from his aunty and £30 from his grandparents.

Oliver would like to save £80 and spend the rest on football cards, which cost £4 a pack each month.

If Oliver does not receive any more money during the next year, can he afford to save £80 and still buy a packet of football cards each month?

20 The running costs of Marc's shop are shown.

Marc's sales for the year were £30 000.

Did his shop make a profit?

Rent	£800 per month
Gas	£40 each week
Electric	£40 each week
Supplies	£200 per month
Wages	£300 per week

21 Rhian owns a small business producing candles.

The costs of her business are shown.

Rhian sells the candles she makes for £25 each.

What is the smallest number of candles that Rhian needs to sell each week to make a weekly profit of at least £100?

Rent	£600 per fortnight
Wages	£500 per week
Wax and other supplies	£10 per candle

22 Mr and Mrs Jones have two children.

Their combined income is £3000 per month.

Their combined expenditures are shown.

Mr Jones wants to save the remaining money in a high-interest savings account.

They can only pay money in multiples of £50 into the savings account.

How much money can they afford to save each week?

Mortgage	£800 per month
Gas, electric and water	a total of £20 per week
Food	£150 per fortnight
Other expenses	£100 per week

23 Bleddyn draws a graph to show how he spends his income each week.

Bleddyn thinks it is sensible to save 15% of his income.

Does Bleddyn achieve this target? If not, how much extra does he need to save each week?

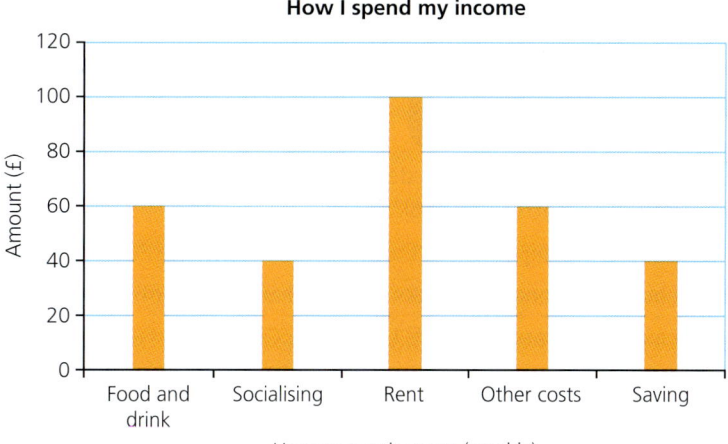

13.2 Foreign currency exchange

Skill checker

Complete each of the calculations below.

① $7 \times 1.2 = \ldots\ldots\ldots$

② $\ldots\ldots\ldots \times 1.4 = 12.6$

③ $15 \div 2.5 = \ldots\ldots\ldots$

④ $22 \div \ldots\ldots\ldots = 20$

Discussion activity
Why is the British pound also known as 'sterling'?

Discussion activity
Which countries use the Euro (€)?

Cross-curricular activity
Why do we need currency?

In your History lessons, research the origin of currency.

Conceptual understanding

When you travel abroad, you need to exchange your money to the currency that is used in the country you visit.

Some popular currencies are the pound (which we use in the United Kingdom), the dollar (which is used in the United States of America) and the Euro (which is used in many of the countries in Europe).

One pound (£) does not equal one dollar ($) or one Euro (€).

To calculate the amount that it costs to convert between one currency and another, exchange rates are used.

Various factors affect exchange rates. This means that the amount of foreign currency which you are able to buy today could be more or less than the foreign currency which you could buy tomorrow.

We can change our money into foreign currencies at a bureau de change. These are found in banks, travel agents, some large supermarkets and online. A bureau de change typically only buys and sells notes of foreign currencies – not coins.

Worked example

Given that £1 = €1.20, find the value of

a £5 in €

b £15 in €

c €3.60 in £

d €26.40 in £

Solution

a £5 × 1.20 = €6.00

b £15 × 1.20 = €18.00

c €3.60 ÷ 1.20 = £3.00

d €26.40 ÷ 1.20 = £22.00

Worked example

Mari visits her local bureau de change to buy some Euros for her trip to Portugal.

She wants to buy as many Euros as possible with £200.

The bureau de change stocks €5, €10 and €20 notes.

The exchange rate is £1 = €1.14.

a How many Euros can Mari buy?

b How much will it cost Mari to buy her Euros?

Solution

a £200 × 1.14 = €228

As the bureau de change stocks only €5, €10 and €20 notes, she can only buy €225.

b €225 in pounds is calculated:

€225 ÷ 1.14 = £197.37 (correct to the nearest penny).

It will cost Mari £197.37 to buy €225.

13   Real-life finance

Discussion activity
Why do banks and bureaux de change offer different rates for buying and selling foreign currency?

Cross-curricular activity
In your Humanities lessons, find out the currency of the countries that you study.

What coins and notes are available?

Are there any images of famous people, buildings or landmarks on these coins and notes? What can you find out about any of these people, buildings or landmarks?

Why do you think they appear on the coins or notes?

13.2 Now try these

Band 1 questions

1. Given that £1 = 1.8 Australian dollars (AUD), find the value of the following in AUD:
   a £4
   b £8
   c £200

2. Given that £1 = 1.7 Canadian dollars (CAD), find the value of the following in CAD:
   a £5
   b £12
   c £300

3. Given that £1 = 1.8 Australian dollars (AUD), find the value of the following in £:
   a 3.60 AUD
   b 18 AUD
   c 270 AUD

4. Given that £1 = 1.7 Canadian dollars (CAD), find the value of the following in £:
   a 6.80 CAD
   b 34 CAD
   c 595 CAD

5. In Paris, a bag costs €100.
   In Cardiff, the same bag costs £70.
   The current exchange rate is £1 = €1.25.
   Where is it cheaper to buy the bag?

6. A pair of jeans costs £80 in Swansea.
   The same pair of jeans costs €110 in Rome.
   The current exchange rate is £1 = €1.25.
   In which city is the pair of jeans cheaper and by how much?

Band 2 questions

7. An identical car is on sale for
   - $15 000 in USA
   - €18 000 in Spain
   - ₩24 000 000 in South Korea
   - £17 000 in UK.

   The exchange rates are shown in the box.

   £1 = £1.25
   £1 = $1.2
   £1 = ₩1600

   In which country is this car on sale for the lowest price?

Strategic competence

8. Ceri needs to exchange some money into Won (₩) for her trip to South Korea.
   Her bank only stocks ₩10 000 notes.
   Ceri wants to buy as many Won (₩) as possible with the £145 she has saved.
   The exchange rate is £1 = ₩1600.
   How much Won (₩) can Ceri buy?

9. Dafydd wants to exchange £600 into Euros.
   His village Post Office only has €50 notes.
   The exchange rate at the Post Office is £1 = €1.10.
   How many Euros can Dafydd buy and how much will it cost him to buy them?

10. Seren, who currently lives in Wales, is applying for a new job, which could be based in Berlin or Washington.
    The job in Berlin pays €45 600 per annum.
    The job in Washington pays $50 050 per annum.
    The current exchange rates are £1 = €1.20 and £1 = $1.30
    Which job pays the better annual salary?

11. Dylan wants to change £200 into Euros (€).
    On Monday, the exchange rate was £1 = €1.20.
    Dylan changes his money on Friday, when the exchange rate is £1 = €1.15.
    How many fewer Euros did Dylan get by getting them on Friday instead of Monday?

Logical reasoning

12. In 2017, Kacper bought £2000 worth of Polish złoty (PLN).
    In 2020, he changed this money back into pounds.
    The exchange rate in each of these years is shown in the table.
    Did Kacper gain or lose money?

Year	Pound (£)	Polish złoty (PLN)
2017	1	4.9
2020	1	5.0

Band 3 questions

13. Mali lives in Australia and is coming to Wales to visit her family.
    The currency in Australia is the Australian dollar (AUD).
    The conversion rate at her local bureau de change is 1 AUD = £0.54.
    The bureau de change only has £20 and £50 notes.
    Mali has 800 AUD saved.
    She wants to get as many pounds as possible.
    To save space in her purse, she asks for the fewest possible number of notes.
    How many £20 and £50 notes will Mali get when she buys her money?
    How much, in Australian dollars, will it cost her to buy them?

Strategic competence

14. £1 buys
    - 8.6 Chinese Yuan
    - 100.8 Indian Rupees
    - 1.2 Swiss Francs.

    How many Chinese Yuan is 100 Swiss Francs worth?

15 Tomos lives in Newport.

His friend Elin lives in Copenhagen.

Tomos and Elin are going on a trip to New York.

The best exchange rate Tomos can get is £1 = $1.20.

The best exchange rate Elin can get is 1 DKK = $0.20.

Using these exchange rates, what is £1 worth in Danish Krone (DKK)?

16 Cerys is going on holiday to Turkey.

Her local bureau de change offers the following rates:

Buy your currency from us:	£1 for 18 Turkish Lira
Sell your currency to us:	£1 for 16 Turkish Lira

The bureau de change will only buy and sell 5, 10 and 20 Lira notes.

Cerys buys as many Lira as possible with £300.

She spends 4000 Lira on holiday.

When she gets back, she sells her remaining 5, 10 and 20 Lira notes to the bureau de change.

How many Lira is she left with after selling these notes?

13.3 Taxation on goods and services

Skill checker

① Calculate the value of
   a   10% of 800
   b   10% of 240
   c   10% of 36.

② Calculate the value of
   a   30% of 400
   b   70% of 120
   c   5% of 280.

③ Increase 42 by 20%.

④ Decrease 64 by 5%.

Taxation transfers wealth from households or businesses to the government. This money is then spent on various things, such as welfare, health, state pensions, defence and education.

There are various types of tax, such as income tax, wealth tax and capital gains tax. In this chapter, we are going to look at Value Added Tax (VAT), which is a tax paid on goods and services.

The rate of VAT is set by the government and varies for different goods and services.

There is
- a standard rate, which applies to most goods and services
- a reduced rate, which applies to items such as children's car seats and home energy
- a zero rate, which applies to most food and children's clothes.

Worked example

The VAT on buying a television in the United Kingdom is 20%.

The price of a television is £800 excluding VAT.

How much will the television cost including VAT?

Solution

20% of £800 is $\frac{20}{100} \times £800 = £160$

Cost including VAT is £800 + £160 = £960

OR

Using a multiplier, cost including VAT is 1.2 × £800 = £960

Discussion activity

Why are some items exempt from VAT?

Cross-curricular activity

It is environmentally friendly to try to reduce the amount of electricity and gas we all use.

In your Science lessons, investigate how much electricity is used by some of your different gadgets or appliances at home.

In your Geography lessons, find out how much electricity and gas are used in the countries you study. Is the consumption of gas or electricity in these countries increasing each year?

Worked example

The VAT on domestic energy in the United Kingdom is 5%.

David pays £210 for his gas and electricity each month. This price includes VAT.

What was the cost of the gas and electricity before VAT was included?

Solution

Think of the cost excluding VAT as 100%.

So, the cost including VAT is 105%.

Using knowledge of reverse percentages, cost excluding VAT is £210 ÷ 1.05 = £200

Note

'Excluding VAT' means 'before VAT was added'.

Cross-curricular activity

In your Science lessons, find out about the units used to measure electricity and gas.

13.3 Now try these

Band 1 questions

1. Carwyn is buying a new car.

   New cars are taxed at the standard rate of 20%.

   His car costs £3000 excluding VAT.

   How much does his car cost including VAT?

13 Real-life finance

2 Bethan is buying a new bike. It costs £550 excluding VAT.
How much does her bike cost including VAT at the standard rate of 20%?

3 Olivia is buying a new child's car seat.
The car seat which she likes costs £40 excluding VAT.
How much does the car seat cost including VAT at 5%?

4 Delyth's car costs £9600, including VAT at the standard rate of 20%.
How much does her car cost excluding VAT?

5 On 1 August 2020, Mr Jones sold £945 of chips at his chip shop.
At this time, hot takeaway food was taxed at a temporary rate of 5%.
How much were his sales of chips excluding VAT?

6 Wyn is buying a new mobile phone.
He pays £150 of VAT at the standard rate of 20%.
How much did his mobile phone cost?

7 Tegan looks at her receipt after buying an item online.
She paid £336 for the item, including £16 tax.
Did she pay standard rate, reduced rate or zero rate VAT on this item?
Explain your reasoning.

Band 2 questions

8 Mrs Evans looks at her electricity bill.
Find the total cost of the units of electricity Mrs Evans used in the period of time between the previous and present meter readings, including VAT.

Present meter reading	7432 units
Previous meter reading	6900 units
Charge per unit	24 pence per unit
VAT	5%

9 Mr Jones looks at the meter readings on his gas bill.
The total cost of his gas is £31.50, including VAT at 5%.
What is the charge per unit of the gas?

Present meter reading	452 units
Previous meter reading	392 units

10 Erin wants to buy 20 new computers for her business.
In shop A, the 20 computers will cost £4960, including VAT at 20%.
In shop B, the 20 computers will cost £4000, excluding VAT.
In shop C, the computers cost £195 each, excluding VAT.
Should Erin buy the computers at shop A, B or C? Explain your reasoning.

11. The pie chart shows the different tax rates that a business in Llanelli paid for its supplies in 2021.

    The business paid £1260 for supplies which have a zero rate of tax.

    How much VAT did it pay in total on its supplies in 2021?

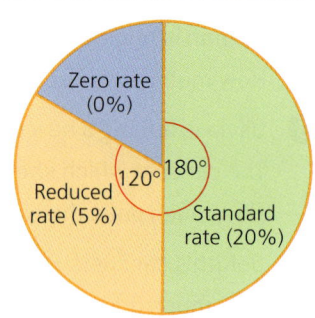

Band 3 questions

12. Prior to January 2011, the standard rate of VAT in the UK was 17.5%.

    If you know 10% of an amount, explain how you could work out 17.5% of the amount without a calculator by doing two divisions and an addition.

13. For a short period of time in 2020, the UK Government reduced the VAT for the hospitality sector from 20% to 5% to encourage spending.

    A meal at a restaurant cost £6 in June 2020, including VAT of 20%.

    How much would a customer save when buying the meal after the reduction in VAT?

14. A mirror is on sale at a shop in Rhyl for £60, including VAT of 20%.

    The shop reduces the cost of the mirror in its Black Friday sale.

    A customer buying the mirror will now pay £6 VAT.

    By what percentage did the shop reduce the cost of the mirror?

15. The standard rate of VAT in France is 20% and in Malta it is 18%.

    Excluding VAT, a table costs the same price in both countries.

    When VAT is calculated for the table, the difference between what is added to the price in France and in Malta is €3.50.

    How much does the table cost in France, including VAT?

16. Mrs Williams received a bill from her energy provider for her electricity use between January and March. The bill was for £120, including VAT, and was based on estimated meter readings taken by the energy company.

    Mrs Williams took two meter readings during this period:
    - 1st January   27 975
    - 31st March   28 275

    Her energy provider charges 25p per unit of electricity, along with a standing charge of £28 per month.

    VAT is charged at 5% on all charges on energy bills in the UK.

    Calculate the difference between what Mrs Williams was billed and what she should pay based on the readings she took.

Key words

Here is a list of the key words you met in this chapter.

Budget	Bureau de change	Exchange rates	Expenditure	Foreign currency exchange
Income	Tax	Taxation	Value added tax	

Use the glossary at the back of this book to check any you are unsure about.

Review exercise: real-life finance

Band 1 questions

1. Carys is paid £950 each month.
   How much is her income each year?

2. Raj is saving to buy a new television.
   The television costs £364.
   How much money will he have to save each week to buy the television in a year?

3. Charlotte is buying a new computer.
   Her computer costs £350 excluding VAT.
   How much does her computer cost including VAT at the standard rate of 20%?

4. In Lisbon, a t-shirt costs €80.
   In Swansea, the same t-shirt costs £60.
   The current exchange rate is £1 = €1.25.
   Where is it cheaper to buy the t-shirt?

5. Ross is buying a new electric bike.
   He pays £230 VAT at the standard rate of 20%.
   How much did his bike cost, including VAT?

6. Ava wants to buy a new pair of trainers that cost £70.
   She is able to save £6 each week to pay for the trainers.
   In how many weeks will Ava be able to afford to buy the trainers?

7. A coat costs
   - £70 in Wrexham
   - €90 in Napoli.

   The current exchange rate is £1 = €1.25.
   In which city is the coat cheaper and by how much?

Band 2 questions

8. A bakery has weekly costs of £343.
   On average, its weekly sales are £587.
   What is the bakery's profit each year?

9. During an average week, Bill the Barber cuts the hair of 35 customers.
   Bill's weekly costs are £250, and he charges £10.50 to cut each customer's hair.
   What is Bill's profit each week?

10. Mr Jones's electricity bill for September to December is shown below.

Present meter reading	2689 units
Previous meter reading	1831 units
Charge per unit	29 pence per unit
VAT	5%

    Find the total cost of the units of electricity Mr Jones used between September and December, including VAT.

Strategic competence

11. Mari wants to exchange £950 to buy Euros.

    Her nearest bureau de change has only €20 and €50 notes.

    The exchange rate at the bureau de change is £1 = €1.13.

    How many Euros can Mari buy and how much will it cost her to buy them?

Logical reasoning

12. Angharad's monthly income is £1500.

    She saves $\frac{3}{5}$ of this money to buy a new car, which will cost £5000.

    How many months will it be before Angharad can afford to buy the car?

13. In 2015, Artyom bought £1500 worth of Russian Roubles.

    In 2017, he changed his money back into pounds.

    The exchange rates in each of these years are shown in the table.

    Did Artyom gain or lose money?

Year	Pound (£)	Russian Roubles
2015	1	109
2017	1	78

14. Jenny wants to buy 50 new laptops for her business.

    In shop A, the 50 laptops will cost £22 500, including VAT at 20%.

    In shop B, the laptops cost £400 each, excluding VAT.

    In shop C, the 50 laptops will cost £18 500, excluding VAT.

    Should Jenny buy the laptops at shop A, B or C? Explain your reasoning.

Band 3 questions

Fluency

15. Lloyd receives £10 pocket money each week.

    He earns £15 per fortnight by delivering newspapers.

    Each week, Lloyd spends £5 per week going to the gym and £6 going to watch his local rugby side. He saves the rest.

    How much will Lloyd have saved in 20 weeks' time?

Strategic competence

16. Tegan lives in Welshpool. Her friend Ifan lives in Warsaw.

    Tegan and Ifan are planning a trip to the USA.

    They each visit their local bureau de change to look at exchange rates between their own currency and US dollars.

    £1 = $1.35   1 PLN = $0.25

    Using these exchange rates, what is £1 worth in Polish złoty (PLN)?

17. A shop in Narbeth sells a television for £456, including VAT of 20%.

    The shop reduces the cost of the television in a sale.

    A customer buying the television will now pay £57 VAT.

    By what fraction did the shop reduce the cost of the television?

Logical reasoning

18. Rhodri owns a small business selling birthday cakes.

    The costs of his business are:
    - **Rent:** £1000 per fortnight
    - **Wages:** £900 per week
    - **Ingredients:** £15 per cake

    Rhodri sells the cakes he makes for £100 each.

    What is the smallest number of cakes that Rhodri needs to sell each week to make a weekly profit of at least £500?

Consolidation 5: Chapters 11–13

Unless the answers are exact, round any lengths to 3 significant figures and any angles to 1 decimal place.

Band 1 questions

1 Calculate the surface area of each of these cuboids.

a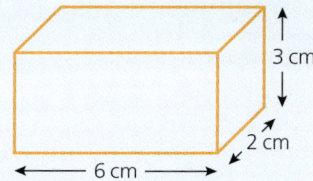

b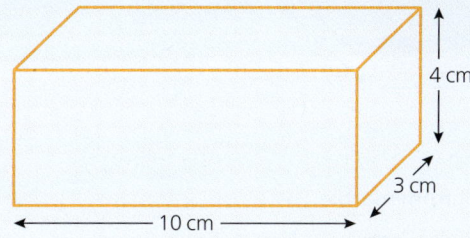

c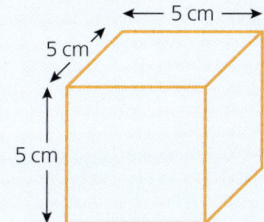

d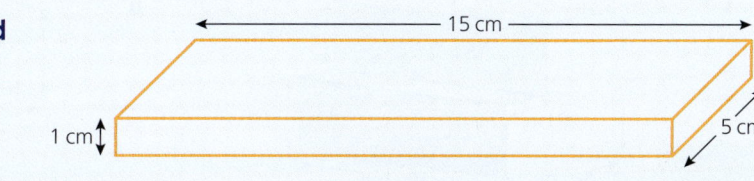

2 Would you expect these two boxes to weigh the same?

Think about the density of the contents as well as the volumes. Explain your answer fully.

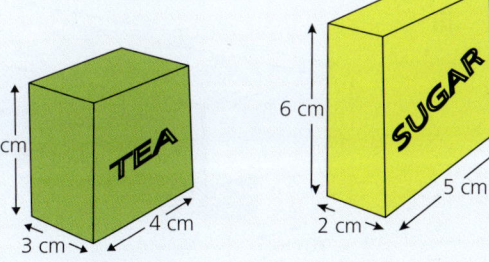

3 This is the swimming pool at Afonffordd Leisure Centre.

  a What shape is the swimming pool?
  b What is the volume of the swimming pool?
  c How many litres of water does the swimming pool hold?
  d The bottom and sides of the inside of the pool are lined with tiles. What area is tiled?

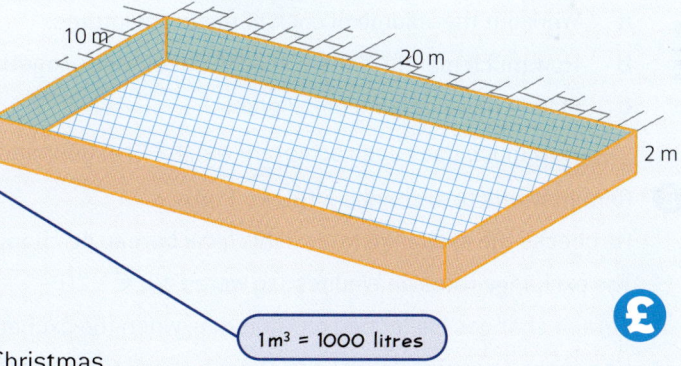

1 m³ = 1000 litres

4 Laura receives £20 pocket money each fortnight.

She also receives £90 for her birthday and £150 for Christmas.

How much is Laura's income over the year?

Band 2 questions

5 Calculate the volume and surface area of this prism.

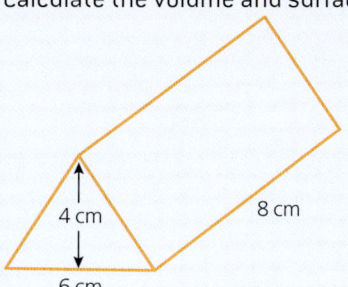

291

6. Find the volume and surface area of each of these solids.

   Give your answers in suitable units.

   a The lengths are in centimetres
   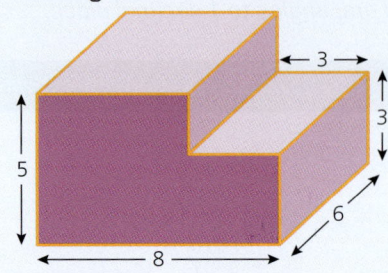

   b The lengths are in millimetres.
   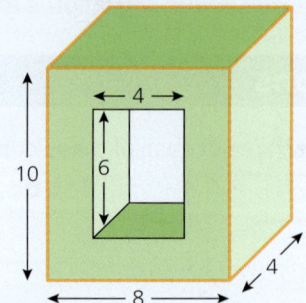

7. Ella is an artist.

   She makes sculptures from marble.

   i 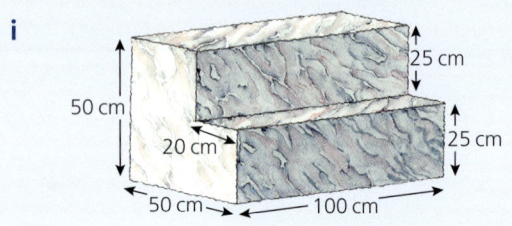   ii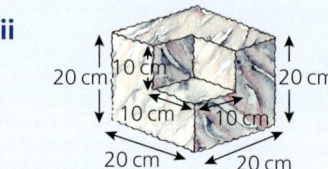

   iii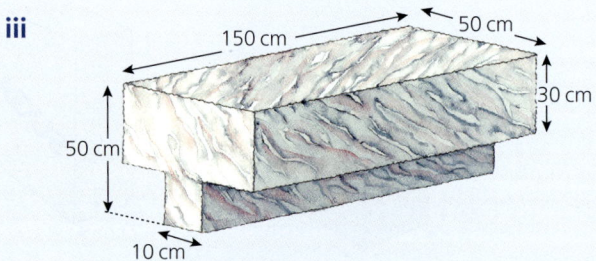

   a Work out the volume of each of these sculptures.
   b How much marble is there in all the sculptures together?
   c Marble has a density of 2.7 g/cm^3.

      How much does each sculpture weigh? Give your answers in kilograms.

8. Geraint wants to exchange £700 for Euros.

   He checks the exchange rates in his local bureau de change.

   The exchange rates on Wednesday were £1 = €1.20.

   Geraint changes his money on Thursday, when the exchange rates were £1 = €1.15.

   How many fewer Euros did Geraint get by getting them on Thursday instead of Wednesday?

9. Abby gets £50 pocket money every 4 weeks.

   Each week, she spends her pocket money on
   - a fruit smoothie, costing £1.99
   - a trip to the gym, costing £5
   - a trip to the cinema, costing £4.75.

   She puts the remaining money into her savings account.

   How much money does Abby save every 4 weeks?

Consolidation 5

10 Mrs Thomas looks at the meter readings on her gas bill.

Present meter reading:	1322 units
Previous meter reading:	1252 units

The total cost of her gas is £30.87, including VAT at 5%.

What is the charge per unit of the gas, including VAT?

11 This right-angled triangular prism has height h, width w and length l.

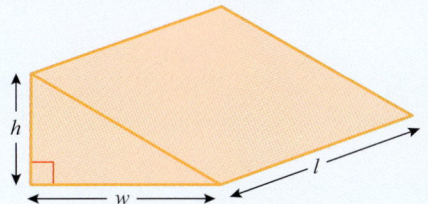

a Write an expression for the volume.

b Write an expression for the surface area.

12 Isobel has saved £5 pocket money each week for the last year.

She is given £100 at Christmas.

Isobel would like to join a gym that costs £30 a month.

The gym has a minimum subscription period of 12 months.

Can Isobel afford to start the subscription after Christmas using just her saved pocket money and the money she received at Christmas?

13 A business in Carmarthen is calculating how much VAT they pay each year.

The pie chart below shows the different tax rates that the business paid for its supplies in 2020.

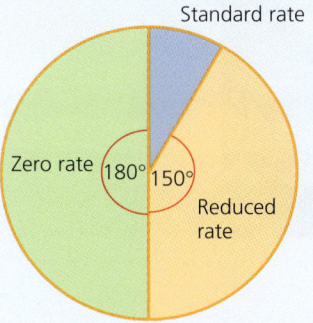

The business paid £2520 for supplies which have a zero rate of VAT.

The standard rate of VAT is 20%.

The reduced rate of VAT is 5%.

How much VAT did the business pay in total on its supplies in 2020?

Band 3 questions

14 Fatima measures rainfall for a Geography project.

The rain is collected from a cylindrical tray which has radius 25 cm.

It then flows into a measuring cylinder underneath, which has diameter 10 cm.

What is the depth of water in the measuring cylinder after 0.4 cm of rain has fallen into the tray?

15 The running costs of Rhiannon's shop are as follows:

Rent	£1000 per month
Gas	£50 each week
Electricity	£50 each week
Supplies	£250 per month
Wages	£400 per week

The sales for 2021 were £40 000.

Did her shop make a profit?

16 Tom is a film star. He sends out souvenir models of his Oscar.

The souvenirs are packed in boxes.

There are two sizes, large and small.

The boxes are similar.

On the front of each box is a photograph of Tom.

Photograph A is an enlargement of photograph B.

Photograph A — 16 cm × 24 cm
Photograph B — 8 cm × height

12 cm, 8 cm, 10 cm

a What is the ratio, width : height, for photograph A?

b What is the height of photograph B?

The small box is 10 cm by 8 cm, and 12 cm high.

The ratio of the heights of the two boxes is 2 : 1.

c What are the dimensions of the larger box?

d What is the surface area of

 i the larger box

 ii the smaller box?

e **i** What is the ratio of the surface areas (large : small) in its simplest form?

 ii What is the connection between this and the ratio of the heights?

f Calculate the volume of each box and write down the ratio of their volumes in its simplest form.

g Find a connection between this ratio and the ratio of the heights.

17 Look at this prism. The end face is an isosceles triangle.

Calculate

a the volume

b the surface area of the prism.

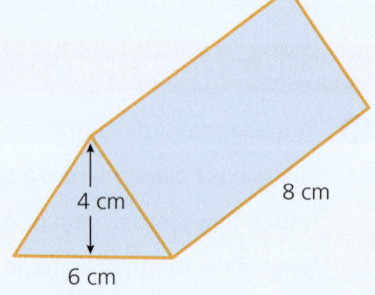

4 cm, 6 cm, 8 cm

18 The diagram shows the quadrilateral WXYZ.

∠WYZ = 50°, ∠XYW = 30°, ∠XWY = ∠WZY = 90° and WX = 12.6 cm.

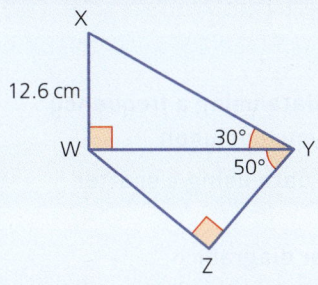

Calculate the area of WXYZ.

Give your answer correct to the nearest square centimetre.

19 This glass display case is of length 60 cm and external diameter 24 cm.

The glass is 1 cm thick.

The case stands on a wooden base 1 cm thick.

  a Find the volume of the wooden base.

  b Find the volume of glass required to make the case.

20 The diagram shows a regular pentagon inscribed inside a circle of radius 10 cm.

Work out the area of the shaded region.

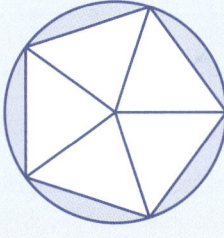

21 Ben is going on a skiing holiday to Switzerland.

His local Post Office offers the following rates on Swiss Franc (CHF).

| Buy your currency: | £1 for 1.3 Swiss Franc (CHF) |
| Sell your currency: | £1 for 1.1 Swiss Franc (CHF) |

The Post Office will only buy and sell 10, 20 and 50 Swiss Franc notes.

Ben buys as many Swiss Franc as possible with £565.

He spends 400 CHF on his holiday.

When he gets back, he sells his remaining notes to the Post Office.

How many pounds does he get back?

14 Working with data

Coming up...

- ▶ Hypothesis testing and questionnaires
- ▶ Calculating mean, median, mode and range from a frequency table
- ▶ Grouping data
- ▶ Estimating the mean from a grouped frequency table
- ▶ Identifying the modal group and the group containing the median from a grouped frequency table
- ▶ Displaying grouped data using a frequency diagram and a frequency polygon
- ▶ Displaying bivariate data using a scatter diagram
- ▶ Interpreting a scatter diagram

Roll the dice!

Here are some graphs. They show the scores for 20 rolls of a dice. Two of them are not correct.

A

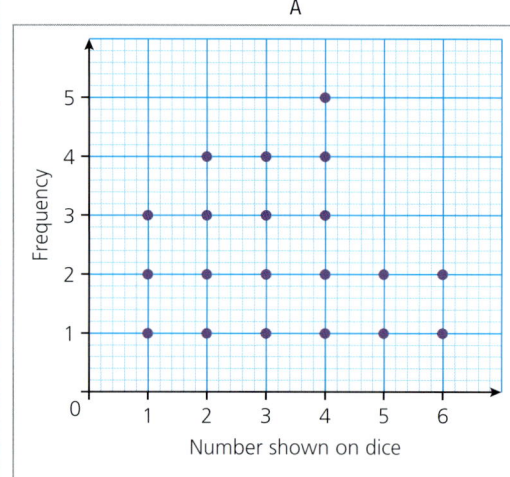

B

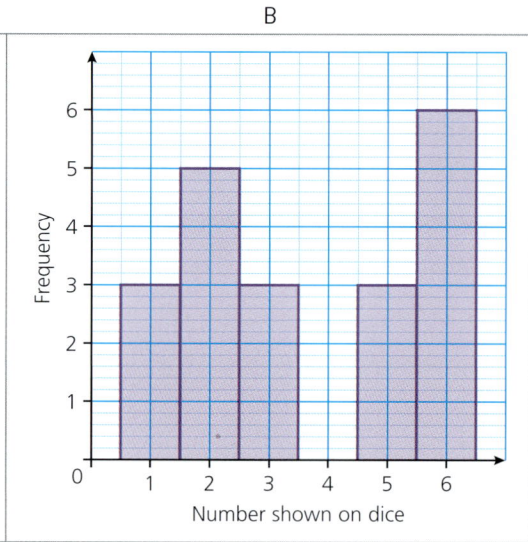

C

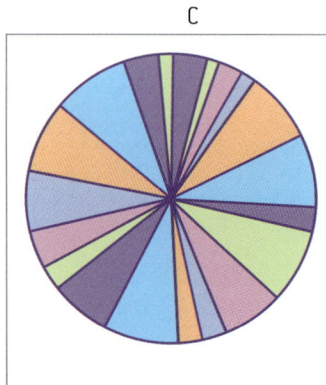

D

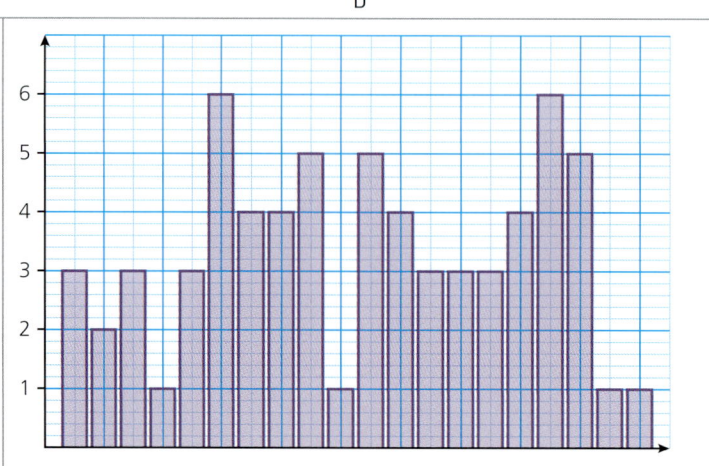

a Which of the graphs are not correct?
b Here are some sets of data. Match them with the graphs.

1	2	3	4
3, 1, 2, 1, 6, 6, 2, 6, 5, 2, 2, 6, 5, 2, 3, 5, 6, 6, 3, 1	3, 2, 3, 1, 3, 6, 4, 4, 5, 1, 5, 4, 3, 3, 3, 4, 6, 5, 1, 1	4, 3, 4, 6, 2, 3, 4, 4, 1, 2, 3, 3, 2, 5, 1, 4, 5, 1, 2, 6	3, 5, 3, 2, 3, 5, 2, 4, 6, 6, 5, 2, 6, 2, 2, 1, 1, 6, 5, 6

c Explain why the two graphs you identified in part **a** are incorrect.
d Explain why the two incorrect graphs are poor representations of the data.
e Explain why the two correct graphs are better representations of the data.

14.1 Hypothesis testing and questionnaires

Skill checker

① Three symbols commonly used in mathematics are $>$, $<$ and $=$.
Use the correct symbol to complete each of the statements below.
a 5 ___ 3
b 4 ___ 2 + 2
c 10 ___ 4 × 3
d 5 × 4 ___ 16 + 3

② Two other symbols used in mathematics are $\geqslant$ and $\leqslant$.
$\geqslant$ means 'greater than or equal to'.
$\leqslant$ means 'less than or equal to'.
For each of the statements below, state whether it is true or false.
If it is false, write the statement correctly using the $\geqslant$ and $\leqslant$ symbols.
a $12 \geqslant 13$
b $11.9 \leqslant 12$
c $15 \leqslant 15$
d $7 + 5 \geqslant 12$

③ In which group, either $3 < x \leqslant 5$ or $5 < x \leqslant 7$, does each of the following lie?

| $x = 4$ | $x = 6$ | $x = 5$ | $x = 6.5$ | $x = 4.5$ | $x = 5.2$ |

$3 < x \leqslant 5$	$5 < x \leqslant 7$

Note

$3 < x \leqslant 5$ is read as '3 is less than x and x is less than or equal to 5'.

For a value to be in this group it must meet both of these conditions.

▶ Hypothesis testing

A hypothesis is a prediction that is supported by an explanation. Hypothesis testing can be used to confirm whether a variable influences another variable.

An example of a hypothesis is that the more you practise your times tables, the more correct answers you will get in a multiplication test. The two variables are the time spent practising and the number of correct answers in the test.

▶ Questionnaires

Questionnaires are used to collect data on people's views. Surveys using questionnaires can be carried out in many ways, such as over the phone, face-to-face, by post and online.

Questionnaires must be easy to understand and unbiased. Being biased favours one view or aspect over another and leads to unreliable results in a questionnaire.

Response boxes are used in questionnaires to make them quicker and easier for people to fill in. Sometimes this leads to the grouping of data. However, you must be careful when designing response boxes, to ensure that they do not overlap and that they cover all possible outcomes.

Questionnaires can be used to collect both discrete and continuous data.

> Discrete data is data which has specific values, e.g. the number of pupils in a class.

> Continuous data is data which can have any value; it does not have any gaps, e.g. height, mass, time.

Worked example

Dewi wants to find out how much money people spend on magazines.

He uses this question:

▶ How much money do you spend on magazines each week?

☐ £1–£2

☐ £2–£3

☐ Over £3

a Write down two criticisms of this question.

b Dewi asks 10 of his classmates this question. Explain why his sample is biased.

Solution

a ▶ There is no option for those who spend less than £1 on magazines.
   ▶ The categories overlap (£2 appears in both £1–£2 and £2–£3).

b Dewi's classmates are all likely to spend a similar amount of money on magazines. He should ask a wider range of people.

14 Working with data

> **Worked example**
>
> Kate wants to find out whether or not people support the building of a new road.
> She uses this question:
> - The new road will cause a lot of problems for our town, don't you agree?
>   - ☐ Yes
>   - ☐ Possibly
>   - ☐ Maybe
>
> **a** Write down two criticisms of her question.
>
> **b** Design a better question and more suitable responses for Kate to use.
>
> **Solution**
>
> **a** ▸ The question is a biased (leading) question – it is clear from the question that Kate thinks the road will cause a lot of problems. This makes it a less reliable question, as people are more likely to say yes in order to be seen to agree with Kate.
>
> ▸ The categories are unclear – 'possibly' and 'maybe' could mean the same thing.
>
> **b** Do you think the new road will cause problems for our town?
>   - ☐ Yes
>   - ☐ No
>   - ☐ Unsure

> **Worked example**
>
> Aled makes the hypothesis that taller people have more brothers and sisters.
>
> What data should Aled collect to test this hypothesis? Design a suitable questionnaire for Aled to use.
>
> **Solution**
>
> To test this hypothesis, Aled needs to collect data on the heights of pupils in his class and the number of brothers and sisters they have.
>
> A suitable questionnaire would be:
> - How tall are you?
>   - ☐ $h < 1.0\,\text{m}$
>   - ☐ $1.0\,\text{m} \leq h < 1.4\,\text{m}$ ← *This includes 1.0 m but doesn't include 1.4 m.*
>   - ☐ $1.4\,\text{m} \leq h < 1.8\,\text{m}$
>   - ☐ $h \geq 1.8\,\text{m}$
> - How many brothers and sisters do you have?
>   - ☐ 0   ☐ 1   ☐ 2   ☐ 3   ☐ 4   ☐ 5+
>
> *5+ means 5 or above.*
>
> *'Or above' or 'other' categories capture values not included in other categories. We don't want many values in these categories, otherwise the questionnaire isn't capturing the data we really want.*

14.1 Now try these

Band 1 questions

Fluency

1. Megan is carrying out a survey of the height and shoe size of pupils in her class.
   Write a hypothesis involving height and shoe size.

2. Tomos asks his friends about the time they spent revising for their Welsh test and the score they got in the test.
   Write a hypothesis involving time spent revising and test score.

Strategic competence

3. Dafydd wants to find out how much pocket money his friends have each week. He asks:
   - How much pocket money do you have each week?
     ☐ £1–£4   ☐ £5–£10   ☐ £10 or more
   Write down two criticisms of his question.

4. Roisin asks her friends how much they each spend on food every day. She asks:
   - How much money do you spend on food every day?
     ☐ £0–£1   ☐ £1–£2   ☐ £2–£3   ☐ £3–£4
   Write down two criticisms of her question.

Band 2 questions

Fluency

5. Mali is carrying out a survey of the distance her friends live from school and the time they spend travelling to school.
   Write a hypothesis involving the distance lived from school and time spent travelling.

6. Aron is creating a questionnaire about exercise.
   Write a hypothesis involving time spent exercising and resting heart rate.

Strategic competence

7. Stef wants to find out what her family thinks about the colour of her mother's new car. She asks:
   - Mum's new car is a really nice colour, don't you think?
     ☐ Yes   ☐ No   ☐ Not really
   a Write down two criticisms of her question.
   b Write a better version of her question, with response boxes.

8. Bleddwyn wants to find out what types of food his friends like eating in school. He asks:
   - What food do you like eating?
     ☐ Pasta   ☐ Panini   ☐ Salad   ☐ Chicken curry
   a Write down two criticisms of his question.
   b Write a better version of his question, with response boxes.

Logical reasoning

9. Mrs Jones wants to open a new café.
   She wants to find out what kind of drinks people like.
   a Write a question, with response boxes, which she could use to find out what drinks people like.
   b Mrs Jones posts her questionnaire to 200 people across Wales.
      Explain why this might not be sensible.
   c Mrs Jones asks three of her friends, 'Isn't coffee better than tea?'
      Give two reasons why this might not be a good way to find out what people in her town like to drink.

Band 3 questions

10 A restaurant manager wants to find out how often people go to his restaurant.

He writes the following question on his questionnaire:
- How many times do you visit my restaurant each year?
  - ☐ All the time   ☐ A lot   ☐ Sometimes

   a Write down what is wrong with his question.

   b Write a better question for the restaurant manager to use.

11 Ivy wants to carry out a survey of movies watched by pupils at her school.

She wants to design a questionnaire.

Write a suitable question she could use to find out what types of movies pupils watch.

12 Oliver wants to find out how much money people spend on birthday presents for their family.

Design a question, with response boxes, for his questionnaire, to find out how much people spend on birthday presents for their family each year.

13 Ffion wants to find out what pupils in her school think about the school uniform.

   a She asks 10 pupils in her class.

   Explain why this is not a good sample.

   b She uses this question:
   - What do you think of the school uniform?
     - ☐ Ok   ☐ Not very nice   ☐ Disgusting

   Explain what is wrong with her question.

14 Rhys wants to test his hypothesis that boys do better in mathematics tests than girls.

He writes a questionnaire, shown below, and gives it to five of his friends to complete.
- Are you a boy or a girl?
  - ☐ Boy   ☐ Girl
- What score did you have in your last mathematics test?
  - ☐ 50%–60%   ☐ 70%–80%   ☐ 90%–100%
  - ☐ 60%–70%   ☐ 80%–90%

   a Explain why Rhys's questionnaire won't allow him to successfully test his hypothesis.

   b Write an improved model to test Rhys's hypothesis.

15 Write a hypothesis involving time spent playing computer games and time spent playing sport.

What data must be captured to test this hypothesis?

Design a suitable questionnaire to capture this data.

14.2 Grouped frequency tables

Communication using symbols

Skill checker

① Decide whether each statement is true or false.

   a $4 > 17$ b $3 \leqslant 3$ c $2 < 2$ d $7 \geqslant 6.5$

② Lowri plays cricket for an under-12s team. These are her scores in eight games.

   2, 4, 10, 9, 20, 5, 30, 2

   a Say whether each statement is true or false for this data set and explain how you know.

   i The mode is 2.   ii The median is 7.5.   iii The range is 16.

③ Cerys and Dewi both choose five cabbage plants.

   They count the numbers of caterpillars on them. Their results are in this table.

   | Cerys | 2, 0, 5, 14, 10 |
   | Dewi | 7, 2, 5, 3, 1 |

   a Work out the range for each set of data.
   b Work out the mean for each set of data.

④ Trishna asks everyone in her class how much pocket money they get each week.

   Here are her results.

Amount	Frequency
£2	9
£2.25	6
£2.50	5
£3	8
£3.50	2

   Work out the mean amount of pocket money for the class.

Activity

Which of the three measures of average – mean, median and mode – is the most appropriate to use with the shoe sizes of pupils in a class? Explain your reasoning.

▶ Grouped frequency tables

Activity

Look at the frequency tables.
What is the same and what is different?

Score	Frequency
10	3
20	8
30	5
40	4

Score	Frequency
$10 \leqslant s < 20$	3
$20 \leqslant s < 30$	8
$30 \leqslant s < 40$	5
$40 \leqslant s < 50$	4

Activity

Grouped frequency tables present data in an organised, easy-to-read format. What are the negatives of grouping data? Discuss with a partner.

The second frequency table contains grouped data. It is a **grouped frequency table**. The data set is sorted into groups defined so that there is exactly one group for each item of data.

The original data was:

10, 12, 14, 20, 22, 24, 24, 25, 26, 27, 28, 30, 31, 34, 36, 38, 42, 46, 48, 49

The groups have been chosen with a width of 10.

There are four groups.

Between 4 and 10 is not too many and not too few.

It is easy to assign each item to its group and to draw a graph of the data.

The groups are described using inequality notation.

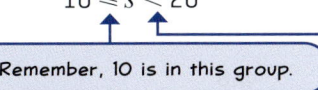

$10 \leq s < 20$

Remember, 10 is in this group.

Remember, 20 is not in this group but is in the next one.

This ensures that each value belongs to exactly one group.

Activity

Use the frequency tables in the previous activity.

a Work out the mean for the first frequency table.

b The scores for the first frequency table are changed to 20, 30, 40 and 50. Work out the mean.

c The scores for the first frequency table are changed to 15, 25, 35 and 45. Work out the mean.

d Which of the answers for **a**, **b** and **c** do you think is closest to the mean of the data in the second frequency table?

Remember

To work out the mean from a frequency table

▶ multiply each score by its frequency
▶ add the results
▶ divide by the sum of the frequencies (20).

The first frequency table in the activity above shows a different set of data:

10, 10, 10, 20, 20, 20, 20, 20, 20, 20, 30, 30, 30, 30, 30, 40, 40, 40, 40

This table retains all the detail of the data so the exact mean can be calculated. The second frequency table does not have that detail so the mean cannot be calculated.

However, the mean can be estimated.

The midpoint of each group is the most representative value in that group. It is used to estimate the mean.

Worked example

The maximum temperature, in °C, is recorded in Buenos Aires for each day in June one year.

| 16.4 | 12.8 | 17.6 | 19.1 | 16.6 | 15.5 | 11.2 | 18.7 | 19.5 | 16.1 | 15.3 | 14.2 | 15.8 | 15.7 | 14.9 |
| 14.4 | 13.4 | 12.1 | 13.9 | 11.9 | 13.1 | 12.6 | 10.9 | 13.5 | 14.2 | 15.4 | 16.6 | 15.9 | 15.6 | 14.3 |

a Present this data in a grouped frequency table.
b Calculate an estimate for the mean temperature.
c Find the group that contains the median.

The data ranges from 10.9 to 19.5 so groups of width 2 are chosen.

It is more efficient to be systematic and use tallies to count the number of values in each group.

Solution

a

Temperature, T °C	Tally	Frequency
$10 \leq T < 12$	|||	3
$12 \leq T < 14$	⋀⋀ ||	7
$14 \leq T < 16$	⋀⋀ ⋀⋀ ||	12
$16 \leq T < 18$	⋀⋀	5
$18 \leq T < 20$	|||	3

b

Temperature, T°C	Midpoint, m	Frequency	$m \times f$
$10 \leq T < 12$	11	3	33
$12 \leq T < 14$	13	7	91
$14 \leq T < 16$	15	12	180
$16 \leq T < 18$	17	5	85
$18 \leq T < 20$	19	3	57
	Totals	30	446

The midpoint of each group is calculated.

The calculation is the same as for ungrouped frequency tables.

Mean = 446 ÷ 30 = 14.866 = 14.9 °C (1 d.p.)

This is an estimate of the mean of the data.

The calculation is an estimate because the midpoints are used instead of the original data.

In this example the estimate could be compared to the mean calculated from the original data.

c To find the median, add 1 to the total frequency and then divide by 2.

$\frac{30 + 1}{2} = 15.5$

Usually the original data is not available.

The median is between the 15th and 16th values.

Working down the frequency column, the 15th and 16th values are both in the $14 \leq T < 16$ group, so this is the median group.

14.2 Now try these

Band 1 questions

1 The Dragons football team has played eight games this season.

The chart below shows the number of people who have attended their games, to the nearest thousand.

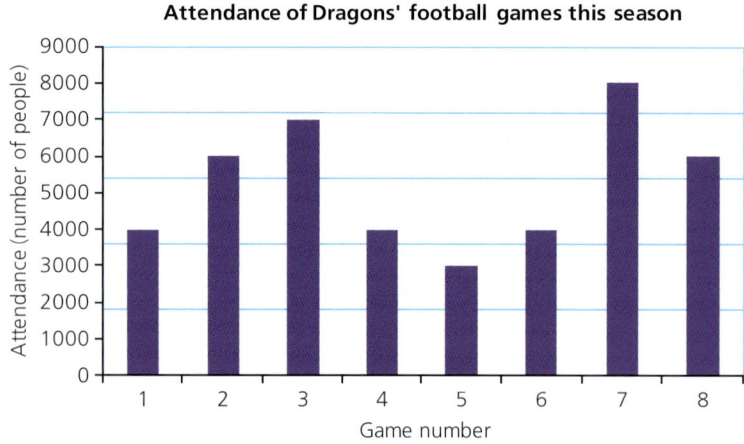

Attendance of Dragons' football games this season

a Which game had the highest attendance?

b What is the range of the attendance figures?

c The club makes a profit if the attendance is over 2500.

In what percentage of their matches so far has the club made a profit?

2. Samir has counted the lengths of the first 50 words in his reading book. Here are his results.
   a Find the mode.
   b Draw a vertical line chart to show Samir's results.

Number of letters	Frequency
1	5
2	7
3	10
4	17
5	5
More than 5	6

3. Three athletes enter the long jump trial for their club.

Athlete	Length of jump (metres)								
Siôn Wells	3.45	4.12	3.56	3.75	2.96	2.77	3.98	3.25	3.78	3.44
Dipesh Raj	2.91	3.46	3.75	2.88	3.24	3.51	3.63	2.99	3.33	3.22
Tristan Trent	4.02	3.55	4.06	4.03	3.70	3.84	3.99	4.03	3.24	3.53

   a How many jumps does each athlete complete?
   b Find the median length of jump for each athlete.
   c Find the range for each athlete.
   d Which athlete has the longest jump?
   e Which athlete would you choose to represent the club? Give a reason for your choice.

Note
Remember, the median is the middle value once data has been ordered.

4. A jogger runs an average of 7 km each day for 16 days.
   She runs an average of 13 km each day for the next 8 days.
   Find the average distance each day over the 24 days.

5. A survey was taken for the number of pets in 20 households.
   The results were:
   1, 2, 1, 3, 2, 3, 4, 3, 3, 22, 5, 4, 3, 2, 0, 1, 0, 3, 4, 4
   a Copy and complete the frequency table.
   b Which of the numbers is an **outlier**?
      Give one reason for including it.
      Give one reason for removing it.

Number of pets						
Frequency						

   Work out the answers to parts **c** and **d** both with the outlier and without it.
   c i What is the mean number of pets?
      ii What is the mode?
      iii What is the median?
   d What is the range of this data?
   e Comment on the effect of the outlier on the answers to parts **c** and **d**.
   f Present the survey results in the grouped frequency table.

Number of pets, P	Frequency
$0 \leqslant P < 4$	
$4 \leqslant P < 8$	
$8 \leqslant P < 12$	
$12 \leqslant P < 16$	
$16 \leqslant P < 20$	
$20 \leqslant P < 24$	

6 Seema goes 10-pin bowling.

The table shows how many pins she knocks down.

Number of pins	0	1	2	3	4	5	6	7	8	9	10
Number of turns	5	0	2	2	1	4	6	5	3	2	0

a   How many turns does Seema have?

b   Work out the total number of pins Seema knocks down.

c   Work out Seema's mean score.

Band 2 questions

7 The list below contains the total time, in minutes, spent at a gym by 30 people during a period of 3 weeks.

30	360	380	320	360	310	260	360	80	510
520	410	270	370	440	440	520	270	40	160
210	280	200	320	450	450	470	170	40	100

a   Complete a copy of this tally chart and frequency table.

Time, t (minutes)	Tally	Frequency
$0 \leqslant t < 99$		
$100 \leqslant t < 199$		
$200 \leqslant t < 299$		
$300 \leqslant t < 399$		
$400 \leqslant t < 499$		
$500 \leqslant t < 599$		

b   Explain why the groups are a good choice for this data.

c   Which is the modal group?

8 Angharad's telephone bill shows the length of her last 20 calls to the nearest minute.

10 min	2 min	5 min	8 min
7 min	6 min	12 min	2 min
15 min	12 min	3 min	4 min
8 min	14 min	2 min	14 min
12 min	10 min	13 min	9 min

a   What is the range of the data?

b   Copy and complete this tally chart and frequency table.

Length of call, t (minutes)	Tally	Frequency
$2 \leqslant t < 4$		
$4 \leqslant t < 6$		
$6 \leqslant t < 8$		
$8 \leqslant t < 10$		
$10 \leqslant t < 12$		
$12 \leqslant t < 14$		
$14 \leqslant t < 16$		

c   How many calls lasted

   i   at least 10 minutes but less than 16 minutes?

   ii 8 minutes at the most?

d   What is the modal group?

9 A newsagent records the number of magazines they sell on each of 15 days.

  a Copy and complete this grouped frequency table.

Number of magazines sold, n	Midpoint, m	Frequency, f	$m \times f$
$10 \leqslant n < 20$		3	
$20 \leqslant n < 30$		7	
$30 \leqslant n < 40$		5	
Totals			

  b In which group does the median value lie? Estimate its value.
  c Use the table to estimate the mean number of magazines sold.

10 A speed camera recorded the speed, v mph, of 60 cars on a busy road.

Speed, v (mph)	Midpoint, m	Frequency, f	$m \times f$
$0 \leqslant v < 10$		1	
$10 \leqslant v < 20$		3	
$20 \leqslant v < 30$		12	
$30 \leqslant v < 40$		38	
$40 \leqslant v < 50$		6	
Totals		60	

  a State the modal class.
  b What do you think the speed limit was on this road?
     Give a reason for your answer.
  c Estimate the mean car speed on this road.

11 The grouped frequency table shows information about the waiting times (in minutes) for a bus.

Waiting time, b (minutes)	Number of people
$0 \leqslant b < 10$	3
$10 \leqslant b < 20$	6
$20 \leqslant b < 30$	10
$30 \leqslant b < 40$	4
$40 \leqslant b < 50$	2

  a The time spent waiting by one person for a bus is 20 minutes.
     In which class interval is this recorded?
  b One person arrived at the bus stop at the same time as the bus.
     In which class interval is their waiting time recorded?
  c Write down the modal class interval.
  d What can you say about the range of the waiting times?
  e Estimate the mean time spent waiting for a bus.

12 Two researchers, Gwyn and Malini, recorded the lengths (in seconds) of birds' songs.

Gwyn's data	Frequency
$10 \leq x < 14$	5
$14 \leq x < 18$	8
$18 \leq x < 22$	4
$22 \leq x < 26$	3

Malini's data	Frequency
$12 \leq x < 18$	3
$18 \leq x < 24$	4
$24 \leq x < 30$	2
$30 \leq x < 36$	1

  a For each set of data
      i estimate the mean
      ii state the modal class.
  b Do you think they were recording the same birds? Explain your answer.

Band 3 questions

13 Hassan is collecting data on the heights of adult men for a clothing company.
He measures the height, in centimetres, of 50 adult men.

  a What is the modal class?
  b Extend your table and calculate an estimate for the mean.
  c Why do you think Hassan collected this information?

Height, h (cm)	Frequency
$155 \leq h < 160$	2
$160 \leq h < 165$	6
$165 \leq h < 170$	8
$170 \leq h < 175$	10
$175 \leq h < 180$	11
$180 \leq h < 185$	7
$185 \leq h < 190$	4
$190 \leq h < 195$	2

14 A safari park keeps a record of the number of cars visiting each day during the month of June.

Number of cars, n	Frequency
$100 \leq n < 150$	1
$150 \leq n < 200$	1
$200 \leq n < 250$	2
$250 \leq n < 300$	4
$300 \leq n < 350$	6
$350 \leq n < 400$	8
$400 \leq n < 450$	3
$450 \leq n < 500$	4
$500 \leq n < 550$	1

  a Use your table to calculate an estimate for the mean.
  b What is the maximum possible value for the range?
  c What is the minimum possible value for the range?

15 The table shows the time, in minutes, that 30 customers had to wait to get a new car tyre.

  a Calculate an estimate for the mean waiting time.
  b Work out an estimate for the range.
  c At another garage, the mean waiting time is 18 minutes with a range of 5 minutes.
      Compare the waiting times of the two garages.

Time, t (minutes)	Frequency
$0 \leq t < 8$	4
$8 \leq t < 16$	16
$16 \leq t < 24$	6
$24 \leq t < 32$	4

16 A consumer organisation wants to compare two different seed composts, A and B.

150 seeds are sown in each compost and the heights of the seedlings that have germinated are measured after three weeks.

This table shows the heights, h mm, of each set of seedlings.

	A	B
$0 \leqslant h < 10$	6	18
$10 \leqslant h < 20$	23	16
$20 \leqslant h < 30$	45	20
$30 \leqslant h < 40$	36	48
$40 \leqslant h < 50$	11	31

a Estimate the mean height for each set of seedlings.

b Which compost do you think is better and why?

17 In a survey, a group of Year 9 students were asked how long they had spent on their Science homework the previous night.

Time, t (minutes)	Frequency
$0 \leqslant t < 10$	2
$10 \leqslant t < 20$	6
$20 \leqslant t < 30$	7
$30 \leqslant t < 40$	8
$40 \leqslant t < 50$	5
$50 \leqslant t < 60$	3
$60 \leqslant t < 70$	4

The results, in minutes, are shown.

a How many students were surveyed?

b Use the table to estimate the mean time spent on Science homework.

c **i** Using 20-minute intervals, draw and complete another grouped frequency table for the data.

   ii Calculate an estimate for the mean using this table.

d Compare the two means you have calculated. Which estimate is more accurate?

18 A fish breeder is testing two new types of food.

Two hatchings of baby fish are fed the different foods.

After two months the weights of the baby fish are measured to the nearest 0.1 gram.

Weight of baby fish, w (grams)	Batch A frequency	Batch B frequency
$0 \leqslant w < 0.5$	0	2
$0.5 \leqslant w < 1.0$	3	2
$1.0 \leqslant w < 1.5$	7	5
$1.5 \leqslant w < 2.0$	7	5
$2.0 \leqslant w < 2.5$	4	4
$2.5 \leqslant w < 3.0$	2	2
$3.0 \leqslant w < 3.5$	0	1
$3.5 \leqslant w < 4.0$	0	2

a Estimate the mean, the median and the range for Batch A.

b Estimate the mean, the median and the range for Batch B.

c Which fish food do you think is more effective?

Give a reason for your answer.

14.3 Displaying grouped data

Skill checker

① Write down the values indicated by the arrows.

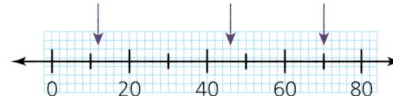

② Llinos has drawn a bar chart to show how the people in her class travel to school.

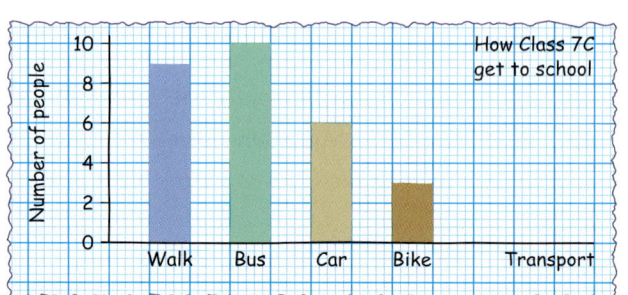

  a What is the most popular way to get to school in Llinos's class? How do you know?
  b How many people walk to school?
  c How many people come to school by bus or car?
  d How many people are there in Llinos's class?
  e Draw a pictogram to show this data.

③ A dice is rolled 21 times. A vertical line chart is drawn of the results.

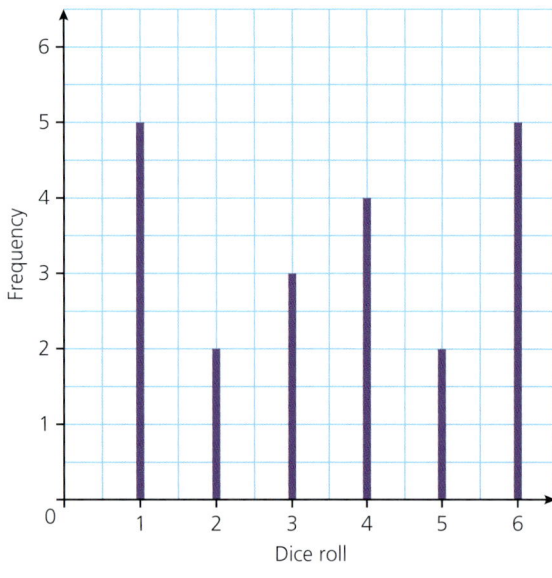

  a There are two modes. What are they?
  b What is the median score?

④ Mrs Patel sets her classes a test and wishes to group the results. They are percentages.

57	82	73	64	67	41	75
33	88	79	48	66	58	71
60	64	73	52	66	35	22
81	93	52	67	70	45	56
94	51	69	49	55	71	81
56	62	50	73	85	81	47
56	64	67	96	48	70	72

Why would $20 \leqslant m < 30$, $30 \leqslant m < 40$ up to $90 \leqslant m < 100$ be suitable groups for the results?

▶ Displaying grouped data

Activity

Sanjay and four of his friends measure the length of their feet, in cm rounded to the nearest mm:

22.5, 24.2, 23.3, 24.1, 25.8

and record their shoe sizes:

5, 7, 6, 7, 9

a Why are the foot lengths rounded?

b The shoe sizes are not rounded. Why not?

c What is the modal shoe size?

d Why is there not a modal foot length?

The shoe sizes are **discrete data**. The UK sizes include $5\frac{1}{2}, 6\frac{1}{2}, \ldots$ as well as the whole numbers. They do not include any other values such as $5\frac{1}{3}$ or 6.235 in between.

The foot lengths are **continuous data**.

> Continuous data does not have gaps in the possible values.

They are both examples of **numerical** data. Numerical data consists of numbers.

24.1 cm and 24.2 cm are both size 7. The lengths in between 24.1 and 24.2, such as 24.102674, are also size 7.

Continuous data is usually rounded before it is recorded.

> Rounded data is still continuous even if it looks as if there are gaps!

Categorical data is mentioned in Book 2 on pie charts.

It is data that is not numerical. It describes a property or characteristic. Sometimes digits are included but they are not used in calculations. They are used to identify something. A phone number is an example of a number used to identify something.

Worked example

Which of these are examples of the following types of data?

| Heights of meerkats | Colours of mobile phone cases | Numbers of people at events |
| Car number plates | Numbers of goals scored | Times to run a race | Makes of car |

a discrete data
b continuous data
c categorical data

Solution

a Numbers of people at events are discrete data.
   Numbers of goals scored are discrete data.

b Heights of meerkats are continuous data.
   Times to run a race are continuous data.

c Colours of mobile phone cases are categorical data.
   Makes of car are categorical data.
   Car number plates are categorical data.

> You can only have a whole number of people.

> These are always whole numbers; you cannot score half a goal.

> Heights take all values within a range although they can only be measured to a certain accuracy.

> Times to run a race can be measured to extreme precision.

> Colours are not numerical.

> Makes of car are not numerical.

> They contain numbers but these are used to identify the car. Bus numbers are also identifiers.

Curriculum for Wales Mastering Mathematics: Book 3

A **frequency diagram** is plotted from a grouped frequency table. It consists of bars placed on a linear scale covering the range of values in the groups. A frequency polygon can also be plotted from a grouped frequency table. It consists of plots — these are plotted at the midpoint of the group against the frequency — which are then joined with straight lines.

Cross-curricular activity

We often measure and record continuous data, such as time and distance, in PE lessons.

Measure the time taken by your friends to run 100 m and their resting heart rate. How could you display the results? Ask your PE teacher to explain the link between fitness and resting heart rate.

Worked example

Plotting a frequency diagram

Alex and Emily measured the heights, h m, of members of their athletics club.

1.45	1.57	1.60	1.48	1.60	1.77	1.56	1.55	1.66	1.66
1.70	1.62	1.60	1.42	1.52	1.55	1.59	1.72	1.52	1.62
1.80	1.52	1.75	1.55	1.70	1.63	1.44	1.73	1.50	1.54
1.36	1.62	1.54	1.64	1.55	1.82	1.47	1.68	1.55	1.70
1.60	1.75	1.63	1.75	1.44	1.60	1.60	1.42	1.58	1.80

a Display this information using a grouped frequency table.
b Use your grouped frequency table to plot a frequency diagram.
c Use your grouped frequency table to plot a frequency polygon.
d Describe the distribution of the data.

The groups are 0.1 m (10 cm) wide. There are six groups.

Solution

a

Height, h (m)	Frequency
$1.30 \leqslant h < 1.40$	1
$1.40 \leqslant h < 1.50$	7
$1.50 \leqslant h < 1.60$	15
$1.60 \leqslant h < 1.70$	15
$1.70 \leqslant h < 1.80$	9
$1.80 \leqslant h < 1.90$	3

b

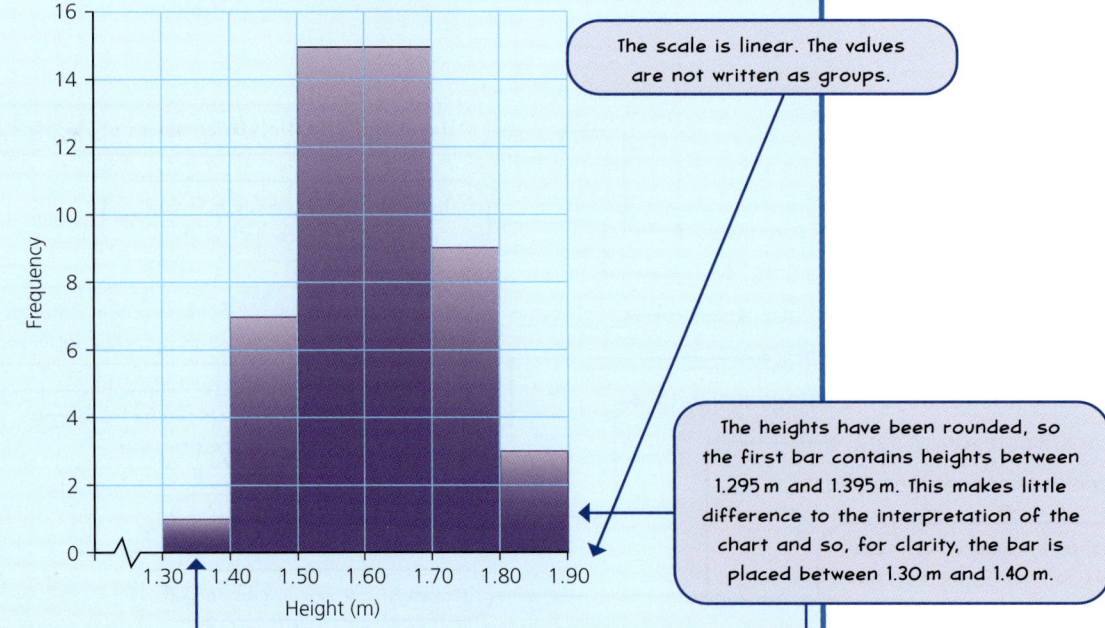

Frequency diagram showing the heights of members of an athletics club

The scale is linear. The values are not written as groups.

The heights have been rounded, so the first bar contains heights between 1.295 m and 1.395 m. This makes little difference to the interpretation of the chart and so, for clarity, the bar is placed between 1.30 m and 1.40 m.

Communication using symbols

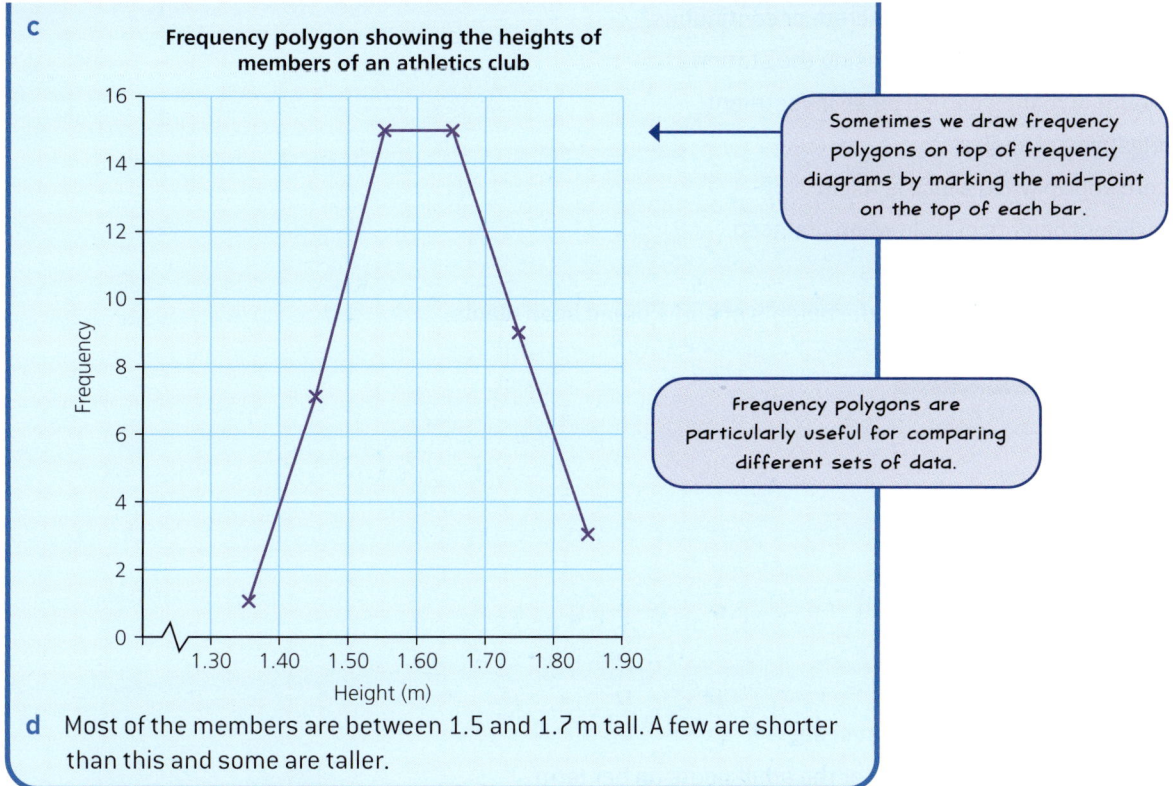

d Most of the members are between 1.5 and 1.7 m tall. A few are shorter than this and some are taller.

Maths in context

Many popular ways of displaying data were invented by a Scottish political economist, William Playfair. Born in 1759, he was also a secret agent.

William Playfair produced the first bar chart, pie chart and area graph.

As we learn more about graphs we have to consider which graph is the most appropriate for the data we want to display.

Use the internet to review and interpret some of William Playfair's graphs.

14.3 Now try these

Band 1 questions

1. State whether each type of data is categorical or numerical.
   a flavours of crisps
   b amount of salt in packets of crisps
   c methods of transport
   d passwords
   e PIN numbers
   f times taken to walk to school

2. Myfanwy writes a table like this.

Time, t (hours)	Tally	Frequency
$0 \leqslant t < 5$		
$5 \leqslant t < 10$		
$10 \leqslant t < 15$		
$15 \leqslant t < 20$		

   a What does $0 \leqslant t < 5$ mean?
   b Which category does 5 hours belong to?
   c Explain why $0 \leqslant t < 5$ is a better group size than $0 \leqslant t < 6$?

3. State whether each type of data is discrete or continuous.
   a The number of people in cars passing the station
   b The amount of money earned by an investment
   c The wingspans of eagles
   d The score shown when a dice is rolled
   e The maximum heights of balls when bounced
   f The weights of new-born dolphins

4. Zelma draws a frequency diagram of the heights of some broad bean plants.

Heights, h (m)	Frequency
$0.8 \leqslant h < 0.9$	7
$0.9 \leqslant h < 1.0$	9
$1.0 \leqslant h < 1.1$	25
$1.1 \leqslant h < 1.2$	9

   The horizontal scale is shown below.

   0.8–0.9   0.9–1.0   1.0–1.1   1.1–1.2

   Explain why Zelma should not use that scale.

5. Tegan has measured the height in cm of the adult goats on her farm.
   Here are her results.

   132   121   146   134   114   125   137   129
   127   130   124   136   141   118   129   136
   123   128   135   132   129   117   140   135
   139   131   127

   a Copy and complete this tally chart.

Height, h (cm)	Tally	Frequency
$110 \leqslant h < 115$		
$115 \leqslant h < 120$		
$120 \leqslant h < 125$		
$125 \leqslant h < 130$		
...		

   b Which class has the highest frequency?

6. This frequency diagram shows the ages of children at a summer camp.
   a How many children are between 6 and 7 years old?
   b How many children are over 8 years old?
   c How many children are between 8 and 9 years old?
   d What is the median age of the children?

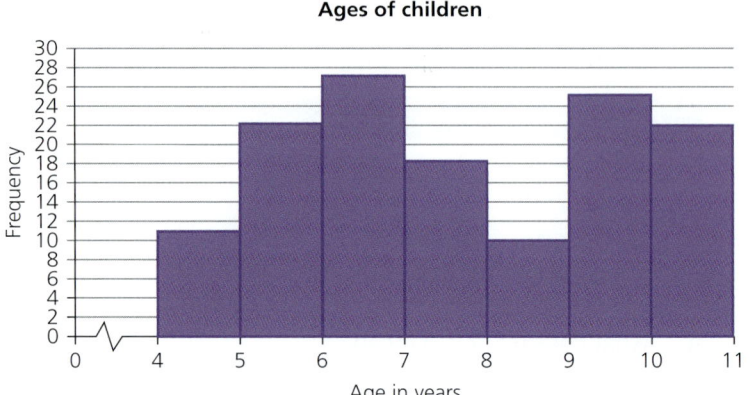

Band 2 questions

7 Medwyn works at a swimming pool.

He is doing a survey to find out what age groups use the swimming pool most.

He writes down the ages, y years, of all the people who visit the pool one morning.

Here are his results.

32	55	41	12	13	13	62	45	23
6	4	3	29	27	56	19	17	34
47	8	7	11	15	70	33	26	14
37	13	14	22	30	45	52	15	14
31	53							

a Make a tally chart using groups $0 \leqslant y < 5$, $5 \leqslant y < 10$, $10 \leqslant y < 15$ and so on.

b Draw a frequency diagram.

c Draw a frequency polygon on top of your frequency diagram.

d How many people visited the pool that morning?

e Which age group used it most?

8 Mr Harris has given his class a test. He has given each student a mark, m, out of 50.

Here are the results for his class.

32	38	43	21	30	46	29	35
37	42	27	25	33	32	16	48
39	37	26	19	40	23	35	24
8	35	39	41	36			

Mr Harris is going to give each student a grade.

He wants to make a tally chart to show his results.

a Copy and complete the table to show what grades Mr Harris's students received.

Grade	Mark	Tally	Frequency
E	$0 < m \leqslant 10$		
D	$10 < m \leqslant 20$		
C	$20 < m \leqslant 30$		
B	$30 < m \leqslant 40$		
A	$40 < m \leqslant 50$		

b What is the modal grade awarded?

c Draw a frequency diagram to show the results.

d Draw a frequency polygon to show the results.

e What grade does a mark of 20.5 get?

9 Siân keeps a record of her scores when she plays her new computer game.

Here are her scores, s points, for the first few weeks.

17	55	82	134	168	143	182	174
240	113	194	98	257	322	286	301
126	264	228	319	160	200	258	320
247	281	177	346				

The maximum possible score is 500.

a Make a tally chart using groups 0–49, 50–99, 100–149 and so on.

b Draw the frequency diagram.

c Draw the frequency polygon.

d What do you think the frequency diagram will look like after a few months?

10 Neila times how long her friends can hold their breath for. The time is t seconds.
She records her data in the table below.

Time t (seconds)	Frequency
$0 \leqslant t < 10$	0
$10 \leqslant t < 20$	1
$20 \leqslant t < 30$	3
$30 \leqslant t < 40$	19
$40 \leqslant t < 50$	12
$50 \leqslant t < 60$	5

a How many people took part?
b Draw a frequency diagram.
c Estimate the median time. Is it in the modal group?
d Draw a frequency polygon.
e One of Neila's friends holds his breath for 30.5 seconds.
Which group does this time go in?

11 The maximum temperature, in °C, is recorded in Cardiff for each day in June.

16.4	12.8	17.6	19.1	16.6	15.5
11.2	18.7	19.5	16.1	15.3	14.2
15.8	15.7	14.9	14.4	13.4	12.1
13.9	11.9	13.1	12.6	10.9	13.5
14.2	15.4	16.6	15.9	15.6	14.3

a Draw a frequency diagram, using five groups of equal width, starting at 10 °C.
b On how many days was the temperature
    i at least 16 °C but less than 18 °C   **ii** less than 14 °C   **iii** at least 12 °C?
c Draw a frequency polygon.
d Describe the distribution.

12 Seren carried out a survey to find out the ages of people using her local swimming pool during one hour.
She draws a frequency diagram to show her results.

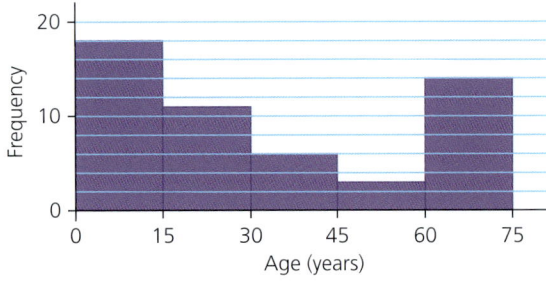

a How many people aged 60 or over used the pool during the survey?
b Do you think Seren carried out her survey in the morning or in the evening?
Explain your answer.
c Draw a frequency polygon of Seren's results.

Band 3 questions

13 Steffan and his sister Anna share a computer.

Steffan and Anna both keep a record of the length of time, in minutes, that they use the computer each day for one month.

Time, t (minutes)	Steffan's frequency	Anna's frequency
$0 \leq t < 15$	0	8
$15 \leq t < 30$	2	4
$30 \leq t < 45$	5	2
$45 \leq t < 60$	11	3
$60 \leq t < 75$	10	2
$75 \leq t < 90$	2	7
$90 \leq t < 105$	1	4
$105 \leq t < 120$	0	1

a   Draw frequency diagrams to show
  i   Steffan's data
  ii Anna's data.
b   Anna thinks Steffan uses the computer more than she does. Do you think that Anna is right?
    Explain your answer.
c   Calculate an estimate of the mean for Steffan and for Anna.

14 A group of people go on a course to maintain a new machine. At the end they are given a test.

These are their marks, as percentages.

a   Draw a frequency diagram to display the data.
b   What is the modal group?
c   Estimate the mean using the grouped frequency table.
d   Which is the more useful, the frequency diagram or the grouped frequency table? Explain why.
e   The pass mark is 60. What percentage of the people pass?

Marks, m	Frequency
$10 \leq m < 20$	2
$20 \leq m < 30$	1
$30 \leq m < 40$	2
$40 \leq m < 50$	4
$50 \leq m < 60$	5
$60 \leq m < 70$	7
$70 \leq m < 80$	5
$80 \leq m < 90$	3
$90 \leq m < 100$	1

15 Zubert is a diver. He holds a competition to find out how far his friends can swim underwater without taking a breath.

He records the results in a frequency table.

Distance (metres)		Frequency
At least	Below	
0	10	2
10	20	11
20	30	4
30	40	2
40	50	1

a   How many people take part in the competition?
b   What is the shortest possible distance swum by any of the participants?
c   Display this information as a frequency diagram.
d   Describe the distribution.
e   Calculate the percentage of people in the modal group.
f   Calculate an estimate of the mean distance.

16 Here is some data about the members of Castell Football Club.

Name	Age (years)	Height (cm)	Goals, g, this season
Nabil	11	138	16
Lee	13	150	11
Laamia	11	123	2
Younis	12	141	7
Lorraine	13	135	0
Chantel	11	152	3
Heidi	13	129	14
Sabrina	13	137	9
David	12	131	0
Jody	11	148	17
Youssu	14	164	0
Patricia	12	149	2
John	11	128	15
Stuart	12	137	1
Crystal	13	145	22

a Display the goals scored in a grouped frequency table.
Use groups of $0 \leqslant g < 5$, $5 \leqslant g < 10$ and so on.
b Display the height data in a grouped frequency table.
Use groups $120 \leqslant h < 130$, $130 \leqslant h < 140$, etc.
c Display the age data in a frequency table.
d Find the median of each set of data.
e Draw frequency diagrams to illustrate the data.
f Which player do you think is most representative of the group. Why?
g Who is most likely to be the goalkeeper? Explain your reasoning.

17 Below are the weights, in kg, of all the babies born in a hospital in one month.

Weights, w (kg)	Boys frequency	Girls frequency
$2.8 \leqslant w < 3.0$	1	3
$3.0 \leqslant w < 3.2$	6	7
$3.2 \leqslant w < 3.4$	8	14
$3.4 \leqslant w < 3.6$	14	14
$3.6 \leqslant w < 3.8$	19	18
$3.8 \leqslant w < 4.0$	20	13
$4.0 \leqslant w < 4.2$	11	10
$4.2 \leqslant w < 4.4$	4	1
$4.4 \leqslant w < 4.6$	1	0

a Draw appropriate diagrams to compare the data.
b Describe the similarities and differences between the weights of the boys and girls.
c Using your diagrams, calculate an estimate of the mean for boys and an estimate of the mean for girls.

18 Take your pulse for one minute. This is your pulse rate.
  a Find the pulse rate for each person in your class. (You will need at least 20 measurements.)
     Record this in a suitable grouped frequency table.
  b Use suitable charts and statistical measures to summarise your data.

19 Lowri measures the mass of pupils using her school gym.
The masses she measured are shown in the frequency diagram below.

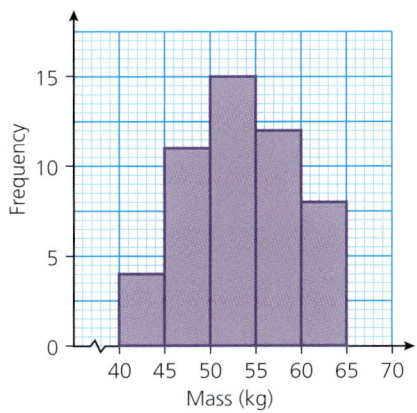

  a How many pupils did Lowri weigh?
  b What is the modal group?
  c What is the median group?
  d Calculate an estimate for the mean.

20 Rhys measures the height of some flowers in a meadow.
The heights he measured are shown in the frequency polygon below.

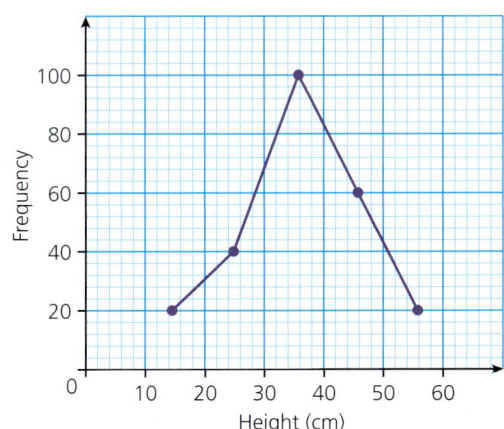

  a How many flowers did Rhys measure?
  b What is the modal group?
  c What is the median group?
  d Calculate an estimate for the mean.

14.4 Scatter diagrams

Skill checker

① Write down the coordinates of A, B, C and D.

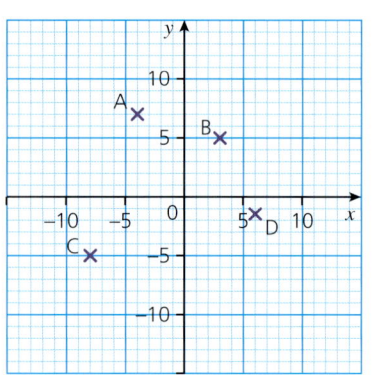

> The x-coordinate is written first.

② Write down the numbers indicated by the arrows.

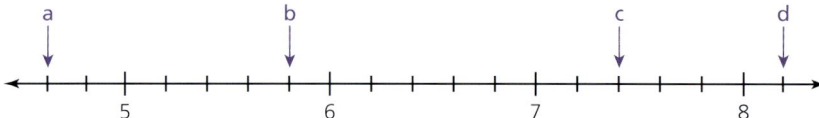

③ Two cars are represented on the graph.

Decide whether each of the following statements is true or false.

▸ The older car is cheaper.
▸ The more expensive car is older.

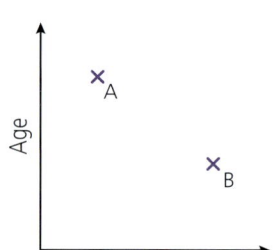

④ Two people are represented on the graph.

Decide whether each of the following statements is true or false.

▸ The taller person is younger.
▸ The older person is shorter.

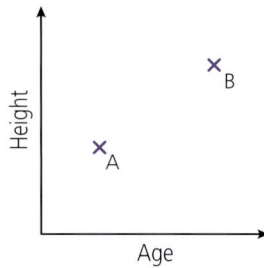

▶ Scatter diagrams

A **scatter diagram** is drawn using pairs of data as coordinates.

Activity

Ms Calligraphy asks 10 students in her class to write their first names, in a line, as many times as they can in 30 seconds. She asks them to measure the length of their writing in cm, rounded to the nearest mm.

Student	Number of times name written	Length of writing (cm)
1	10	20.2
2	11	30.3
3	9	20.4
4	12	28.7
5	11	24
6	14	21.9
7	11	31.5
8	6	16.8
9	13	25.1
10	11	31.3

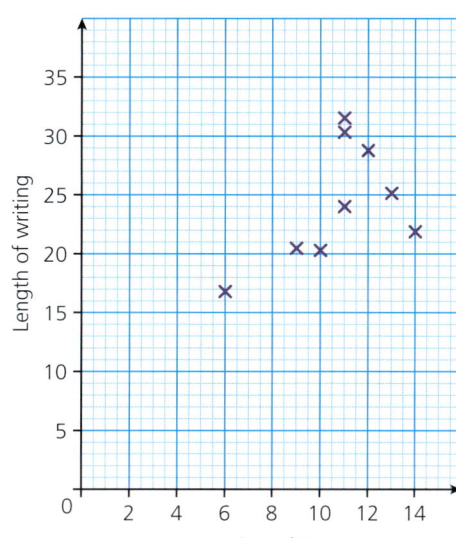

(6, 16.8) represents the data from one of the students.

a   Mai has a very short name. Where do you think her result is?
b   Marc writes slowly. Where do you think his result is?
c   Angharad writes slowly. Where do you think her result is?
d   Where do you think Angharad's result is compared with Marc's?
e   What can you say about the person whose result is at (14, 21.9)?

The results in the activity are mostly grouped in one area of the graph.

The number of repetitions of the names is not related to the length of the writing.

Correlation is when there appears to be a link between how one measure changes with how another measure changes.

There is no correlation between the number of repetitions and the length of the writing.

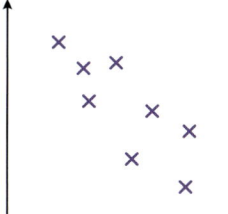 

| Positive correlation: Points are roughly in a line from bottom left to top right. As one quantity increases, the other one does too. | Negative correlation: Points are roughly in a line from top left to bottom right. As one quantity increases, the other one decreases. | No correlation: Points are in no particular pattern. As one quantity increases, it is not possible to say what happens to the other one. |

A **line of best fit** can be drawn when there is correlation.

It is a straight line that follows the trend of the data.

Worked example

Oliver says, 'I think that people with long legs can jump further than people with short legs.'

Susai says, 'Rubbish! The length of a person's legs does not affect how far they can jump.'

They decide to collect data from their friends to find out who is right.

	Alan	Barry	Claire	Dipak	Ernie	Flora	Gurance	Habib	Ivan
Inside leg measurement (cm)	60	70	50	65	65	70	55	75	60
Standing jump distance (cm)	85	90	65	90	80	100	80	95	70

a Draw a scatter diagram.

b i Describe the correlation in the scatter diagram for Oliver and Susai's data.

   ii Does the scatter diagram support Oliver or Susai?

c i Draw a line of best fit.

   ii Jemima has an inside leg measurement of 70 cm. Use your graph to estimate what distance she is likely to jump.

d Have Oliver and Susai got enough data to be certain about their findings?

Solution

a

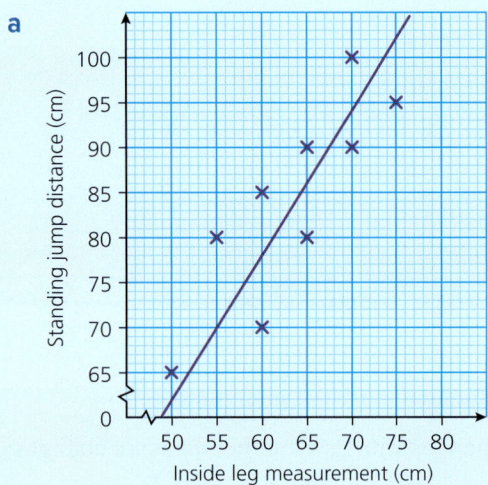

b i The graph shows positive correlation: as the inside leg measurement increases, so does the distance jumped.

   ii This supports Oliver's claim.

c i A line of best fit is drawn on the graph.

      The line of best fit leaves an even distribution of points on either side of the line.

   ii From the graph we can see that Jemima should jump about 94 cm.

d They do not have enough data to be certain. More points on the graph would help.

Cross-curricular activity

Scatter diagrams are used in Science to look at the relationship between two variables.

Investigate the correlation between the length and the width of leaves from a species of tree. Does the same correlation exist for other trees?

14 Working with data

14.4 Now try these

Band 1 questions

1. Here are some examples of scatter diagrams.

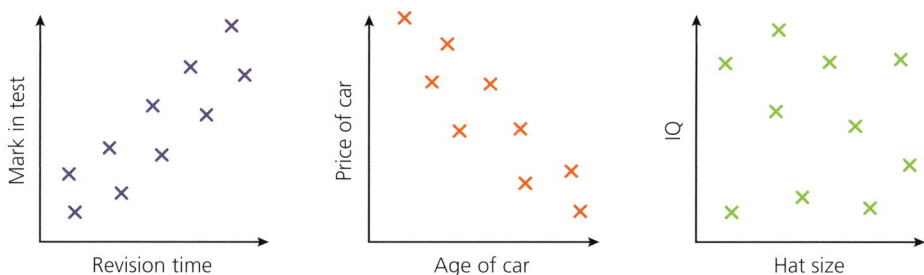

Copy and complete each sentence using the words in the box.

| increases | decreases | positive | negative | no |

a   As the revision time ____ the test mark ____, so there is ____ correlation between revision time and test mark.
b   As the age of the car ____ the price ____, so there is ____ correlation between age and price of car.
c   Hat size does not appear to be related to IQ, so there is ____ correlation between hat size and IQ.

2. A scatter graph is drawn to show how much five students enjoy reading and swimming.
Match the points with the students.

Nia quite likes doing both.
Osian enjoys reading very much but is not keen on swimming.
Sioned enjoys both reading and swimming very much.
Tomos dislikes both.
Mali swims a lot but seldom reads.

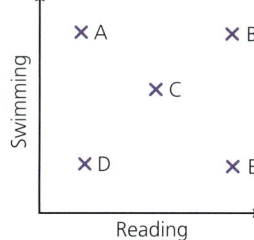

3. The data in the graph shows positive correlation. The horizontal axis is the final height of a tomato plant.

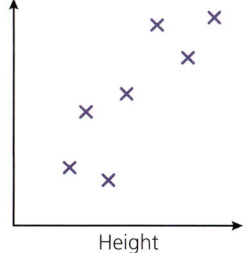

Explain why the amount of fertiliser used could be on the vertical axis.

4. The data in the graph show negative correlation. The horizontal axis is the cost of a car.

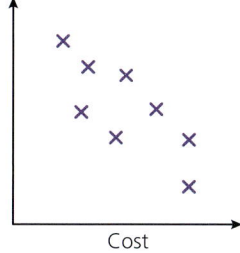

Explain why the vertical axis could be the number of miles on the mileometer at the time it is sold.

5 The results on two papers in a Science exam are plotted on a scatter diagram.
  a Describe the correlation.
  b i   What do the zig-zags across each axis mean?
     ii Why are they used in this graph?
  c Jac scored 54 on paper 1. What mark did he get on paper 2?

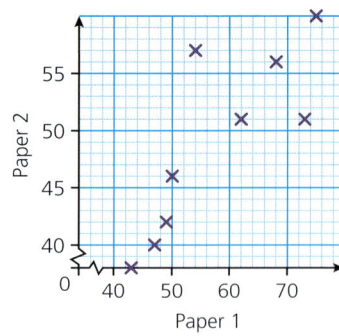

6 Owain says, 'The scales on a graph have to start at 0.'
   Explain why Owain is wrong.

Band 2 questions

7 Describe the correlation you would expect to find between the following data sets.
  a Score in History test and score in Science test
  b Leg length and arm span
  c Typing speed and length of grass in garden
  d Death rate per 1000 and average salary per capita

8 This scatter diagram shows the relationship between the price of some one-bedroom flats and their distance from the centre of Valletta, the capital city of Malta.

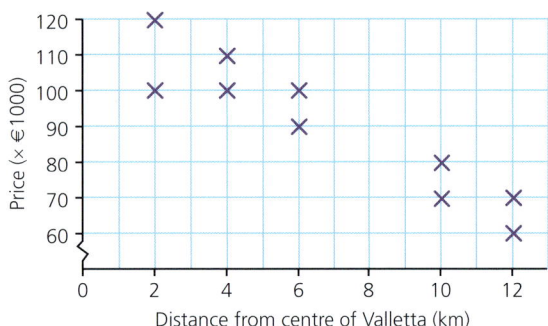

  a Copy the scatter diagram and draw a line of best fit.
  b Use the scatter diagram to estimate
     i   the price of a flat 8 km from the centre of Valletta
     ii the distance from the centre of Valletta of a flat costing €105 000.

9 A group of friends take part in a sponsored swim.

Name	Abdul	Sonia	Lizzy	Geoff	Angey	Carrie	Cherry
Age	20	38	60	51	28	33	26
Number of lengths	30	19	14	11	26	34	42

   a Draw a scatter diagram to show this data.
   b Describe the correlation between the ages of these swimmers and the number of lengths they swim.
   c Terry is 46 and Clara is 31. They also swim.
      i Who do you expect to swim further?
      ii Can you be sure?

10 Catrin took ten sets of tests for a coaching certificate.

Each test had a theory section and a practical section. Her results are shown below.

Theory (%)	64	45	56	72	65	78	32	85	76	90
Practical (%)	30	40	50	65	68	70	22	81	73	86

   a Draw a scatter diagram showing her results on the theory and practical sections.
   b Describe the correlation between Catrin's results on the two sections.
   c Draw a line of best fit on your scatter diagram.
   d Catrin took another test.

     She was absent for the practical section but got 60% on the theory.

     Use the line of best fit to suggest a suitable mark for the practical section.

     Can you be certain that Catrin would get this mark? Explain your reasoning.

11 Eleri thinks that the more football matches you go to, the less you read.

She collected the data below for the last month.

Name	Amy	Ben	Des	Ella	Mo	Sid
Number of books read	3	7	2	6	3	0
Number of football matches attended	4	5	2	0	3	5

   a Plot the data on a scatter diagram.
   b What correlation does the graph show?
   c Is Eleri correct? Explain your answer by referring to your graph.

12 Skylar says, 'A line of best fit must go through the point (0, 0).' Explain why Skylar is wrong.

13 Tristan measures the floor area of a number of classrooms in his school.

He also counts the number of desks in them.

Area (m^2)	25	32	26	38	40	42	48	50	55	65
Number of desks	10	13	16	17	11	15	20	19	17	20

   a Draw a scatter diagram and add a line of best fit if appropriate.
   b Another classroom has floor area 46 m^2.

     Use your scatter diagram to estimate how many desks it can have.

   c Comment on the correlation between the area and the number of desks.

Band 3 questions

14 Look at this scatter diagram.

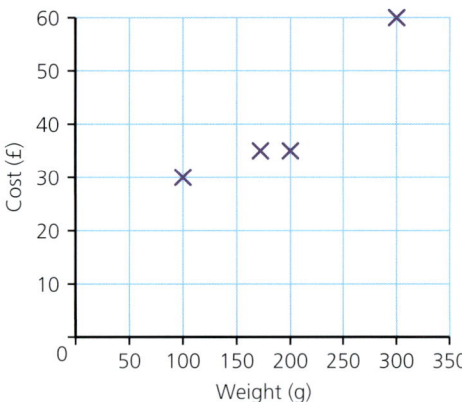

It shows the weight and cost of four books taken off a shelf at an antique book shop.

Rhiannon says, 'You can see that heavier books cost more.'

a Why does Rhiannon say this?

Another seven books are taken off the shelf at random.

Weight (g)	100	250	55	200	205	300	50
Cost (£)	45	15	50	8	10	7	36

b Copy the scatter diagram and plot the new points.

Tecwyn says, 'There is no correlation between a book's weight and its cost.'

c Which of the two opinions is more reliable, Rhiannon's or Tecwyn's? Explain your reasoning.

15 Ellie thinks that there is a connection between the number of reported crimes in a town and the number of police officers on the beat.

She collects information for 15 towns in her county for one month.

Town	1	2	3	4	5	6	7	8	9	10	11	12	13	14	15
Number of police officers	6	16	15	28	30	22	12	23	11	20	17	16	19	25	27
Number of reported crimes	180	150	111	70	30	64	80	95	98	36	83	58	121	146	42

a Draw a scatter diagram for this data set.

b Describe the correlation.

c What other factors might affect the number of reported crimes?

16 In a fishing competition, the judges measure the length and mass (weight) of each fish caught.

These are the results for the first six fish caught.

Fish	1	2	3	4	5	6
Length (cm)	24	50	29	19	42	45
Mass (kg)	1.5	13.2	5.5	0.9	10.6	7.8

a Draw a scatter diagram for this data set.

b Describe the correlation.

c Another fish is caught which has a mass of 8 kg.

Use the line of best fit to estimate the length of this fish.

17 Shamicka thinks that people who are good at sprinting are also good at long jump.
    She collects some data.

Time to run 100 m (s)	Distance jumped (m)
14.3	1.81
12.7	1.94
13.9	1.84
15.4	1.72
14.0	1.88
17.3	1.52
12.8	2.03

Time to run 100 m (s)	Distance jumped (m)
16.0	1.68
15.3	1.75
13.4	1.88
13.6	1.92
14.7	1.78
12.5	2.08
16.4	1.58

a Draw a scatter diagram to show Shamicka's data.
b Is Shamicka right?
c Find the median running time.
d Find the median distance jumped. Is this the same person as in part **c**?
e Shamicka thinks the quickest 25% of people in her sample should be classified as good.
   Do the same people feature in the top 25% of jumpers?

18 Choose two of the following body measurements and investigate whether or not they are connected.

   Length of arm   Head circumference   Handspan   Wrist circumference   Length of middle finger

a Draw a table to collect your data and think carefully about the number of items of data you will collect and how you will choose them.
b Collect your data and display your results on a scatter graph.
c Describe the correlation between your chosen measurements.

19 Wil says both of these graphs show positive correlation. Explain why Wil is wrong.

a
b

20 Investigate whether some people have quicker reaction times regardless of whether they are using their dominant hand or their other hand.

Key words

Here is a list of the key words you met in this chapter.

Categorical data	Continuous data	Correlation	Discrete data	Frequency diagram
Grouped frequency table	Hypothesis testing	Line of best fit	Modal group	Negative correlation
Numerical	Positive correlation	Questionnaire	Scatter diagram	

Use the glossary at the back of this book to check any you are unsure about.

Curriculum for Wales Mastering Mathematics: Book 3

Review exercise: working with data

Band 1 questions

1 Lowri is carrying out a survey of the height and mass of pupils in her class.
Write a hypothesis involving height and mass.

2 Meic wants to find out how much money his friends spend on football cards each month.
He uses this question:
- How much money do you spend on football cards each month?
  ☐ £1–£2   ☐ £3–£4   ☐ £5 or more

Write down two criticisms of his question.

3 A group of 40 people took an aptitude test for a flying school.
The table shows their results.

Score, s	Frequency, f	Midpoint, m	$m \times f$
$40 \leq s < 50$	5		
$50 \leq s < 60$	10		
$60 \leq s < 70$	12		
$70 \leq s < 80$	8		
$80 \leq s < 90$	5		
Totals			

a Complete a copy of the table and estimate the mean mark.
b The pass mark for entry to the flying school is 65.
Estimate how many people passed.

4 The table summarises the heights of 25 football players in a team.

Height, h (cm)	Number of football players
$140 < h \leq 144$	2
$144 < h \leq 148$	6
$148 < h \leq 152$	10
$152 < h \leq 156$	3
$156 < h \leq 160$	3
$160 < h \leq 164$	1

Calculate an estimate for the mean height.

5 The ages of the people on a package holiday are as follows.

45	38	27	32	30	7	4	28	48	42	15	31
13	10	56	58	37	24	25	69	34	26	44	46
32	35	64	51	60	49	46	17	38	62	57	52
52	53	29	24	27	23	57	45	43	26	37	31

Morgan says that groups of width five, $0 \leq a < 5$ and so on, are best in this case. Dom says that groups of width 20, $0 \leq a < 20$ and so on, are best.
Tegan says that groups of width 10, $0 \leq a < 10$ and so on are best.
Explain why Tegan is right.

14 Working with data

6 Are the following data sets discrete or continuous?
   a The points scored by a group of friends on a computer game.
   b The numbers of students in each class at school.
   c The weights of moon rock specimens.
   d The distances travelled by snails in a day.
   e The numbers of letters arriving each day at your house.

7 This table shows the marks (out of 10) given by judges at a local vegetable show.

Name	Marks from Judge 1	Marks from Judge 2
Mrs Griffiths	1	2
Mr Hands	6	8
Mr Smith	3	3
Mr Taylor	5	5
Mr Thomas	7	8
Ms Barrett	2	3
Mrs Hogg	8	10
Mr Jones	9	7
Mr Evans	4	6

   a Draw a scatter diagram showing the marks of the two judges.
   b Is there any correlation between the marks of the two judges? If so, what kind?
   c If your graph shows correlation, add in a line of best fit.

8 State the type of correlation, if any, each graph shows.

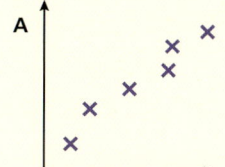 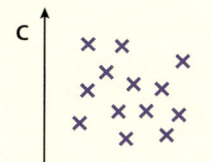 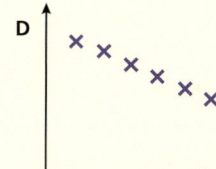

9 Are the following data sets numerical or categorical?
   a The rugby teams supported by the students in your class.
   b The shoe sizes for the students in your class.
   c The favourite TV programmes of the students in your class.
   d The lengths of the feet of the students in your class.
   e The mobile phone numbers in your contacts list.

Band 2 questions

10 Nia is carrying out a survey of the number of brothers and sisters her friends have, and the number of bedrooms in their houses.

Write a hypothesis involving numbers of brothers and sisters, and house size.

11 Aled wants to find out whether his friends like his new hair style.

He uses this question on a questionnaire:

- My new hair style is really nice, don't you think?
  - ☐ Yes
  - ☐ Yes, I love it
  - ☐ No

a Write down two criticisms of his question.

b Write a better version of his question, with response boxes.

12 A teacher records the times taken by 25 children to read a story.

Time, t (minutes)	Number of children
$10 \leq t < 20$	4
$20 \leq t < 30$	8
$30 \leq t < 40$	7
$40 \leq t < 50$	5
$50 \leq t < 60$	1

a Calculate an estimate for the mean time taken to read the story.

b Which time interval contains the median time to read the story?

c A lesson is 45 minutes long. Estimate how many children do not finish the story.

13 Shamicka asks her school friends to record the number of minutes they spend playing computer games during May.

Here are her results.

25	150	262	30	143	0	55	320	260	60
65	140	40	170	74	130	45	125	300	220
96	132	90	185	89	167	68	50	160	82

a Construct a grouped frequency table using groups of $0 \leq t < 50$, $50 \leq t < 100$ and so on.

b Draw a frequency diagram and a frequency polygon to show this information.

c Shamicka plans to ask the same question in August.

Predict the shape of the new diagram.

14 The time, in seconds, for 25 fireworks in a display to burn out was recorded.

10	11	12	27	20
16	27	31	14	18
29	15	32	35	33
32	26	34	19	29
35	20	11	28	23

a Draw and complete a grouped frequency table using the class intervals $10 \leq t < 15$, $15 \leq t < 20$ and so on.

b Draw a frequency diagram and a frequency polygon. What can you say about the organiser's choice of fireworks?

15 Feroza collects data about the wingspans, w cm, of a type of bird.

   a   The first bird measured had a wingspan of 145 cm. In which class interval is this recorded?
   b   Write down the modal class.
   c   Estimate the mean wingspan.

   Feroza goes to another location.
   She measures the wingspans (in cm) of five more birds of the same type.
   They are 162, 160, 165, 175, 168
   d   Calculate the mean of the new birds' wingspans.
       Give a possible explanation for the difference from the original group.

Wingspan, w (cm)	Number of birds
$125 \leq w < 130$	3
$130 \leq w < 135$	6
$135 \leq w < 140$	10
$140 \leq w < 145$	4
$145 \leq w < 150$	2

16 The table shows the maximum temperature and the number of hours of sunshine in 12 British cities on one day in August.

City	Max temp (°C)	Hours of sunshine
London	24	10
Birmingham	22	9
Manchester	20	8
Edinburgh	17	9
Glasgow	18	6
Bangor	23	7

City	Max temp (°C)	Hours of sunshine
Southampton	25	9
Cardiff	23	11
Liverpool	22	8
Newport	25	10
Newcastle	20	9
Swansea	21	8

   a   Draw a scatter diagram to illustrate this data.
   b   Draw a line of best fit on your diagram.
   c   York is not included in the data.
       It had a maximum temperature of 19 °C on the day the data was recorded.
       Estimate the number of hours of sunshine in York that day.
   d   St David's had 12 hours of sunshine on the same day.
       Estimate the maximum temperature in St David's.

17 Sara thinks that older people watch more television.
   She carries out a survey of 20 people of different ages.
   They each keep a record of how much TV they watch during one week.
   Here are her results.

Name	Age	Hours of TV
Fatima	12	25
Evan	82	6
Harold	45	18
Susan	27	18
Gerald	62	12
Spike	18	26
Sunil	21	19

Name	Age	Hours of TV
Goulu	56	12
Tara	6	20
Comfort	36	14
Leroy	38	11
Malini	45	9
Robert	72	10
Temba	14	21

Name	Age	Hours of TV
Sally	16	23
George	8	32
Ali	6	22
Paris	16	18
Meena	28	16
Rick	62	10

   a   Draw a scatter diagram of her results.
   b   Is there any correlation between age and the amount of time spent watching TV? What does this mean?
   c   Draw a line of best fit on your scatter diagram.
   d   Bethan Jones is 58 years old. Estimate how much television she watches.
   e   Scott McKenzie watches 8 hours of television per week. Estimate his age.

Band 3 questions

18 Max wants to find out how much money people spend on Christmas presents each year.
Design a question, with response boxes, for his questionnaire to find out this information.

19 Seren wants to find out what the pupils in her school think of the school dinners.
She asks five of her friends.

a Explain why this is not a good sample.

She uses this question:
- What do you think of school dinners?
  ☐ Very nice   ☐ Nice   ☐ Quite nice   ☐ Ok

b Explain what is wrong with her question and write an improved version.

20 A gardener plants 40 seed potatoes and records the mass of potatoes obtained from each plant.

Weight, w (grams)	$750 \leqslant w < 800$	$800 \leqslant w < 850$	$850 \leqslant w < 900$	$900 \leqslant w < 950$
Frequency	8	12	16	4

a Calculate an estimate for the mean mass of potatoes per plant.

b Find the range.

The mean mass of a seed potato is 80 g.

c Estimate the quantity $\dfrac{\text{mass of crop}}{\text{mass of seeds}}$.

21 Here are the times a group of students took to run 100 metres.

Time, t (seconds)	Frequency
$14 \leqslant t < 15$	5
$15 \leqslant t < 16$	9
$16 \leqslant t < 17$	10
$17 \leqslant t < 18$	6
$18 \leqslant t < 19$	8
$19 \leqslant t < 20$	4
$20 \leqslant t < 21$	1
$21 \leqslant t < 22$	2

a Use your table to identify the median time taken by the students.

b What is the modal group?

c Draw a frequency diagram and a frequency polygon to show the data.

d Calculate an estimate for the mean time.

e If you were the coach, what would you do to improve the mean time of the group?

22 Joachim collects data about the reaction times of students in his class. He asks each student to catch a dropped ruler and record how far the ruler falls in cm. The students do this once with their dominant hand and once with their other hand.

The results for 10 of the students are displayed on the scatter diagram.

a A student has results in the bottom left-hand corner of the diagram. What can you say about their reactions?

b Describe the correlation.

c What conclusion could Joachim draw from the correlation?

d Draw the line of best fit on a copy of the graph.

e What line represents the same reaction time with both hands?

f Compare your answers to **d** and **e**. What does it suggest about the reaction times of the students?

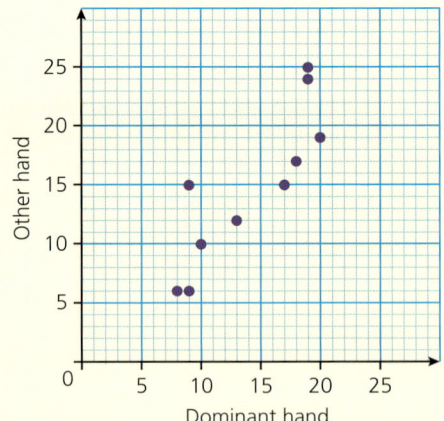

23 190 people take a general knowledge test with 100 questions.
They are hoping to go on a television show.
The table below shows the number of correct answers.

Number of correct answers, q	Frequency
$30 \leqslant q < 40$	80
$40 \leqslant q < 50$	40
$50 \leqslant q < 60$	30
$60 \leqslant q < 70$	20
$70 \leqslant q < 80$	10
$80 \leqslant q < 90$	8
$90 \leqslant q < 100$	2
Total	

a Calculate an estimate for the mean number of correct answers.
b What is the modal group?
c In which group does the median fall?
d Estimate the range.
e People with 75 or more correct answers will appear on the television show.
   Estimate how many people this is.
The producer decides that 75 or more is too generous and the qualifying mark should be 82.
f Draw a frequency diagram to show the data.
g Use the frequency diagram to estimate how many people qualify now.

24 The pupils in Jac's class all plant sunflowers. The heights of the sunflowers are measured after one month and the results are displayed in the frequency polygon below.

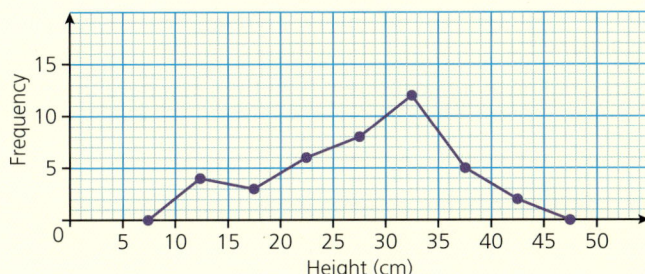

a How many pupils are there in Jac's class?
b What is the shortest possible height of any of the pupils' sunflowers?
c What is the modal group?
d Calculate an estimate for the mean.

Strategic competence

25 Nia asks her friends to sing a musical note for as long as possible without taking a breath.
The results are shown in the frequency diagram.
  a How many people took part in the experiment?
  b What is the shortest possible note sung by any of the participants?
  c What is the modal group?
  d Estimate the mean.
  e Use the diagram to complete a grouped frequency table of the data.

Dylan arrives late so his note is not included in the original data.
He sings a note for 35 seconds exactly.
  f Explain why Dylan's result cannot be recorded in the table.
  g Rewrite the table using the groups $10 \leqslant t < 20$, $20 \leqslant t < 30$, $30 \leqslant t < 40$.
    Include Dylan's result.
  h What is the modal group now?
    Is this what you expected?

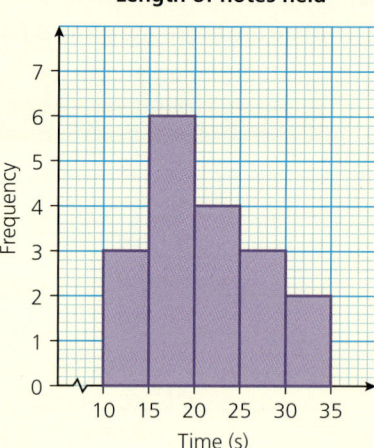

26 Suja thinks that some people are good at remembering things when they see them. Other people are good at remembering things they have seen written down.

She does a test with 10 of her students. She gives them 2 minutes to remember as many as possible out of 15 pictures of everyday foods. They then write down as many of the foods as they can remember. The test is repeated with a list of everyday sights, not including food.

The table shows the results.

Student	1	2	3	4	5	6	7	8	9	10
First test	8	7	8	10	8	9	7	7	11	10
Second test	7	11	8	13	12	9	9	12	7	11

  a Draw a scatter diagram to show the results.
  b Describe the correlation. What conclusion can Suja make?

27 Investigate whether people estimate the length of a straight line more accurately than they estimate the length of a curved line.

15 Probability

Coming up...

- Listing all the outcomes for two or more events
- Using tables to list all the outcomes for two events
- Using grids to enumerate all of the outcomes for two events
- Using probability space diagrams to solve problems involving two events
- Using Venn diagrams to enumerate all of the outcomes for several events
- Using Venn diagrams to solve problems involving several events
- Generating sample spaces for equally likely outcomes for mutually exclusive events
- Using theoretical sample spaces for several events to solve problems in probability

Two events are 'mutually exclusive' if they both cannot happen at the same time.

Heads or Tails?

Anouk and Mustapha play a game.

Each player flips a coin and records the outcome.

Anouk's target is to get TH (tails/heads) on two consecutive throws.

Mustapha's target is to get HH on two consecutive throws.

The first player to reach their target wins.

a Play the game several times. What do you notice?

b Explain your results.

Change the rules so that the game is fairer.

c Play the new game several times. Is it fairer? How do you know?

15.1 Probability space diagrams

Skill checker

1. What is the probability of
    a rolling a six with a single dice
    b getting a head when flipping a coin
    c getting a P when picking a letter at random from the word PROBABILITY
    d getting an I when picking a letter at random from the word PROBABILITY?

② 100 people are asked whether they like strawberry jam and whether they are fans of science fiction films. The table shows the results.

	Likes strawberry jam	Does not like strawberry jam
Likes science fiction films	45	23
Does not like science fiction films	14	18

    a   How many people like both strawberry jam and science fiction films?
    b   How many people like strawberry jam?
    c   How many people do not like science fiction films?
    d   What percentage of the people like strawberry jam but do not like science fiction films?

③ Simplify

    a $\dfrac{12}{24}$ b $\dfrac{12}{18}$ c $\dfrac{3}{18}$ d $\dfrac{15}{20}$

④   a   What is the probability of an impossible event?
    b   What is the probability of a certain event?

Discussion activity

When you roll two dice and add the numbers together, which two totals only have a $\dfrac{1}{36}$ chance of occurring? Why are these totals so difficult to obtain?

Cross-curricular activity

In your Geography lessons, investigate how probability is used in weather forecasting.

▶ Probability space diagrams

Activity

Gwenda is choosing a meal at a restaurant. She has a choice of mains and dessert.
She chooses from:

Mains   **Dessert**

Carbonara   Fruit salad

Burger   Slice of watermelon

Pizza   Apple crumble

Steak   Cheesecake

Fish and chips

Lasagne

  a   List all of the meals that Gwenda could choose.
  b   Have you listed them in the most efficient way? Comment on your answer.

15 Probability

Worked example

Sara is choosing her breakfast. She can choose one cereal and one drink.

Cereals: Wheatamix, Cornflakes or Sugarloops

Drinks: tea or coffee

a Draw a diagram to show all the possibilities for Sara's breakfast.

b How would the diagram change if Sara had three options for drinks, for example, tea, coffee and orange juice?

Solution

a Create a **two-way table** to show the combinations.

	Wheatamix	Cornflakes	Sugarloops
Tea	T&W	T&C	T&S
Coffee	C&W	C&C	C&S

> A two-way table has one set of options as the rows and the second set of options as the columns.

> Listing them in this way avoids missing any combinations.

> It is clear what the letters stand for, so the choices are abbreviated in the table.

b There will be an extra row in the table in answer **a** that lists the options involving orange juice.

There are other ways of listing the combinations in tables or grids.

They are referred to as **probability space diagrams**.

They present the combinations as equally likely outcomes.

It is easy to determine the probability of the outcomes using a probability space diagram.

Worked example

Hannah throws two dice, one red and one green.

a Draw and complete a table to show all the possibilities for the total scores.

b Draw and complete a coordinate grid to show all the possibilities for the total scores.

c What is the probability that the total score is exactly 5?

d Would the answers be the same with two red dice?

Solution

a Create a **two-way table** to show the combinations.

> The scores on the red dice are the rows and the scores on the green dice are the columns.

	1	2	3	4	5	6
1	2	3	4	5	6	7
2	3	4	5	6	7	8
3	4	5	6	7	8	9
4	5	6	7	8	9	10
5	6	7	8	9	10	11
6	7	8	9	10	11	12

> The total score has been calculated for each combination.

337

b A coordinate grid can also be used to show all the outcomes.

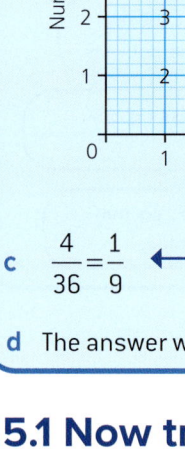

An advantage of using the grid is there is no need to draw out a table.

c $\dfrac{4}{36} = \dfrac{1}{9}$

There are 36 outcomes altogether and four of them are 5s.

d The answer would be the same. The colour of the dice makes no difference.

15.1 Now try these

Band 1 questions

1 Delyth needs to choose one top and one pair of trousers or shorts to pack in her overnight bag.
She can choose from these items.

Tops	Trousers/shorts
Blue sun top	Green shorts
Red t-shirt	Navy shorts
Cropped top	Cream trousers
White shirt	Black trousers
Pink t-shirt	

  a Make a list of all the different combinations of tops and trousers or shorts Delyth can choose.
  b How can you work out the number of different possibilities without listing them all?

2 Elis chooses from a menu with a choice of starters and mains. He does not want a dessert.

Starters	Mains
Soup	Spaghetti bolognese
Crispy chicken wings	Curry
Garlic flatbread	Lemon chicken
	Sausages and chips

  a How many possible meals could he choose?
  b List the outcomes systematically.

3 Harri throws a dice and flips a coin.
  a How many possible outcomes are there?
  b List the outcomes systematically.

15 Probability

4 Samira throws two 4-sided dice. Each dice has its faces labelled 1, 2, 3 and 4.
  a How many possible outcomes are there?
  b List the outcomes systematically.

5 In a game, two fair 8-sided dice with faces numbered from 1 to 8 are thrown and the scores shown added together.
  a What is the smallest possible total?
  b What is the largest possible total?
  c Find the probability that the total is 16.

6 Mari is deciding what to have for lunch. She wants two courses.

Lunch Menu
First course
Cod and chips
Lasagne
Cheese salad
Chicken curry

Second course
Apple pie
Cheesecake
Trifle

First course	Second course
Cod and chips	Apple pie

  a Copy and complete the table of all the possible lunches Mari could choose. Note that you might need to create more rows than in the example above.
  b How do you work out the number of possible lunches?

Band 2 questions

7 The table shows the totals when these spinners are spun.

		Blue							
		1	2	3	4	5	6	7	8
Red	1	2	3						
	2								
	3								
	4								

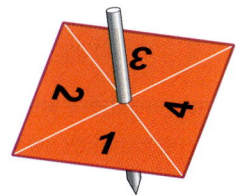 

  a Copy and complete the table. The first two entries have been done for you.
  b There are 32 numbers in the table. How many of them are 12?
  c What is the probability that the total is 12?
  d Write a list of all the possible totals.

339

8 Some children are playing a game using two dice labelled 0, 1, 2, 3, 4, 5.

   a Copy and complete this table to show all the possible totals when the two dice are thrown.

	0	1	2	3	4	5
0	0					
1					5	
2						
3						
4						
5						10

   b Alwyn has the next turn and needs 10 to win. What is the probability he wins on this turn?

9 A spinner has five equal sections, numbered 0 to 4.

   Nerys spins it twice and adds up her two scores.

   a Make a list or table to show all the possible outcomes.
   b What is the probability that Nerys gets 1?

10 In a game, two fair dice are used.

   One has four faces, labelled 2, 4, 6 and 8.

   The other has eight faces, labelled 1 to 8.

   The score is the difference between the numbers on the dice.

   a Make a table or list showing all the possible outcomes.
   b Find the probability of scoring 3.

11 Seren has two bags.

   Each bag contains four balls: one red, one green, one blue and one yellow.

   She picks a ball at random from her first bag.

   Then, she picks a ball at random from her second bag.

   a Copy and complete this table to show all possible pairs of balls.

		First ball			
		R	G	B	Y
Second ball	R	RR	GR		
	G	RG			
	B				
	Y				

   b How many possible outcomes are there altogether?
   c What is the probability that she chooses two red balls?

12 Baz says, 'I cannot see the point in drawing a probability space table. It's too much like hard work.'
   Give two reasons why Baz is wrong.

15 Probability

Band 3 questions

13 Max has two bags.
One bag contains five red discs numbered 1 to 5.
The other bag contains six blue discs numbered 1 to 6.
Max picks one disc from each bag at random.

a Copy and complete the table showing the possible results.

		Blue disc					
		1	2	3	4	5	6
Red disc	1	(1, 1)	(1, 2)			(1, 5)	
	2	(2, 1)			(2, 4)		
	3						
	4		(4, 2)				
	5	(5, 1)			(5, 4)		

b Explain the difference between the outcomes (1, 2) and (2, 1).

c Calculate P(total of 3).

> 'P(total of 3)' means calculate the probability that the total is 3.

14 In a game, a five-sided spinner and a six-sided spinner are spun and their scores are added together. The spinners are fair.

a Copy and complete the table to show all the possible outcomes.

		Spinner 2					
	+	1	2	3	4	5	6
Spinner 1	1	2					
	2				6		
	3						
	4		6				
	5				9		

b Work out the probability of scoring 3.
c Give one advantage of this table compared to the one in question **13**.
d Give one disadvantage of this table compared to the one in question **13**.

15 Loki throws a red dice and a blue dice and multiplies the scores.

a Copy and complete the probability space diagram.
b How many times does 6 appear in the diagram?
c How many times does 7 appear in the diagram?
d What is the modal number in the diagram?

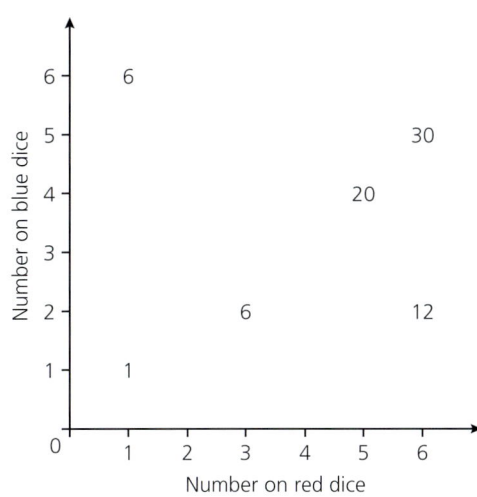

16. Here are nine heart cards numbered 2 to 10 and nine spade cards numbered 2 to 10.

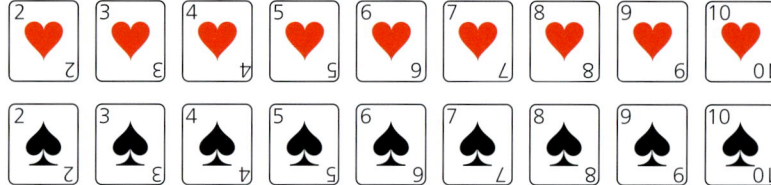

Abdul shuffles the heart cards and places them face down in a row. He does the same with the spade cards.

Abdul chooses one card from each row at random.

   a   Make a table to show all the possible outcomes.

   b   How many different possible outcomes are there?

   c   Work out the probability that he picks two tens.

17. Hywel and Kate decide to play a game with three coins.

   Hywel wins if he gets three heads or three tails, otherwise Kate wins.

   a   Make a list of all the possible outcomes of flipping three coins.

   b   Explain, by referring to the list in **a**, why this game is not fair.

   c   Devise a fairer game that still favours one player. Justify your answer.

18. Kamala and Hilary play a game. They throw two dice, each numbered $-2, -1, 0, 1, 2, 3$.

   The total of the two numbers on the dice gives the score for each turn.

   a   Draw a probability space diagram to show the outcomes.

   b   Devise a game that you would expect Kamala to win one-sixth of the time and Hilary to win the rest of the time.

   c   Devise a game that you would expect Kamala and Hilary to win in the ratio 7 : 5.

15.2 Venn diagrams

Skill checker

1. Write down all of the factors of 20.
2. Write down the first five multiples of 7.
3. Write down a prime number that is a factor of 20 and a factor of 35.
4. Write down a number that is a multiple of 7 and a multiple of 8.

▶ Venn diagrams

We have used Venn diagrams in Chapter 1 with prime factorisation.

Activity

The numbers 1 to 10 are sorted into the regions of this Venn diagram.

Set A is the set of square numbers $\leqslant 10$.

Set B is the set of odd numbers $\leqslant 10$.

a   Explain why 1 and 9 are in the overlap between the two sets.
b   Why are 2, 6, 8, 10 not in either of the circles?
c   Why is 4 the only number in its region of the diagram?
d   11 is added to the diagram. Where should it go?

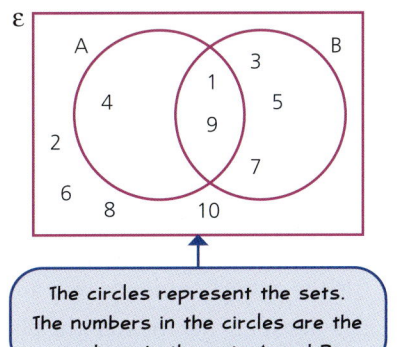

The circles represent the sets. The numbers in the circles are the numbers in the sets A and B.

Venn diagrams are a powerful representation of combinations of events.

They can represent two events or three events.

Worked example

Aled has some cards with different shapes printed on them.
This Venn diagram describes the shapes printed on the cards.

a   How many cards are there in Aled's pack?
b   Aled picks a card at random. What is the probability that his card is
   i   a quadrilateral
   ii   a green quadrilateral
   iii   not green?
c   Draw the shape in the central crossover.
    Write a sentence about the probability of picking it.

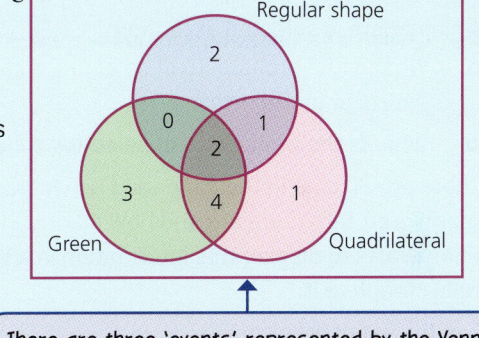

There are three 'events' represented by the Venn diagram: Regular shape, Green and Quadrilateral.

Solution

a   There are 13 cards. ← Add the numbers of cards in all of the regions. The outside region is empty.

b   i $P(\text{quadrilateral}) = \dfrac{8}{13}$ ← Add the numbers of cards in the regions in the pink circle.

   ii $P(\text{green quadrilateral}) = \dfrac{6}{13}$ ← Add the numbers of cards in the overlapping regions of the pink and green circles.

   iii $P(\text{not green}) = \dfrac{4}{13}$ ← Add the numbers of cards in the regions that are not in the green circle.

c   The probability of picking a card with a green square at random is $\dfrac{2}{13}$.

Communication using symbols

343

15.2 Now try these

Band 1 questions

1. There are 10 people on a bus.
   A is the set of people with names beginning with a J.
   B is the set of people with four letters in their name.
   Someone from the bus is selected at random.
   a What is the probability of selecting someone whose name begins with a J?
   b What is the probability of selecting someone whose name is not four letters long or does not being with a J?

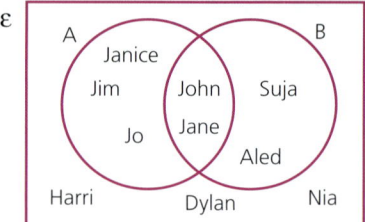

2. Cerys creates a Venn diagram to display information about her four friends.
   C is the set of her friends who are left-handed.
   D is the set of her friends who play musical instruments.
   One of Cerys's friends is selected at random.
   a What is the probability of selecting someone who is not left-handed?
   b What is the probability of selecting someone who is left-handed but doesn't play a musical instrument?

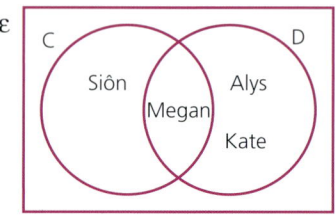

3. Medwyn carries out a survey to find out the favourite sports of his classmates. He displays the results in a Venn diagram.
   E is the set of classmates who like team sports.
   F is the set of classmates who like ball sports.
   A classmate is selected at random.
   a What is the probability of selecting a classmate who doesn't like team sports or ball sports?
   b What is the probability of selecting a classmate who likes ball sports?

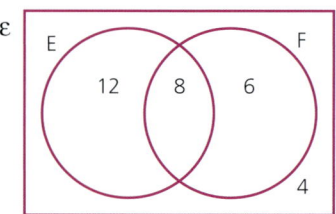

4. There are 20 cars in the car park. Ffion looks at the cars and records how many grey cars there are and how many of the cars have five doors. Ffion starts to display this information in a Venn diagram but does not fully complete it.
   G is the set of grey cars.
   H is the set of five-door cars.
   Ffion selects a car at random from the car park.
   What is the probability of selecting a car which isn't grey and doesn't have five doors?

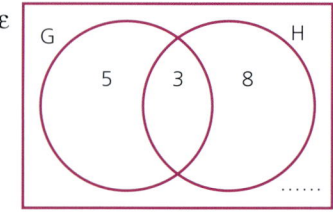

5. Castell Zoo has 30 animals, including 10 birds. The zoo creates a Venn diagram to display information about its animals but does not complete it fully.
   X is the set of birds.
   Y is the set of animals born at the zoo.
   An animal is selected at random from the zoo.
   What is the probability of selecting an animal which wasn't born at the zoo, or isn't a bird?

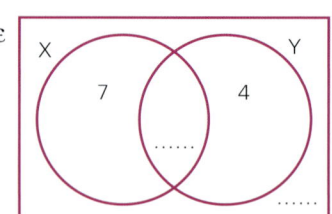

6. There are 28 students in a class.
   P is the set of students who own pets.
   Q is the set of students who are in a sports team.
   a How many students own pets?
   b How many students are in a sports team?
   c How many students do not have pets **and** are not in a sports team?

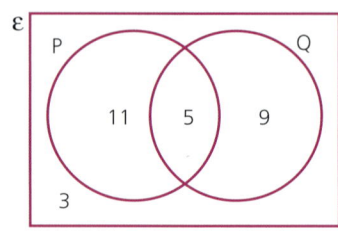

15 Probability

Band 2 questions

7 The Venn diagram shows the students studying GCSE PE. Some like watching cricket (C), and some like watching darts (D).

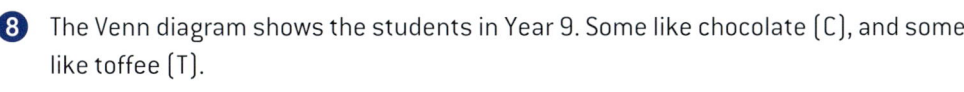

  a How many students like watching both darts and cricket?
  b What is the probability that a student, chosen at random, likes watching both darts and cricket?
  c What is the probability that a student, chosen at random, likes watching cricket?
  d What is the probability that a student, chosen at random, does not like to watch either darts or cricket?

8 The Venn diagram shows the students in Year 9. Some like chocolate (C), and some like toffee (T).

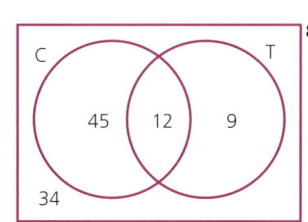

  a How many students like at least one out of toffee and chocolate?
  b What is the probability that a student, chosen at random,
   i likes chocolate   ii does not like toffee   iii likes toffee but does not like chocolate?

9 The Venn diagram shows the membership of a club.
Some members play squash (S), others play tennis (T).
Some play both games and some neither.

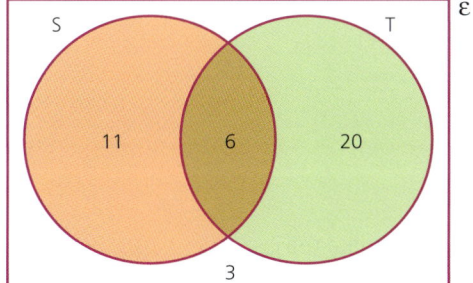

A member is selected at random.

  a Explain why the probability that the member plays squash is $\frac{17}{40}$.
  b Find the probability that the member plays both squash and tennis.
  c Find the probability that the member does not play squash.

10 This Venn diagram shows the prime factors of 70 and 60.
Use the Venn diagram to find

  a the highest common factor of 60 and 70. Justify your answer.
  b the lowest common multiple of 60 and 70. Justify your answer.

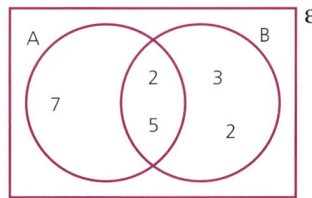

11 a Draw a Venn diagram to show the prime factors of 20 and 72.
  b Use the Venn diagram to find
   i the highest common factor of 20 and 72   ii the lowest common multiple of 20 and 72.
  c Use the comparing prime factorisations method to check your answers for
   i the highest common factor of 20 and 72   ii the lowest common multiple of 20 and 72.

345

12 There are 30 students in a class. Some are learning French (F) and some are learning German (G).

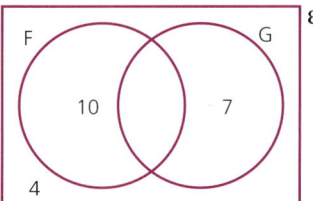

   a   How many students are not learning French nor German?
   b   How many students are learning both French and German?
   c   What is the probability that a student, chosen at random, is
      i   learning both French and German   **ii**   learning just French   **iii**   learning French?

Band 3 questions

13 Out of 80 members of a cricket club, 52 like coffee (C) and 37 like tea (T).

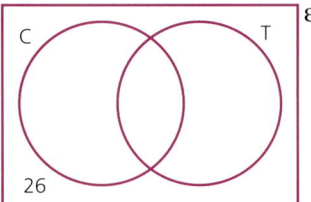

   a   Complete the Venn diagram.
   b   How many do not like either coffee or tea?
   c   How many like both coffee **and** tea?
   d   What is the probability that a club member, chosen at random, does not like tea?

14 Set A is the set of factors of 360.

Which of the following could **not** be a description of set B? Justify your answer.

   i   1, 2, 3, 4, 5, 6
   ii   the factors of 72
   iii   the multiples of 60
   iv   the factors of 45

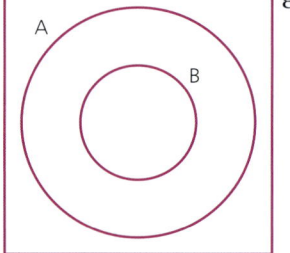

15 The diagram shows the percentage of students studying Physics, History and Geography at GCSE.

What is the probability that a student, chosen at random

   a   studies Geography
   b   does not study Physics
   c   studies History and Geography but not Physics?

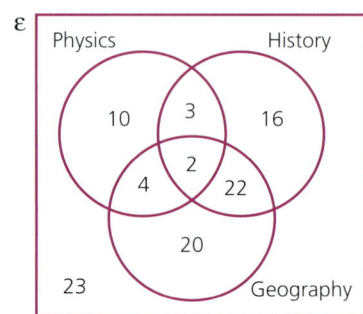

16 The Venn diagram shows the prime factors of 18, 54 and 60.

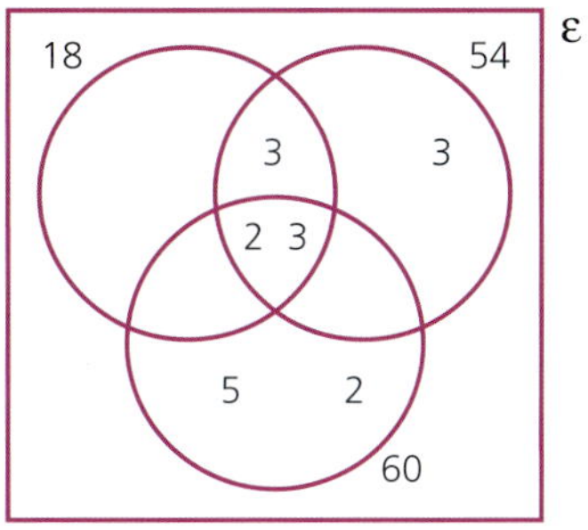

a What is the highest common factor of 18, 54 and 60? Explain how you can use the Venn diagram to find it.
b Explain why the top left-hand region is empty.
c How can you tell from the Venn diagram that 18 is a factor of 54?
d What is the lowest common multiple of 18, 54 and 60? Explain how you can use the Venn diagram to find it.

17 **a** Draw a Venn diagram to show the prime factors of 12, 42 and 105.
b What is the highest common factor of 12, 42 and 105?
c What is the lowest common multiple of 12, 42 and 105?
d Use the comparing prime factorisations method to check your answers for
 i the highest common factor of 12, 42 and 105
 ii the lowest common multiple of 12, 42 and 105.

18 Cai is eating biscuits from a box with four different types of biscuit.
8 biscuits have chocolate and fruit and no nuts.
7 biscuits have both chocolate and nuts, but no fruit.
6 biscuits have just nuts.
4 biscuits contain just fruit.
a Draw a Venn diagram for the biscuits in the box.
Cai chooses a biscuit at random from the box. He likes biscuits with chocolate.
b Find the probability
 i that Cai likes the biscuit
 ii that Cai does not like the biscuit.

15.3 Combined events

Skill checker

1. Simplify:
 a $\dfrac{12}{16}$
 b $\dfrac{30}{45}$

2. Work out:
 a $1 - \dfrac{1}{4}$
 b $1 - \dfrac{3}{8}$

3. Two 8-sided dice are thrown. How many outcomes are there?

4. A normal dice is rolled once. What is the probability of rolling a square number?

Discussion activity

When you flip a fair coin twice, are you more likely to get two tails, two heads or one of each?

Activity

Throw two dice and add the numbers to give a score. If that score is on your bingo card you cross it out.

Make up your own 3 by 3 bingo card with nine numbers on it.

You may repeat numbers on the card but only cross them off one at a time.

Play the game once.

a   What is your opinion of the numbers you chose? What were good choices and what were bad choices?

b   Make up a new bingo card and play again.

c   Were the numbers a better choice this time? What worked well? What can be improved?

d   Make up a new bingo card and play again.

> This game is best played as a class with the teacher rolling the dice. Each student has their own bingo card. The aim is to cross off all of the numbers on your card.

Worked example

Hannie throws two dice, one red and one green.

	1	2	3	4	5	6
1	2	3	4	5	6	7
2	3	4	5	6	7	8
3	4	5	6	7	8	9
4	5	6	7	8	9	10
5	6	7	8	9	10	11
6	7	8	9	10	11	12

What is the probability the total score is

a   exactly 3   b   3 or less   c   greater than 12   d   a prime number?

Solution

There are 36 possible equally likely outcomes.

a   Probability of exactly 3 $= \dfrac{2}{36} = \dfrac{1}{18}$ — There are 2 ways of getting 3, 1 + 2 and 2 + 1, so 2 favourable outcomes.

b   Probability of 3 or less $= \dfrac{3}{36} = \dfrac{1}{12}$ — There are 3 ways of getting 3 or less.

c   Probability of greater than 12 $= 0$ — It is impossible to get more than 12.

d   Probability of a prime number $= \dfrac{15}{36} = \dfrac{5}{12}$ — Prime numbers are 2 (1 way), 3 (2 ways), 5 (4 ways), 7 (6 ways) and 11 (2 ways).

Communication using symbols

15.3 Now try these

Band 1 questions

1 Eleri chooses a t-shirt and shorts.

T-shirt	Shorts
white	red
white	green
blue	red
blue	green
yellow	red
yellow	green

  a How many different outfits are there?

Eleri chooses an outfit at random.

  b Find the probability that she chooses
- **i** red shorts with a blue t-shirt
- **ii** green shorts with a white t-shirt
- **iii** an outfit with a yellow t-shirt.

2 Benny chooses a meal.

	Ice cream	Fruit salad
Curry	Curry, Ice cream	Curry, Fruit salad
Pizza	Pizza, Ice cream	Pizza, Fruit salad
Fish	Fish, Ice cream	Fish, Fruit salad
Chicken	Chicken, Ice cream	Chicken, Fruit salad

  a How many different meals can he choose?

Benny chooses a meal at random.

  b Find the probability that he chooses
- **i** fish and fruit salad
- **ii** a meal that includes pizza
- **iii** a meal that includes fruit salad.

3 Alex throws two coins. The table shows the possible outcomes.

		Second coin	
		Heads	Tails
First coin	Heads	HH	HT
	Tails	TH	TT

  a How many equally likely outcomes are there?

  b Find the probability that he throws
- **i** two heads
- **ii** one of each
- **iii** no heads.

4 Kris throws a dice and flips a coin.

The table shows the possible outcomes.

	1	2	3	4	5	6
H	H1	H2	H3	H4	H5	H6
T	T1	T2	T3	T4	T5	T6

- **a** Explain how you can work out that there are 12 different outcomes without counting them.
- **b** Find the probability that he gets
  - **i** a tail and an even number
  - **ii** a head
  - **iii** a tail and a square number
  - **iv** a head and an 8.

5 Mia throws two 4-sided dice. Each dice has its faces labelled 1, 2, 3 and 4.

		Second dice			
		1	2	3	4
First dice	1	2	3	4	5
	2	3	4	5	6
	3	4	5	6	7
	4	5	6	7	8

- **a** Explain how the numbers in the table have been calculated.
- **b** Find the probability of getting a score of
  - **i** 8
  - **ii** 5
  - **iii** 1.
- **c** Find the probability of getting a score between 1 and 10.

6 In a game, two fair 8-sided dice numbered from 1 to 8 are thrown and the scores shown added together. The table shows the different totals possible.

		Dice 2							
	+	1	2	3	4	5	6	7	8
Dice 1	1	2	3	4	5	6	7	8	9
	2	3	4	5	6	7	8	9	10
	3	4	5	6	7	8	9	10	11
	4	5	6	7	8	9	10	11	12
	5	6	7	8	9	10	11	12	13
	6	7	8	9	10	11	12	13	14
	7	8	9	10	11	12	13	14	15
	8	9	10	11	12	13	14	15	16

- **a** How many different equally likely outcomes are there?
- **b** How many different totals are there?
- **c** Find the probability that the total is
  - **i** an even number
  - **ii** a prime number
  - **iii** a square number
  - **iv** a multiple of 5
  - **v** a triangle number.

> **Remember**
> A triangle number has a pattern of dots in the shape of a triangle.
> 1, 3, 6, 10, …

Band 2 questions

7 The table shows the totals when the spinners are spun.

		Blue							
		1	2	3	4	5	6	7	8
Red	1	2	3	4	5	6	7	8	9
	2	3	4	5	6	7	8	9	10
	3	4	5	6	7	8	9	10	11
	4	5	6	7	8	9	10	11	12

  a Write a list of all the possible totals.
  b Find the probability of
    i getting a total of 7
    ii getting a total of less than 5
    iii getting a total of at least 9.
  c An event has a probability of $\frac{3}{32}$. Give two examples of what it could be.

8 Some children are playing Snakes and Ladders using two dice labelled 0, 1, 2, 3, 4, 5.
The table shows all the possible totals when the two dice are thrown.

	0	1	2	3	4	5
0	0	1	2	3	4	5
1	1	2	3	4	5	6
2	2	3	4	5	6	7
3	3	4	5	6	7	8
4	4	5	6	7	8	9
5	5	6	7	8	9	10

  a Siôn has the next turn and needs 5 to win. What is the probability that he wins on this turn?
  b Mark will land on a snake if he gets 10.
    What is the probability that he lands on a snake?
  c Ffion will land on a ladder if she gets 4 or 11.
    What is the probability that she lands on a ladder?

9 A spinner has five equal sections, numbered 0 to 4.
Tristan spins it twice and multiplies the two numbers.
  a Copy and complete the table to show all the possible outcomes.
  b How many different equally likely outcomes are there?
  c How many different scores are there?
  d Find the probability that Tristan scores
    i 16
    ii an odd number
    iii a square number
    iv 0.

	0	1	2	3	4
0	0	0			
1					
2			2		
3					
4					16

Fluency

10 Sanjit has two bags.

Each bag contains four balls: one white, one green, one black and one yellow.

He picks a ball at random from his first bag.

Then, he picks a ball at random from his second bag.

a Copy and complete this table to show all possible outcomes.

b What is the probability that he chooses two balls of the same colour?

c What is the probability that he chooses at least one green ball?

d What is the probability that he chooses one yellow ball and one of a different colour?

		First ball			
		W	G	B	Y
Second ball	W	WW			
	G				
	B				
	Y				

Strategic competence

11 Two dice are used in a game.

One has four faces, labelled 1, 3, 5 and 7.

The other has eight faces, labelled 1 to 8.

The score is the **difference** between the numbers on the two dice.

a Make a table showing all the possible outcomes.

b Find the probability of scoring

  i 0   ii 1   iii 2   iv 5 or more.

c What score has a probability of $\frac{1}{32}$?

12 A class is playing bingo.

Numbers are generated by rolling two 6-sided dice, with faces numbered 1 to 6, and adding the scores.

Each of the children in the class chooses four numbers.

They win when all of their numbers have been generated.

What numbers would you have on your card? (You can have the same number more than once if you wish.)

Explain your choices.

Fluency

13 Owen and Efa are playing a game with one red dice and one blue dice.

The red dice is numbered 1, 1, 3, 4, 5, 5.

The blue dice is numbered 2, 2, 3, 4, 5, 6.

a Copy and complete the table of the possible outcomes.

		Blue dice					
		2	2	3	4	5	6
Red dice	1	1, 2	1, 2				1, 6
	1	1, 2					
	3						
	4						
	5	5, 2					
	5						5, 6

b Find the probability of

  i a 1 on the red dice and a 2 on the blue dice

  ii an even number on both dice

  iii an odd number on both dice

  iv a 2 on both dice.

Band 3 questions

14 Karl buys two raffle tickets. He draws a probability diagram.

	Win	Lose
Win	WW	WL
Lose	LW	LL

Karl says, 'Three of the outcomes have me winning a prize. That means I have a 75% chance of winning.'

Explain why Karl is wrong.

15 Mari has two bags.

One bag contains five red discs numbered 1 to 5.

The other bag contains six blue discs numbered 1 to 6.

Mari picks one disc from each bag at random.

Calculate

   a   P(same score on both discs)

   b   P(score on the red disc is greater than the score on the blue disc)

   c   P(score on the red disc is less than the score on the blue disc)

   d   P(scores add up to 7).

16 Catrin has two bags.

One bag contains five black marbles numbered 1 to 5.

The second bag contains five white marbles numbered 1 to 5.

Catrin picks one black marble and one white marble at random.

   a   Draw a probability space diagram to show the possible outcomes.

   b   Find the probability that Catrin gets

       i   two even numbers

       ii two odd numbers.

   c   What is the probability that Catrin gets

       i   identical numbers

       ii a bigger number on the white

       iii a bigger number on the black?

   d   Add together your answers from part **c**. Explain the result.

17 Here are nine heart cards numbered 2 to 10 and nine spade cards numbered 2 to 10.

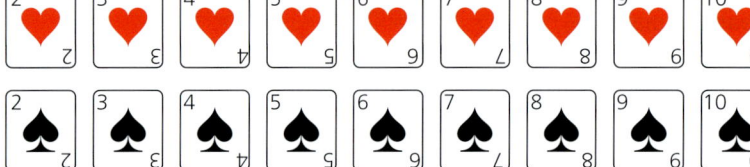

Suja shuffles the heart cards and places them face down in a row. She does the same with the spade cards.

Suja chooses one card from each row at random.

Work out the probability that

a the two numbers are the same
b the two numbers are different
c the number on the heart card is bigger
d the two numbers are both even
e the two numbers are both prime
f the two numbers are both square numbers.

18 In a game, a five-sided spinner and a six-sided spinner are spun, and their scores are added together.

Work out the probability of scoring

a a total of 6
b a total of 7
c a total of 6 or 7
d any total other than 6 or 7
e an even total
f an odd total
g a total that is a square number
h a total that is not a square number.

19 Two identical dice have six faces. They are rolled and the numbers are added.

What numbers are on their faces if

- P(even number) = 1
- there is at least one odd number on the dice
- $P(2) = P(22) = \frac{1}{36}$
- the numbers on the dice form an arithmetic sequence
- 12 is the modal score.

Key words

Here is a list of the key words you met in this chapter.

Probability space diagram   Two-way table   Venn diagram

Use the glossary at the back of this book to check any you are unsure about.

15 Probability

Review exercise: probability

Band 1 questions

1. Caryl takes five tops and three pairs of shorts on holiday. She chooses one top and one pair of shorts at random one day.

   a Make a list of all the possible combinations she could choose.

   b What is the probability that she chooses a top and a pair of shorts which are the same colour?

2. The prime factors of 30 are 2, 3 and 5 because $30 = 2 \times 3 \times 5$.

   The prime factors of 55 are 5 and 11 because $55 = 5 \times 11$.

   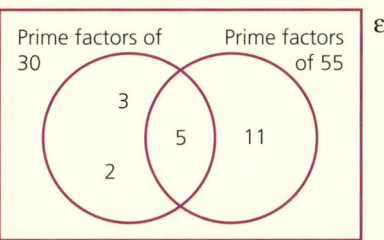

   a What number is a prime factor of both 30 and 55?

   b Work out the product of all four of the numbers in the circles.

   c Explain why the answer to **b** is a multiple of 30.

3. Sadiq chooses a drink and a snack. The drinks available are tea, coffee and orange juice.

   The snacks available are pastries, muffins and cheese scones.

   a Make a list of the combinations of drink and snack that Sadiq can have.

   b Explain the system you used in **a** to make sure you listed all of the possibilities.

4. The numbers from 1 to 10 are shown in the diagram.

   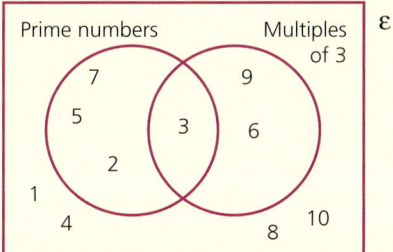

   a How many of the numbers are prime numbers?

   b How many of the numbers are **not** multiples of 3?

   c A number from 1 to 10 is chosen at random. Find the probability that it is not a prime number.

5 A bag contains one red cube, one green cube, one blue cube and one white cube. One cube is removed from the bag, replaced and another cube taken from the bag.

The probability space diagram shows the possible outcomes.

		Second cube			
		R	G	B	W
First cube	R	RR	RG	RB	RW
	G	GR	GG	GB	GW
	B	BR	BG	BB	BW
	W	WR	WG	WB	WW

a How many of the outcomes have at least one green cube?

b Find the probability of picking

  i two blue cubes

  ii two cubes of the same colour

  iii two cubes with different colours.

c Explain why the answers to **ii** and **iii** add to 1.

6 30 people working in an office are asked if they like to watch football and tennis.

The diagram shows the results.

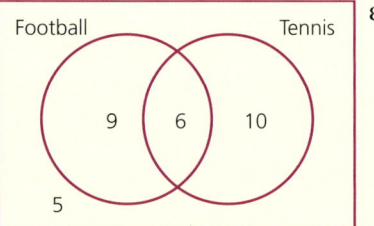

a How many people like to watch football?

b How many people like to watch tennis?

c One of the people, who said they liked watching both football and tennis, changes their mind and says they only like watching football. What changes in the diagram?

7 Siân throws two 4-sided dice. Each dice has its faces labelled 1, 1, 2 and 4. The numbers showing on the two dice are added to get her score.

She draws a probability space diagram to show the outcomes.

		Second dice			
		1	1	2	4
First dice	1	2	2	3	5
	1	2	2	3	5
	2	3	3	4	6
	4	5	5	6	8

a How many equally likely outcomes are there?

b How many different scores are there?

c Find the probability that

  i the score is 2

  ii the score is 5 or more

  iii the score is **not** 3.

15 Probability

Band 2 questions

8 Sophie has two spinners with sides numbered 1, 2 and 3.

She spins both spinners and adds the results.

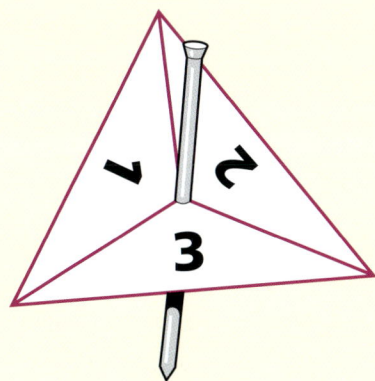

a Copy and complete the table to show all the possible outcomes.

		Second spinner		
		1	2	3
First spinner	1		3	
	2			5
	3			

b What is
  i the most likely total
  ii the least likely total?

c Work out the probability that the total is
  i 5
  ii less than 4
  iii a square number
  iv a prime number.

9 In a game, two fair 6-sided dice numbered from 1 to 6 are thrown. The score is the higher number out of the two.
  a Draw a probability space diagram to show all the possible outcomes.
  b What is the least likely outcome?
  c Find the probability of
    i scoring 4
    ii scoring more than 3
    iii scoring at least 3.

10 a Write these as products of their prime factors.
    i 28
    ii 48
  b Copy and complete the Venn diagram to show the prime factors of 28 and 48.

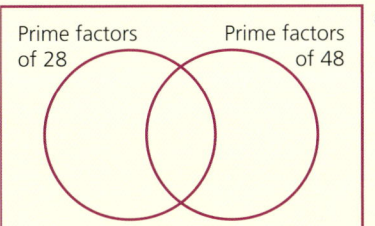

  c Use the Venn diagram to find
    i the highest common factor of 28 and 48
    ii the lowest common multiple of 28 and 48.
  d Use the comparing prime factorisations method to check your answers for
    i the highest common factor of 28 and 48
    ii the lowest common multiple of 28 and 48.

11. At Ysgol Maes, 18% of students study GCSE History but not Geography and 33% of them study GCSE Geography but not History. 30% of the students do not study either History or Geography.

   a What percentage of students study both History and Geography?
   b Copy and complete the Venn diagram.

   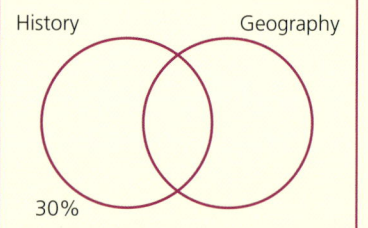

   c Find the probability that a student chosen at random
      i   studies History
      ii does not study Geography
      iii studies History but not Geography.

12. A bag contains two red balls and two blue balls. Fflur removes one ball at random and then replaces it before choosing another at random.
   a Draw a probability space diagram to show the 16 equally likely outcomes.
   b Find the probability of choosing
      i   two balls of the same colour
      ii at least one red ball
      iii a red ball as the first ball.

Band 3 questions

13. Anna collects information on the pets owned by students in Year 9 at her school. She presents the information in a Venn diagram.

    There are 120 students in Year 9. 48 students have cats as pets and 62 students have dogs as pets. There are 33 students who do not have a cat or a dog.

    Draw a Venn diagram to show this information.

14. Janek runs a tourist information website in his local town.

    He wants to put accommodation information on his site in the easiest way for people to understand.

    He has information about the hotels they use in a Venn diagram.

    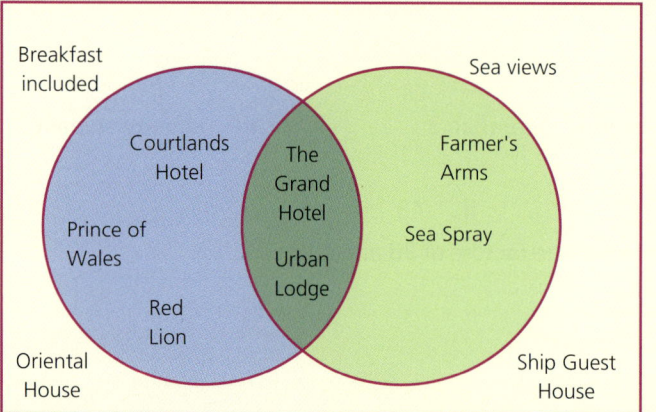

    However, Janek thinks it would be better to show the information in a two-way table instead.
    a Suggest suitable headings for the rows and columns of the two-way table.
    b Complete a two-way table for Janek.
    c What other information would Janek need to give customers coming to the website who wanted to book accommodation in the local area?

15 Rudi throws two dice, each labelled with the numbers 1 to 6. The score is the highest common factor of the two numbers.
   a Draw a probability space diagram to show all the outcomes.
   b What is the most likely score? Explain why that is the case.
   c Find the probability of
      i scoring more than 1
      ii scoring a number greater than 6
      iii scoring an even number.

16 Bag A contains 3 white balls and 2 red balls. Bag B contains 2 white balls and 3 red balls. A ball is taken at random from each bag.
Find the probability that they are different colours.

17 Lili has a set of cards with the numbers from 2 to 20 printed on one side.
   a Mark each one in its correct position on a copy of this Venn diagram.

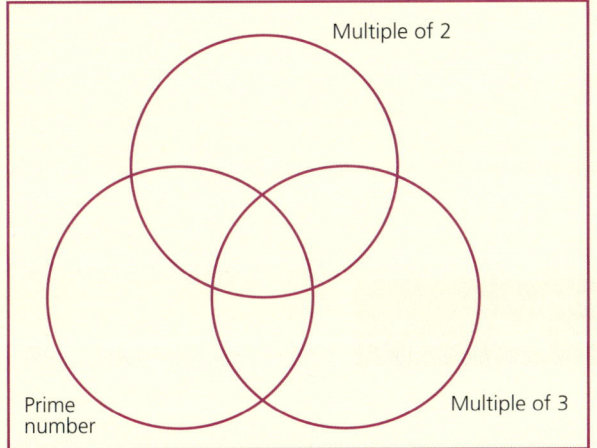

   b A card is picked at random. What is the probability that it is
      i a multiple of 2   **ii** a multiple of 3   **iii** a multiple of both 2 and 3
      iv a prime number   **v** a prime multiple of 3?
   c Why is the centre intersection empty?

18 **a** Draw a Venn diagram to find the lowest common multiple of 18, 21 and 70.
   b What is the highest common factor of 18, 21 and 70?

19 Design two spinners so that
   • there are 15 equally likely outcomes
   • the score is the sum of the numbers on the spinners
   • the highest score is 11
   • the lowest score is 3
   • the probability of scoring 7 is 0.2.

20 Two bags contain identical numbers of red and blue balls. There are 5 balls in each bag.
There are more red balls than blue balls in each bag.
A ball is removed, at random, from each bag.
The probability that the balls are different colours is 0.32.
How many blue balls are in each bag?

Consolidation 6: Chapters 14–15

Band 1 questions

1. Decide whether the following sets of data are categorical, discrete or continuous.
   a   the number of eggs laid by garden birds
   b   the colours of garden birds
   c   the wingspans of garden birds
   d   species of garden birds
   e   weights of garden birds.

2. A bus company is interested in the distance travelled to work by local residents.
   They asked a random sample of 100 workers and recorded the results in this table.

Distance, x miles	$1 \leq x < 3$	$3 \leq x < 5$	$5 \leq x < 7$	$7 \leq x < 9$	$9 \leq x < 11$
Frequency	10	25	50	5	10

   a   What is the modal class?
   b   Estimate
      i   the mean
      ii the range.
   c   What use is information like this to a bus company?

3. Shani throws two 4-sided dice. The dice are labelled with the numbers 1 to 4.
   The score is the difference between the numbers.
   a   Copy and complete the probability space diagram.

		Second dice			
		1	2	3	4
First dice	1				
	2				
	3				
	4				

   b   What is the mode of the scores?
   c   Find the probability of the modal score.

4. Dafydd is carrying out a survey of the number of football training sessions his teammates attend and the number of goals they score in games.
   Write a hypothesis involving the number of training sessions attended and the number of goals scored in games.

5. Megan wants to find out how much money her friends save each week.
   She includes the question below on her questionnaire.
   How much money do you save each week?
   ☐ £1–£2
   ☐ £3–£4
   ☐ £5 or more
   Write down two criticisms of her question.

6. 20 students in Year 9 are asked whether they like fruit or vegetables.
   The Venn diagrams shows their responses.
   a   How many students like both fruit and vegetables?
   b   Find the probability that a student chosen at random does not like vegetables.

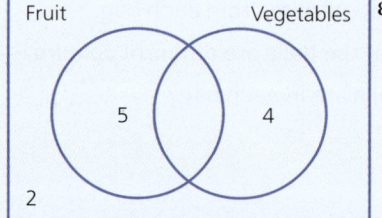

7 Pedr and Rhian wonder whether people with long arms also have long legs.

They collect data from 10 of their friends.

Person	1	2	3	4	5	6	7	8	9	10
Arm length (cm)	42	45	43	42	43	45	36	48	43	44
Leg length (cm)	66	71	65	64	68	68	64	75	66	71

a Plot the data in a scatter diagram.
b Describe the correlation.
c Do people with long arms also have long legs?
d Which point is an outlier? Should it be included in the data?

8 Tomi measured the handspans of 30 students in Year 9.

a Find the probability that a student, chosen at random, has a handspan between 16 and 17 cm.
b Find the probability that a student, chosen at random, has a handspan greater than 16 cm.

Handspan, h cm	Frequency
$14 \leqslant h < 15$	1
$15 \leqslant h < 16$	4
$16 \leqslant h < 17$	7
$17 \leqslant h < 18$	14
$18 \leqslant h < 19$	4

Band 2 questions

9 The grouped frequency table gives information about the distance 100 students travel to school.

Distance travelled, d km	Frequency
$0 \leqslant d < 8$	40
$8 \leqslant d < 16$	25
$16 \leqslant d < 24$	15
$24 \leqslant d < 32$	8
$32 \leqslant d < 40$	12

a What percentage of the 100 students travel at least 24 km to school?
b Calculate an estimate for the mean distance travelled to school by the students.
c What is the probability that a student, chosen at random, travels at least 24 km to school?

10 Ffion wants to find out the views of pupils in her class about the new school badge.
She uses this question:

- The new badge is really nice, isn't it?
  ☐ Yes   ☐ It's ok   ☐ No

a Write down two criticisms of her question.
b Write a better version of her question, with response boxes.

11 Mr Jones wants to open a new restaurant in Holyhead.
He wants to find out what types of food people like.

a Write a question, with response boxes, which he could use to find out what types of food people like.
b Mr Jones posts his questionnaire to 150 people across Wales.
Explain why this might not be sensible.
c Mr Jones asks his friends, 'Aren't pizzas better than burgers?'
Give two reasons why this might not be a good way for Mr Jones to find out what people in his town like to eat.

12. Two 6-sided dice are labelled with the numbers 0 to 5.
    The dice are thrown, and the score is calculated by multiplying the numbers.
    a Copy and complete the probability space diagram for the outcomes.

		Second dice					
		0	1	2	3	4	5
First dice	0						
	1		1				
	2	0					
	3						15
	4						
	5						

   b Find the probability of
      i a score of 0   ii a score greater than 1   iii a score greater than 30.

13. Emily collects the ages of 60 people taking their driving test over one week at the local test centre.
    40, 24, 32, 17, 18, 21, 29, 18, 39, 17, 20, 30, 17, 19, 17, 18, 21, 30, 46, 17,
    26, 19, 17, 21, 18, 19, 28, 18, 21, 37, 19, 31, 40, 18, 21, 18, 25, 18, 40, 18,
    33, 18, 21, 18, 44, 19, 21, 19, 19, 25, 19, 20, 21, 22, 23, 27, 30, 31, 39, 50
    a Construct a grouped frequency table for this data set.
    b Why aren't any of the ages less than 17?
    c What is the modal age group? Why do you think this might be?
    d Find the group with the median age. Do you think this is a useful measure for this data set?
    e Draw a frequency diagram of this data.
    f What do you notice about the shape of the frequency diagram? Give a reason for this.
    g What percentage of people who took their test were over 20?

14. Rani planted 15 packets of rare seeds. She recorded the number, n, that germinated from each packet. Her results are given in this table.

Number germinating, n	Frequency
$10 \leqslant n < 15$	3
$15 \leqslant n < 20$	6
$20 \leqslant n < 25$	4
$25 \leqslant n < 30$	2

   a Calculate the mean number of seeds germinating from each packet.
   b A packet of seeds is chosen at random.
      What is the probability that more than 19 of the seeds will germinate?

15. A random sample of 100 people are tested for immunity against two diseases, P and Q.
    20 are immune to P only and 30 are immune to Q only.
    15 are immune to both P and Q.
    The rest are not immune to either disease.
    a Copy the Venn diagram and fill in the missing numbers.
    b A person is chosen at random. Calculate the probability that the person is immune to
       i both diseases   ii neither disease   iii disease Q.

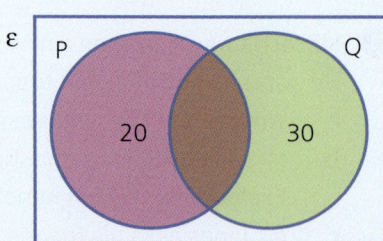

16 The table shows the child mortality and income for nine countries.
The child mortality measure is the number of deaths before the age of five, per 1000 live births.
The income is GDP per person in international dollars.

Country	Botswana	Brazil	Columbia	Egypt	Iran	Mexico	NZ	Philippines	Togo
Child mortality	51.3	63	35.2	85.8	53.7	44.8	11.2	56.7	145
Income	8500	10 300	7730	5900	11 600	13 600	24 000	4010	1300

a Plot a scatter diagram to show this information.
b Describe the correlation and interpret it in context.
c Draw a line of best fit.
d Belarus has a child mortality rate of 15.2. Use the line of best fit to estimate its income.

17 Tegan uses a Venn diagram to find the highest common factor of 30 and 48.

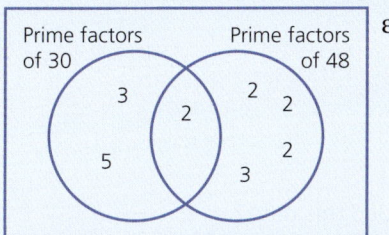

She says that the highest common factor of 30 and 48 is 2.
Explain why Tegan is wrong and what mistake she has made.

18 Dylan measures the length of time it takes for people to do a puzzle.
The results are shown in the frequency polygon.
a How many people did Dylan time doing the puzzle?
b What is the modal group?
c What is the median group?
d Calculate an estimate for the mean.

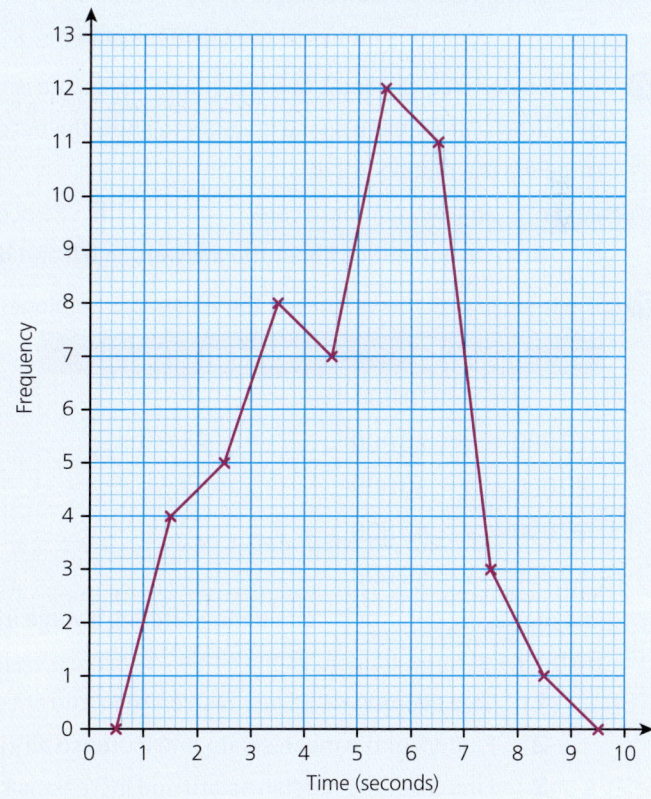

Curriculum for Wales Mastering Mathematics: Book 3

Band 3 questions

19 Use a Venn diagram to find the highest common factor and lowest common multiple of 108, 140 and 300.

20 An ice-cream seller kept a daily record of the highest temperature and the number of ice-creams sold.
The results are shown below.

Temperature (°C)	32	30	31	27	20	25	23	14	27	21	19	18	21
Number of ice-creams sold	190	180	188	156	36	150	150	100	160	158	136	186	156

a Show this data on a scatter diagram.
b Describe any correlation between the two variables.
c How is this information useful to the ice-cream seller?
d There is an outlier in the data. State which it is and suggest a reason for it.
e What is the maximum number of ice-creams sold in any one day?
Why might this information be useful to the ice-cream seller?
f What is the temperature range over the time the ice-cream seller recorded this data?
g The next day the temperature is forecast to be around 17 °C.
How many ice-creams do you estimate will be sold?
h The bank asks for a business plan.
Calculate the mean number of ice-creams sold per day.
How would this influence the business plan?

21 Two 6-sided dice labelled 1 to 6 are thrown. The score is the lowest common multiple of the two numbers on the dice.
a Draw a probability space diagram to show the outcomes.
b Find the probability of
  i an even number score
  ii a score of less than 3
  iii a score of at least 3.
c What do you notice about the answers to **b ii** and **b iii**? Explain why that is so.

22 The table shows the salaries of the 25 people employed by a small firm.

Salary, £s	Number of people
$0 \leqslant s < 10\,000$	6
$10\,000 \leqslant s < 20\,000$	11
$20\,000 \leqslant s < 30\,000$	5
$30\,000 \leqslant s < 40\,000$	2
$40\,000 \leqslant s < 50\,000$	1

The managing director is considering offering a wage increase of 5%.
He thinks of three ways to achieve this:
  1 Give everyone a 5% increase on their current salary.
  2 Find 5% of the mean salary and increase all salaries by this amount.
  3 Find 5% of the median salary and increase all salaries by this amount.
a Which members of staff would benefit most from each scheme?
b Which scheme do you think he should choose? Give a reason for your answer.

Consolidation 6

23 Dewi is collecting data on the heights of adult men for a clothing company.

He measures the heights of 50 adult men.

Maya collects similar data from 50 women.

a Draw frequency diagrams for the men's and women's heights.

b Describe what your graphs show.

c What are the modal heights?

d i Which gender has the greater range of heights?

ii What effect will this have on the garments produced by the clothing company?

e What other data might it be useful for the clothing company to collect?

f What is the probability that an adult, chosen at random, is between 165 and 175 cm tall?

Height, h cm	Women frequency	Men frequency
$135 \leq h < 140$	1	0
$140 \leq h < 145$	1	0
$145 \leq h < 150$	1	0
$150 \leq h < 155$	2	1
$155 \leq h < 160$	3	2
$160 \leq h < 165$	11	5
$165 \leq h < 170$	14	8
$170 \leq h < 175$	9	9
$175 \leq h < 180$	5	12
$180 \leq h < 185$	2	7
$185 \leq h < 190$	1	4
$190 \leq h < 195$	0	2

24 Use the information below to complete the Venn diagram representing the percentage of students studying Mathematics, English and Biology in Sixth Form.

20% of students do not study any of the three subjects.

41% of students study Biology.

40% of students study English.

47% of students study Mathematics.

7% of students study all three subjects.

18% of students study Mathematics and English.

20% of students study Mathematics and Biology.

11% of students study Biology but not Mathematics or English.

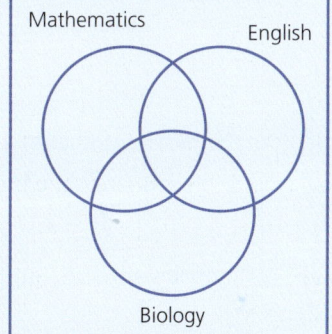

25 Write a hypothesis involving time spent watching television and time spent doing homework. What data must be collected to test this hypothesis? Design a suitable questionnaire to collect this data.

26 Two 6-sided dice labelled 1 to 6 are thrown. The score is the mean of the two numbers on the dice.

a Draw a probability space diagram to show the outcomes.

b Find the probability of

i a score that is not a whole number

ii a score of 3.5

iii a score of at least 6.5.

c What do you notice about the numbers in the diagram? Explain why that is so.

Glossary

3D shape A solid shape with three dimensions: length, width and depth.

Accuracy The accuracy of a measurement tells you how close it is to the actual value.

Acute An acute angle lies between 0° and 90°.

Addition (add) Addition or adding is finding the total of two or more amounts. For example, 3 + 4 = 7.

Adjacent Adjacent means next to. In a right-angled triangle the adjacent is the side which is next to the angle.

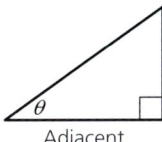
Adjacent

Allied angles A pair of angles on the inside of a pair of parallel lines and on the same side of a transversal. Allied angles are sometimes referred to as C angles because of the shape they make.

Alternate angles A pair of angles on the inside of a pair of parallel lines, but on opposite sides of a transversal.

Angle An angle is a measure of turning. Angles are measured in degrees. For example, a full turn is 360°.

Angle of depression When you look down at an object then the angle of depression is the angle between the horizontal and the line of sight.

Angle of elevation When you look up at an object then the angle of elevation is the angle between the horizontal and the line of sight.

Appreciation The increase in value of a product over time.

Approximate The answer to a calculation which has been found by rounding the numbers in the calculation. An approximate answer is close to the solution but not the exact solution.

Area The area of a shape is the amount of surface that it covers. Area is measured in square units such as mm^2, cm^2, m^2 and km^2.

Arithmetic sequence An arithmetic sequence is a sequence where the difference between any two consecutive terms is a constant. For example, 7, 11, 15, 19, …

Arrowhead An arrowhead is a kite with a reflex angle.

Average An average is a measure of the typical value in a data set. Common averages are the mean, the mode and the median.

Axes Axes is the plural of axis. An axis is a fixed reference line for the measurement of coordinates. The x-axis is horizontal and the y-axis is vertical.

Bar chart A bar chart is a chart that uses rectangular bars to display data. The height of each bar represents the frequency.

Base The base of a shape is the line or surface on which it appears to be standing.

Bank A bank is a financial institution licensed to receive deposits and make loans. Banks may also provide other financial services, such as foreign currency exchange.

Bearing An angle measured clockwise from North.

Best buy The cheaper or cheapest option when a decision must be made between two or more similar items for sale. The best buy can often be decided using the unitary method.

Biased Data is biased if some outcomes occur more or less than would be expected. For example, a dice is biased if the number 6 hardly ever comes up.

BIDMAS BIDMAS is a way of remembering the order in which you carry out the different operations in a calculation: Brackets, Indices, Division, Multiplication, Addition and Subtraction.

Bisector A bisector is a line which cuts an angle, line or shape in half.

Budget A budget is a spending plan based on income and expenses.

Bureau de change A bureau de change is a business where people can exchange one currency for another.

Brackets Brackets are a way of grouping numbers or algebraic terms together. For example, (3 + 5) ÷ 2 means the same as 8 ÷ 2.

Cancel To cancel a fraction is to simplify it by dividing the numerator and denominator by a common factor.

Capacity The capacity of a 3D shape is the volume it can hold. It is measured in cubic units such as mm^3 and cm^3, or for liquids litre (l) and millilitre (ml).

Categorical Categorical data is non-numerical data that can be put into groups such as favourite colour or types of pet.

Centilitre 1 centilitre (1 cl) is one hundredth of a litre.

Centimetre 1 centimetre (1 cm) is one hundredth of a metre.

Centre of rotation The centre of rotation is the 'pivot' point about which an object is rotated.

Certain If an event is certain, its probability is 1. For example, it is certain that the Sun will rise tomorrow.

Chance The chance of something happening is how likely it is to happen. Words can be used – it is unlikely that Sam will get up before midday. Or numbers can be used – the probability of Sam getting up before midday is about 0.09.

Circle A circle is a shape made up of all the points that are a specific distance from the centre.

Circumference The circumference is the perimeter of a circle.

Clockwise The direction followed by the hands of a clock.

Coefficient The number written in front of a letter.

Co-interior angles See allied angles.

Column method Digits are written with units in the same column, tens in the same column, hundreds in the same column, etc.

Common denominator The common denominator of two or more fractions is a common multiple of all of the denominators. It is used when adding or subtracting fractions.

Common factor A common factor of two or more numbers is a number that divides exactly into all the numbers. You simplify fractions by cancelling by a common factor.

Commutative An operation is commutative if the order in which you do it doesn't matter. Addition and multiplication are commutative operations.

Compound interest Compound interest is when interest is compounded. This means that when interest is added to an amount in one time period, the added interest is then part of the amount on which the next period's interest is calculated.

Compound measure A quantity that is made up from more than one other quantity. Examples include density (which uses mass and volume) and speed (which uses distance and time).

Concave All sides are oriented inwards.

Concentric Concentric circles have the same centre but different radii (sizes). They look like ripples after a stone is dropped into water.

Cone A cone is a 3D shape that has two faces: a flat circular base and a single curved face forming a single vertex. It could be described as a circular based pyramid.

Congruent Congruent shapes are exactly the same shape and size – they are identical.

Construct Draw accurately using only a compass and a straight edge.

Continuous data Continuous data does not have gaps in the possible values.

Conversion graph A conversion graph is used to change one unit into another. This could be changing between miles and kilometres, pounds to a foreign currency, or the cost of a journey based on the number of miles travelled.

Convex At least one side is oriented outwards.

Coordinates Coordinates are a way of showing position on a pair of axes or graph. For example, the point $(3, -5)$ is 3 to the right and 5 down.

Correlation Correlation is when there appears to be a link between how one measure changes and how another measure changes.

Corresponding angles A pair of angles in identical positions relative to a transversal.

Corresponding sides Equivalent sides on a pair of shapes (the shapes are either congruent or one is an enlargement of the other).

Cosine The cosine of an angle, cos θ, in a right-angled triangle is the ratio of the side adjacent to the angle to the hypotenuse: $\cos \theta = \dfrac{\text{adjacent}}{\text{hypotenuse}}$

Cube (number) To find the cube of a number you multiply the number by itself twice. For example, the cube of 5.6 (5.6^3) is $5.6 \times 5.6 \times 5.6$ or 175.616.

Cube (shape) A cube is a 3D shape with six identical square faces.

Cube root The cube root of a number is the number that, when multiplied by itself twice, gives the original number. The inverse of cubing is cube rooting. For example, the cube root of 27 is 3 (as $3 \times 3 \times 3 = 27$). The symbol $\sqrt[3]{}$ is used for the cube root of a number, so $\sqrt[3]{27} = 3$.

Cuboid A 3D shape whose six faces are all rectangular or square. A rectangular prism.

Currency The money used in a country, for example the currency of the UK is pounds.

Currency conversion The process of changing one currency to another, for example pounds to euros. A currency conversion can be done using a conversion graph.

Curve For example, the circumference of a circle. A curve can also be the shape of a graph described by the relationship between two variables, for example $y = x^2$.

Cylinder A cylinder is a 3D shape with a constant circular cross-section.

Data A group of facts or statistics which can be numerical (for example heights) or non-numerical (for example favourite crisp flavour).

Decagon A ten-sided polygon.

Decimal A decimal is a number written using a decimal point. For example, 82.17. The digits after the decimal point represent a value less than one.

Decimal place The number of decimal places of a decimal is the number of digits after the decimal point. For example, 3.2 is written to one decimal place and 5.678 is written to three decimal places.

Glossary

Decimal point The decimal point is the dot in a decimal number. The digits before the decimal point represent whole numbers, the digits after the decimal point represent fractions. For example, the number 4.37 means four units plus three tenths plus seven hundredths.

Denominator The denominator is the bottom integer of a fraction. It tells you how many equal parts the whole is divided into. For example, $\frac{3}{8}$ has a denominator of 8, so the 'whole' has been divided into eight equal parts.

Density A type of compound measure measuring the mass per m³ or per cm³, etc.
Density = Mass ÷ Volume.

Depreciation The decrease in value of a product over time.

Diameter A diameter is a line that passes through the centre of a circle and joins two points on the circumference.

Difference The difference between two numbers is the result of subtracting the smaller from the larger. For example, the difference between 15 and 6 is $15 - 6 = 9$.

Digit A digit is one of the symbols 0, 1, 2, 3, 4, 5, 6, 7, 8 or 9. Digits are used to write numbers.

Dimension A dimension is a measure of the size of something. For example, height, length or width.

Directed A number with a positive or a negative sign in front of it. If a sign is missing, then we assume the number is positive.

Direct proportion Two quantities are in direct proportion if, when one increases, the other also increases. An example is the number of items bought and their total cost.

Discrete data Discrete data does not include any in between values. It has gaps.

Distance-time graph See Travel graph.

Divide To divide two numbers you share the first number into the number of equal parts given by the second number. For example, 24 divided by 3 is 8 ($24 ÷ 3 = 8$).

Division Division is the process of dividing two numbers.

Divisor In a division the divisor is the number you divide by. For example, in the division $24 ÷ 3$ the divisor is 3.

Edge In a 3D shape, an edge forms the boundary between two faces. In a cube, for example, there are 12 edges that form the boundaries between the 6 faces.

Enlargement An enlargement changes the size of an object. When a shape is enlarged, the image is mathematically similar to the object but of a different size.
Note: the image may be smaller than the object.

Equal If two quantities or expressions are equal, they have the same value. The symbol = is used to show equality. For example, $14 + 8 = 22$.

Equally likely In probability, if two events are equally likely they have the same probability of happening. For example, when you roll a fair six-sided dice the possible outcomes of 1, 2, 3, 4, 5 or 6 are equally likely. They each have a probability of $\frac{1}{6}$.

Equation An equation is where one expression is equal to another. An equation always contains an equals sign. For example, $6 + 4 = 16 - 6$. When an equation contains an unknown it can be solved. For example, the solution to the equation $x + 4 = 16 - x$ is $x = 6$.

Equilateral An equilateral triangle has three equal angles (all 60°) and three sides of equal length.

Equivalent Equivalent fractions represent the same value. For example, $\frac{3}{5}$ and $\frac{9}{15}$ are equivalent.

Estimate To estimate the answer to a calculation is to use rounding to find an approximate answer.

Evaluate Evaluate means 'work out the value of'.

Even An even number is an integer that is a multiple of 2. For example, 2, 4, 6, ..., 48, ..., etc.

Evens or even chance Evens or even chance are ways of saying the probability is 0.5.

Event An event is any of the possible outcomes from an experiment. For example, I roll two dice and add the numbers together. What is the probability of the event 'my total is 10'?

Exchange rates An exchange rate is the rate at which one currency may be exchanged for another currency.

Expenditure Expenditure is the total amount of money that a person or organisation spends.

Expand To multiply out or remove the brackets from an expression.

Experiment An experiment or trial is a procedure in probability that can be repeated over and over again and where we know what the possible outcomes are. For example, rolling two dice and recording the total of the two numbers.

Exponential function An exponential function is in the form $y = a^x$ where a is a positive constant.

Exponential graph A graph showing an exponential function.

Expression An expression is numbers and symbols and operators (for example subtraction) grouped together. For example, $3x - 8$. An expression does not contain an equals sign.

Exterior The exterior angles of a polygon add up to 360°.

Face In a 3D shape, a face is a surface that forms a part of the boundary of the shape. For example, a cube has six faces.

Factor A factor of a number divides into that number exactly. For example, the factors of 18 are 1, 2, 3, 6, 9 and 18.

Factorise Opposite of expand.

Fair In probability a dice or coin is fair if all the possible outcomes are equally likely.

Fibonacci sequence The Fibonacci sequence begins 1, 1, 2, 3, 5, 8, …
Each term is the sum of the previous two terms.

Fibonacci-type sequence Any sequence in which each term is the sum of the previous two terms, for example 3, 4, 7, 11, 18, …

Foreign currency exchange Foreign currency exchange is the trading of one currency for another. This often takes place at a bank or bureau de change.

Formula A formula is a rule or relationship connecting two or more variables that are represented by letters. For example, the formula for the area of a triangle is $A = \frac{1}{2} bh$.

Fraction A fraction consists of a numerator and a denominator. A fraction can represent part of a whole or represent a decimal number.

Frequency The frequency of a data value is the number of times it occurs.

Frequency diagram A frequency diagram is plotted from a grouped frequency table. It consists of bars placed on a linear scale covering the range of values in the groups.

Frequency table A frequency table shows the frequency of each data value in a data set.

Function machine A flow diagram that takes an input value, applies a set of mathematical operations and outputs the answer.

Geometric sequence A geometric sequence is a sequence where the ratio between any two consecutive terms is a constant. For example 5, 10, 20, 40, … (in this sequence the ratio is 2).

Gradient The gradient is a measure of how steep a line is. The gradient of the line joining two points can be calculated as the change in the y values divided by the change in the x values.

Gram Gram (g) is a metric unit of mass. 25 g is approximately the mass of an ordinary bag of crisps.

Graph A graph is the depiction of a relationship between (usually) two variables. For example, it is possible to draw a graph of the relationship $y = x^2$.

Greater than The symbol > indicates the value to the left is bigger than the one to the right.

Grid method A method for multiplying numbers which involves dealing with units, tens, hundreds, etc. separately.

Grouped frequency table A grouped frequency table is obtained when the data set is sorted into groups defined so that there is exactly one group for each item of data.

Height The height is the distance from the base to the top.

Heptagon A heptagon is a seven-sided polygon.

Hexagon A hexagon is a six-sided polygon.

Highest common factor (HCF) The highest common factor (HCF) of two numbers is the greatest integer that divides exactly into both numbers. For example, the highest common factor of 6 and 15 is 3.

Horizontal A horizontal line is parallel to the horizon. It runs from left to right.

Hundreds In place value, hundreds is the place value of the first digit in a three-digit number. In the number 714, the 7 represents seven hundreds.

Hundredths In place value, hundredths is the place value of the second digit after the decimal point. In the number 7.14, the 4 represents four hundredths.

Hypotenuse The hypotenuse is the longest side of a right-angled triangle.

Hypothesis testing Hypothesis testing uses sample data to test whether a hypothesis is true.

Image When a shape undergoes a transformation, the resulting shape is the image.

Impossible An event with probability 0 is impossible.

Improper fraction In an improper fraction, the numerator is greater than the denominator. For example, $\frac{8}{3}$ is an improper fraction.

Included angle The angle between two given sides.

Included side The side between two given angles.

Income Income is the total amount of money that a person or organisation receives.

Indices The index is the power to which a number is raised. For example, in 4^3 the power (or index) is 3 and so $4^3 = 4 \times 4 \times 4$. The plural of index is indices.

Inequality Compares the size of two values, showing if one is less than, or greater than, the other.

Glossary

Input A starting value in a function machine.

Integer An integer is a positive or negative whole number (including zero).

Interest Interest can be: 1) a payment made to an investor or 2) money charged for borrowing. See also simple interest and compound interest.

Interior Inside a shape.

Internal The sum of the interior or internal angles of an n-sided polygon is $180° \times (n - 2)$.

Inverse Inverse means opposite. Subtraction is the inverse of addition. Division is the inverse of multiplication.

Inverse proportion Two quantities, x and y, are in inverse proportion when an increase in one quantity causes a decrease in the second quantity. For example, when the number of men building a wall is doubled then the time to build the wall is halved.

Irregular (shape) A shape where all angles are not an equal size or all sides are not an equal length.

Isosceles An isosceles triangle has two equal angles and two sides of equal length.

Key A key is used on some diagrams such as pictograms to show what a symbol represents. For example, Key: ☐ represents four people.

Kilogram Kilogram (kg) is a metric unit of mass equal to 1000 grams. 1 kg is approximately the mass of a bag of sugar.

Kilometre Kilometre (km) is a metric unit of distance equal to 1000 metres. 1 km is $2\frac{1}{2}$ times round an ordinary athletics track.

Kite A kite is a quadrilateral which has two pairs of adjacent, equal sides. Other properties are one pair of equal angles and diagonals that cross at right angles.

Length The length of an object is a way of measuring its size. For example, the length of this rectangle is 13 cm.

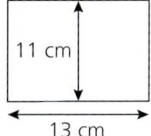

Less than The symbol < indicates the value to the left is less than the one to the right.

Likely An event with probability greater than 0.5 but less than 1 is likely.

Line A line is a straight one-dimensional figure with no thickness and extending infinitely in both directions. It is often called a straight line.

Line of best fit A line of best fit can be drawn when there is correlation. It is a straight line that follows the trend of the data.

Line segment A line segment is part of a line. It has a beginning and an end.

Linear Relating to a straight line. An arithmetic sequence can be described as linear, since on a graph of the terms of the sequence, the points would lie on a straight line.

Litre Litre (l) is a metric unit of capacity equal to 1000 millilitres. 1 litre is usually the capacity of a carton of fruit juice.

Long division A method for dividing numbers which clearly shows how each remainder is calculated.

Long multiplication A method for multiplying two numbers.

Loss If an item is sold for less than the buying price, the loss is the difference between these two prices. If the selling price is greater than the buying price, there is a profit.

Lower bound Measurement is only approximate, so the actual value of a measurement could be half the rounded unit below the given value. The lower bound is the least possible value that the true measurement could be. For example, the measurement of the length of a pencil is 15.5 cm to the nearest millimetre. So the lower bound is 15.5 cm $- \frac{1}{2}$ mm = 15.45 cm. The actual length, l, of the pencil is greater than or equal to 15.45 cm, so $15.45 \leq l$.

Lowest common multiple (LCM) The lowest common multiple of two numbers is the least integer that is a multiple of both numbers. For example, the lowest common multiple of 6 and 15 is 30.

Lowest terms A fraction is in its lowest terms (or simplest form) when it can't be simplified or cancelled any further.

Mass The mass of an object is a measure of how much matter it contains. Mass is measured, for example, in kilograms. In Maths at this level, mass and weight are considered to be the same thing.

Mean The mean is found by adding together all of the data values and then dividing this total by the number of data values. The mean is one of the three main ways to measure an average.

Median The median is the middle value when a data set is organised in order of size. The median is one of the three main ways to measure an average.

Metre Metre (m) is a metric unit of length equal to 1000 millimetres. 2 m is roughly the height of a doorway.

Metric unit Examples of metric units are:
- For length: millimetres, centimetres, metres, kilometres
- For mass: milligrams, grams, kilograms, tonnes
- For volume: millilitres, litres.

Milligram Milligram (mg) is a metric unit of mass equal to one thousandth of a gram. Milligrams are used to measure out doses of drugs. For example, a small aspirin might contain 75 mg.

Millilitre Millilitre (ml) is a metric unit of capacity equal to one thousandth of a litre. For example, a teaspoon holds 5 ml.

Millimetre Millimetre (mm) is a metric unit of length equal to one thousandth of a metre. For example, a small ruler is 150 mm long.

Millions In place value, millions is the place value of the first digit in a seven-digit number. In the number 2 714 806, the 2 represents two millions.

Mirror line The line in which a shape is reflected is known as the mirror line.

Mixed number A mixed number is made up of a whole number and a fraction. For example, $2\frac{3}{4}$ is a mixed number.

Modal The mode of a set of data can be called the modal value. In grouped data, the modal group is the group with the highest frequency.

Mode The mode is the value occurring most often in a data set. The mode is one of the three main ways to measure an average.

Multiple The multiple of a number is the result when you multiply that number by a positive integer. For example, the multiples of 6 are 6, 12, 18, 24, 30, …

Multiplication (multiply) Multiplication comes from repeated addition, so 7×3 is the same as $7 + 7 + 7$ or 21.

Multiplier The multiplier is the number you are multiplying by. For example, in 3×7 the multiplier is 7.

Mutually exclusive events Two events are mutually exclusive if they both cannot happen at the same time. For example, when rolling a dice, getting a 3 and getting a 5 are mutually exclusive as you can't get both a 3 and a 5 from rolling one dice.

Negative A value which is less than zero.

Negative correlation Negative correlation occurs when points are roughly in a line from top left to bottom right of a scatter diagram.

As one quantity increases, the other one decreases.

Nonagon A nonagon is a nine-sided polygon.

Number line A line with numbers indicating their size relative to each other.

Numerator The numerator is the top number in a fraction. For example, $\frac{3}{8}$ has a numerator of 3.

Numerical data Numerical data consists of numbers only.

Object The object is the shape that a transformation is applied to.

Obtuse An obtuse angle lies between 90° and 180°.

Octagon An octagon is an eight-sided polygon.

Odd An odd number is an integer that is not a multiple of 2. For example, 1, 3, 5, …, 57, …, etc.

Ones In place value, ones is the place value of the digit in a single-digit number. In the number 8, the 8 represents eight ones. In the number 7246, the 6 represents six ones. Ones are also known as units.

Operations An operation is a procedure that is carried out to a number, such as dividing or raising to a power. See BIDMAS for the order in which to carry out operations.

Outlier In a set of data an outlier is a value that lies outside the range of the rest of the data — it is much larger or much smaller. It also refers to a value that does not fit the same pattern as the rest of the data.

Pair of compasses An instrument for drawing a circle, or an arc of a circle. Sometimes incorrectly referred to as a compass.

Parallel A pair of parallel lines can be continued to infinity in either direction without meeting.

Parallelogram A parallelogram is a quadrilateral that has two pairs of parallel sides. Other properties are that opposite sides are equal and opposite angles are equal.

Partitioning Breaking numbers into smaller parts to make them easier to work with, for example into units, tens, hundreds, and so on.

Pentagon A pentagon is a five-sided polygon.

Percentage A percentage (%) is a number expressed as a fraction of 100. For example, 38% is equivalent to $\frac{38}{100}$ or 0.38.

Percentage change A change in a quantity expressed as a percentage of the original quantity. For example, if the price of an item falls from £4 to £3, the decrease is £1 and the percentage decrease is 25% (since $\frac{1}{4} = 25\%$).

Percentage decrease The decrease in a quantity expressed as a percentage of the original quantity. For example, if the price of an item falls from £4 to £3, the decrease is £1 and the percentage decrease is 25% (since $\frac{1}{4} = 25\%$).

Glossary

Percentage error The error in the measurement of a quantity expressed as a percentage of the true quantity. For example, if the length of an object is measured as 49 cm but its true length is 50 cm, the error is 1 cm and the percentage error is 2% (since $\frac{1}{50} = 2\%$).

Percentage increase The increase in a quantity expressed as a percentage of the original quantity. For example, if the price of an item rises from £4 to £5, the increase is £1 and the percentage increase is 25% (since $\frac{1}{4} = 25\%$).

Percentage loss If an item is sold for less than it was bought, the percentage loss is the difference between the buying price and the selling price expressed as a percentage of the buying price. For example, if the buying price of an item is £5 and the selling price is £4, the loss is £1 and the percentage loss is 20% (since $\frac{1}{5} = 20\%$).

Percentage profit If an item is sold for more than it was bought, the percentage profit is the difference between the buying price and the selling price expressed as a percentage of the buying price. For example, if the buying price of an item is £4 and the selling price is £5, the profit is £1 and the percentage profit is 25% (since $\frac{1}{4} = 25\%$).

Perimeter The perimeter of a shape is the distance around the outside edge of the shape. Perimeter can be measured in mm, cm, m or km.

Perpendicular Two lines are perpendicular if they meet at right angles.

Pi, π The Greek letter pi is the number 3.14159..., it is equal to the circumference of a circle divided by its diameter.

Pie chart A circular chart that is divided into sectors to represent different groups.

Piece-wise linear A function comprising of more than one straight line segment.

Pictogram A pictogram is a chart that uses pictures or symbols to display data.

Place value The place value of a digit is the number that a particular digit in a number represents. For example, in the number 4137.59, the 4 represents four thousands and the 9 represents nine hundredths.

Polygon A polygon is a closed 2D shape made up of straight lines.

Position-to-term rule The position-to-term rule (or nth term) is a way of describing the terms of a sequence using their position in the sequence. For example, the position-to-term rule for the sequence 7, 13, 19, 25, 31, ... is $6n + 1$.

Positive A value which is greater than zero.

Positive correlation Positive correlation is when points are roughly in a line from bottom left to top right of a scatter diagram. As one quantity increases, the other one does too.

Powers For example, in 4^3 the power is 3 and so $4^3 = 4 \times 4 \times 4$.

Power of 10 10 raised to an integer power. For example, $10^3 = 1000$.

Prime A prime number is a number with exactly two factors: 1 and itself. The prime numbers are 2, 3, 5, 7, 11, ...
Note: 1 is not a prime number as it has only one factor.

Prism A 3D shape that has a constant cross-section and that does not have any curved faces.

Probability Probability is the study of chance. The probability of an event happening is a measure of how likely that event is to happen.

$$\text{Probability} = \frac{\text{number of favourable outcomes}}{\text{total number of equally likely outcomes}}$$

Probability scale The probability scale goes from 0 (an impossible event) to 1 (a certain event).

Product The product of two numbers is the result of multiplying them together. For example, the product of 8 and 3 is 24 ($8 \times 3 = 24$).

Profit The profit is the difference between the selling price and buying price of an item. If the selling price is lower than the buying price there is a loss.

Proportion A proportion is a part of a whole. Proportions can be given as fractions, decimals or ratios.

Protractor A protractor is an instrument used for measuring angles.

Pythagoras' theorem A theorem that says that the square of the hypotenuse (longest side) of a right-angled triangle is equal to the sum of the squares of the two shorter sides. It is written as $a^2 + b^2 = c^2$.

Quadrant A quadrant is any of the four regions of a graph divided by the x and y axes.

Quadrilateral A quadrilateral is a four-sided polygon (a four-sided plane shape).

Questionnaire A questionnaire is a research tool featuring a series of questions used to collect useful information from those who complete it.

Radius A radius is a line that joins the centre of a circle to a point on the circumference. The plural is radii.

Random An event is random if it is done without method or conscious decision.

Range Range is a measure of the spread of a data set. The range is the difference between the highest and lowest data values.

Ratio A ratio compares two quantities.

Real-life graph Any graph that arises from a real-life situation rather than just an equation. Real-life graphs include travel graphs and currency conversion graphs.

Reciprocal The product of a number and its reciprocal is 1. So $4 \times \frac{1}{4} = 1$ and $5 \times \frac{1}{5} = 1$.

Reciprocal function A function of the form $y = \frac{k}{x}$ where k is a constant.

Reciprocal graph A graph of a reciprocal function.

Rectangle A rectangle is a quadrilateral with the following properties: two pairs of parallel sides; four equal angles (90°). It also has opposite sides of equal length.

Reflection A reflection is a 'flip' movement about a mirror line. The mirror line is the line of symmetry between the object and its image.

Reflex A reflex angle is an angle between 180° and 360°.

Regular polygon A shape where all angles are equal sizes and all sides are an equal length.

Reverse percentage A calculation that allows you to work out the original cost, size, etc, after an increase or decrease.

Rhombus A rhombus is a quadrilateral with four equal sides. It also has two pairs of parallel sides, equal opposite angles and diagonals that cross at right angles.

Right angle A right angle is 90°.

Right-angled triangle A triangle where one of the internal angles is 90°.

Rotation A rotation is a 'turning' movement about a specific point known as the centre of rotation.

Rounding (round) Rounding is a way of rewriting a number so it is simpler than the original number. A rounded number should be approximately equal to the unrounded (exact) number. For example, 86 rounded to the nearest ten is 90.

Scale In a scale drawing, the scale is the ratio or multiplier that tells you how much bigger (or smaller) the image is compared to the original drawing.

Scalene A scalene triangle has three sides of different lengths and three different angles.

Scatter diagram A scatter diagram is drawn using pairs of data values as coordinates.

Sector A sector is part of a circle that looks like a piece of pie, it is enclosed by two radii and part of the circumference.

Semi circle A semi circle is half a circle.

Sequence A sequence is a collection of terms arranged in a specific order.

Short division An abbreviated version of long division where the remainders are evaluated without writing down the calculation.

Sign A symbol written in front of a number to indicate whether it is positive or negative.

Significant figure The first significant figure of a number is the first non-zero digit in the number. The second significant figure is the next digit in the number and so on. For example, in the numbers 78046 and 0.0078046 the 1st significant figure is 7, the 2nd significant figure is 8 and the 3rd significant figure is 0.

Similar Two shapes are similar if corresponding sides are the same proportion and corresponding angles are equal; the shapes are enlargements of each other.

Simple interest Simple interest is when a percentage is worked out for an amount and then multiplied by the number of time periods the money is invested for.

Simplify To simplify an expression you write it in an equivalent way but using smaller numbers or fewer terms. You can simplify a fraction, a ratio or an algebraic expression. For example, you can simplify $\frac{9}{15}$ to $\frac{3}{5}$ or $x + x + y + y + y$ to $2x + 3y$ or $21 : 35$ to $3 : 5$.

Sine The sine of an angle, sin θ, in a right-angled triangle is the ratio of the side opposite the angle to the hypotenuse: $\sin \theta = \frac{\text{opposite}}{\text{hypotenuse}}$

Solid Another word for a 3D shape.

Solve To solve an equation you find the value of the unknown that satisfies the equation. For example, the solution of $7x = 35$ is $x = 5$.

Speed A type of compound measure measuring the distance travelled per second or per hour, etc. Average speed = Distance travelled ÷ Time taken

Sphere A 3D shape that is perfectly round, like a ball.

Square (shape) A square is a quadrilateral with four equal sides and four equal angles (90°). It also has two pairs of parallel sides and diagonals that cross at right angles.

Square metre A square metre (m^2) is a unit of area. 1 m^2 is equivalent to the area of a square of side 1 m.

Square number A square number is the result when an integer is multiplied by itself. The square numbers are 1, 4, 9, 16, 25, …

Square root The square root of a number is the number that, when multiplied by itself, gives the original number. The reverse of squaring is square rooting. Every positive number has two square roots. For example, the square

Glossary

root of 9 is 3 (as $3 \times 3 = 9$) or -3 (as $-3 \times -3 = 9$). The symbol $\sqrt{}$ is used for the positive square root of a number, so $\sqrt{9} = 3$.

Square-based pyramid A 3D shape whose base is a square. It has four other faces, all of which are isosceles triangles, meeting at a single point.

Standard form Standard form is a way of writing very large or very small numbers. A number in standard form is written as $A \times 10^n$ where $1 \leq A < 10$ and n is an integer. For example, $0.000\,034 = 3.4 \times 10^{-5}$ and $590\,000\,000 = 5.9 \times 10^8$.

Straight line graph A straight line graph, or linear graph, is a graph in which the points are arranged in a straight line.

Subject The subject of a formula is the variable (letter) which appears on its own on one side of the formula.

Substitute To substitute is to replace the variables (letter symbols) in an expression or formula with numbers.

Subtraction Subtraction or taking away is finding the difference between two or more amounts. For example, $7 - 4 = 3$.

Sum The sum of two numbers is the result of adding them together. For example, the sum of 8 and 3 is 11 ($8 + 3 = 11$).

Surface area The total area of all the faces of a 3D shape. For example, the surface area of a cube is $6 \times$ the area of one face.

Symmetry A shape has symmetry if it does not change under a transformation such as reflection or rotation.

Tally chart A tally chart is a way of collecting or representing data using tally marks. For example:

Score	Tally	Frequency
0		
1		
2							5
3												10
4													11
5					3

Tangent The ratio of the opposite side to the adjacent side in a right-angled triangle: $\tan \theta = \dfrac{\text{opposite}}{\text{adjacent}}$.

Tax A tax is a compulsory financial charge, or some other type of levy, imposed on a taxpayer by a governmental organisation.

Taxation Taxation is the system by which a government takes money from people and spends it on things such as education, health and defence.

Tens In place value, tens is the place value of the first digit in a two-digit number. In the number 57, the 5 represents five tens. In the number 7246, the 4 represents four tens.

Tenths In place value, tenths is the place value of the first digit after the decimal point. In the number 0.83, the 8 represents eight tenths.

Term Each number in a sequence is called a term.

Term number The term number gives the position of a term in a sequence. For example, in the sequence 3, 13, 23, 33, 43, ..., the term number of 13 is 2 as it is second in the sequence. The term number is often referred to as n.

Term-to-term rule A term-to-term rule describes how to use one term in a sequence to find the next term.

Terminating decimal A terminating decimal has digits that do not continue forever. For example, 0.123 and 0.987 654 321.

Thousands In place value, thousands is the place value of the first digit in a four-digit number. In the number 7246, the 7 represents seven thousands.

Thousandths In place value, thousandths is the place value of the third digit after the decimal point. In the number 0.691, the 1 represents one thousandth.

Tonne Tonne (t) is a metric unit of mass equal to 1000 kilograms. 1 t is approximately the mass of a small car.

Total The total is the result of adding numbers together.

Transformation A transformation changes the size, shape, position or orientation of an object. Examples of transformations are reflection, rotation, enlargement and translation.

Translation A translation is a sliding movement.

Trapezium A trapezium is a quadrilateral with one pair of parallel sides.

Travel graph A travel graph is a diagram showing the journey of one or more objects on the same pair of axes.

Triangle A triangle is a three-sided polygon.

Triangle numbers The sequence of numbers 1, 3, 6, 10, 15, ...

Trigonometry An area of maths that involves calculating unknown angles and lengths in triangles.

Turn To turn is to rotate about a point. A full turn is a rotation of 360°.

Two-way table A two-way table has one set of options as the rows and the second set of options as the columns.

375

Unitary method A method used to determine a best buy. The unitary method finds the cost of one item, when two or more purchase options are available. For example, if 6 eggs cost £1.20 the unit price is 20p; if 10 eggs cost £1.50 the unit price is 15p and the pack of 10 eggs is the better buy.

Units (number) In place value, units is the place value of the digit in a single-digit number. In the number 8, the 8 represents eight units. In the number 7246, the 6 represents six units.

Units (shape) The standard units used for measurement, such as kilogram, metre and second.

Unknown The unknown is the letter you are trying to find a value for in an equation. The unknown is most commonly given the symbol x.

Unlikely An event with probability greater than 0 but less than 0.5 is unlikely.

Upper bound Measurement is only approximate; the actual value of a measurement could be half the rounded unit above the given value. The upper bound is the greatest possible value that the true measurement could be. For example, the measurement of the length of a pencil is 15.5 cm to the nearest millimetre. So the upper bound is 15.5 cm + $\frac{1}{2}$ mm = 15.55 cm. The actual length, l, of the pencil is less than 15.55 cm, so $l < 15.55$.

Variable A variable is a quantity that may change in maths problems, such as the area of a triangle. A variable is usually represented by a letter.

VAT VAT stands for 'value added tax'. This is a tax that is added to most goods that we buy. The Government sets the percentage VAT that is charged.

Vector A vector is a quantity with both magnitude (size) and direction. A vector can be written using two components: the first denoting movement in the x-direction and the second the movement in the y-direction. In this form a vector is often used to describe a translation.

Venn diagram A Venn diagram is a diagram made from overlapping circles, which are used to display sets.

Vertex A vertex of a shape is a point where two sides meet.

Vertical A vertical line is perpendicular to the horizon. It runs up and down.

Vertical line graph A vertical line graph is a graph that uses vertical lines to display data. The height of each line represents the frequency.

Vertically opposite Vertically opposite angles are formed when two lines cross. Vertically opposite angles are equal.

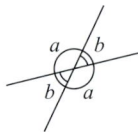

Volume Volume is the amount of space that a 3D shape takes up.

Weight Weight is an everyday term that means the same as mass. When Science and Maths get more precise, a different definition in used.

Width The width of an object is a way of measuring its size. For example, the width of this rectangle is 11 cm.

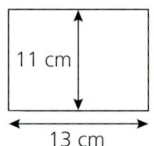

x-coordinate The x-coordinate is the horizontal value in a pair of coordinates. It is how far left or right the point is. In the coordinate (7, 2) the x-coordinate is 7.

y-coordinate The y-coordinate is the vertical value in a pair of coordinates. It is how far up or down the point is. In the coordinate (7, 2) the y-coordinate is 2.

y-intercept The point at which a straight line meets the y-axis.